Edited by
Andreas Seidel-Morgenstern

Membrane Reactors

Related Titles

Peinemann, K.-V., Pereira Nunes, S. (eds.)

Membrane Technology

6 Volume Set

ISBN: 978-3-527-31479-9

Pereira Nunes, S., Peinemann, K.-V. (eds.)

Membrane Technology

in the Chemical Industry

2006

ISBN: 978-3-527-31316-7

Drioli, E., Giorno, L. (eds.)

Membrane Operations

Innovative Separations and Transformations

2009

ISBN: 978-3-527-32038-7

Sammells, A. F., Mundschau, M. V. (eds.)

Nonporous Inorganic Membranes

for Chemical Processing

2006

ISBN: 978-3-527-31342-6

Güell, C., Ferrando, M., López, F. (eds.)

Monitoring and Visualizing Membrane-Based Processes

2009

ISBN: 978-3-527-32006-6

Freeman, B., Yampolskii, Y., Pinnau, I. (eds.)

Materials Science of Membranes for Gas and Vapor Separation

2006

ISBN: 978-0-470-85345-0

Koltuniewicz, A., Drioli, E.

Membranes in Clean Technologies

Theory and Practice

2008

ISBN: 978-3-527-32007-3

Edited by Andreas Seidel-Morgenstern

Membrane Reactors

Distributing Reactants to Improve Selectivity and Yield

WILEY-VCH Verlag GmbH & Co. KGaA

The Editor

Andreas Seidel-Morgenstern
Otto-von-Guericke-Universität
Institut für Verfahrenstechnik
Universitätsplatz 2
39106 Magdeburg
Germany

and
Max-Planck-Institut für Dynamik
komplexer technischer Systeme
Sandtorstr. 1
39106 Magdeburg
Germany

Library of Congress Card No.: applied for

British Library Cataloguing-in-Publication Data
A catalogue record for this book is available from the British Library.

Bibliographic information published by the Deutsche Nationalbibliothek
The Deutsche Nationalbibliothek lists this publication in the Deutsche Nationalbibliografie; detailed bibliographic data are available on the Internet at <http://dnb.d-nb.de>.

Composition Toppan Best-set Premedia Limited, Hong Kong
Printing and Binding betz-druck GmbH, Darmstadt
Cover Design Schulz Grafik-Design, Fußgönheim

Printed in the Federal Republic of Germany
Printed on acid-free paper

ISBN: 978-3-527-32039-4

Contents

Membrane Reactors: Distributing Reactants to Improve Selectivity and Yield.
Edited by Andreas Seidel-Morgenstern

ISBN: 978-3-527-32039-4

Preface

The chemical and pharmaceutical industries are characterized by the fact that a very large spectrum of different target compounds is synthesized. Due to the variety of the physical and chemical properties of the reactants and products involved a wide spectrum of operating conditions and reactor concepts is applied (Moulijn *et al.*, 2001).

Despite the long history of chemical reaction engineering there are still many problems which are not solved in a satisfactory manner (Levenspiel, 1999). One of the most difficult problems motivated the research leading to this book. This problem lies in the fact that during the synthesis of a certain target component typically undesired parallel or consecutive reactions occur which reduce the achievable yields. Formed side products have to be separated from the target product at the reactor outlets, which is a demanding and expensive task. Thus, there is considerable interest in developing technologies which allow increasing the selectivity and yield with which a certain target product can be generated.

It is well known that the selectivity in reaction networks towards a target compound can be increased following various concepts. First, a careful selection of the reaction temperature can be made to favor the formation of the target. A second, more versatile and very important direction is connected with the intensive activities devoted to developing and applying dedicated catalysts which accelerate specifically the desired reaction (Ertl *et al.*, 2008). The third approach, which is studied in this book, exploits the fact that the selectivity with respect to a certain desired product can be increased by properly adjusting the local concentrations of the reactants involved.

Innovative distributing and dosing concepts have for some time been an objective of research in chemical reaction engineering (Levenspiel, 1999). Besides adding certain reactants in a discrete manner into chemical reactors, various possibilities have been suggested for using different porous or non-porous membranes in order to arrange different ways of contacting the reactants (Coronas, Menendez, and Santamaria,1994; Lu *et al.*, 1997; Seidel-Morgenstern, 2005). Although the membrane reactor concept has been identified as interesting and promising, there are still several difficult problems that must be solved prior to an industrial application.

Membrane Reactors: Distributing Reactants to Improve Selectivity and Yield.
Edited by Andreas Seidel-Morgenstern

ISBN: 978-3-527-32039-4

This book describes results obtained within the research project "Membrane supported reaction engineering", which was funded by the German Research Foundation (DFG) and was carried out between 2001 and 2008 at the Otto von Guericke University and the Max Planck Institute for Dynamics of Complex Technical Systems in Magdeburg (Germany). Chemists, chemical engineers and mathematicians worked in this project together in order to investigate various options of the so-called distributor type of membrane reactors (Dittmeier and Caro, 2008). The projects focused on a single and important class of reactions, namely the selective oxidations of gaseous alkenes. These reactions require typically elevated temperatures and suffer from severe selectivity limitations (Hodnett, 2000). Different types of membranes and reactor configurations were studied, concentrating on the oxidative dehydrogenation of ethane to ethylene. In order to allow a comparison of the results of different project partners working on packed-bed membrane reactors, fluidized membrane reactors and electrochemical membrane reactors, the same type of vanadia-based catalyst was always used. To implement a range of dosing concepts various high-temperature-resistant membranes were applied. On the one hand, this book was written to summarize the large amount of experimental material generated during the project. On the other hand, it was also the goal of the authors to provide theoretical concepts which allow quantitative descriptions and evaluations of the various types of membrane reactors investigated.

The book starts in Chapter 1 with an introduction into some basics of chemical reaction engineering. The equations presented are helpful to evaluate in a quantitative manner the potential of dosing reactants via reactor walls, as applied in membrane reactors of the distributor type, in order to enhance the selectivity of producing a certain target component in a network of reactions.

Chapter 2 gives a summary of the current state of the art of modeling packed-bed reactors. This chapter serves to introduce the overall notation and provides a frame for modeling mass and heat transfer processes relevant in the different reactors studied. Appropriate model reduction methods, discretization techniques and suitable solvers for the typically large systems of algebraic equations are also discussed in this chapter.

Chapter 3 introduces the heterogeneously catalyzed oxidative dehydrogenation (ODH) of ethane which was studied as a model reaction in various projects. Although this reaction is currently far from a wide industrial application, it was identified to be a suitable object of investigation for the purpose of the overall project. Chapter 3 provides further an important basis for the experimentally oriented projects described later. It summarizes the preparation, characterization and properties of the vanadia catalysts used, the reaction network and a model capable of quantifying the reaction kinetics.

The aim of Chapter 4 is the analysis of the relevant transport phenomena, in particular the superposition of convection and diffusion processes. Another point addressed in this chapter is the experimental and theoretical analysis of transport processes in membranes. Finally, an analysis of the influence of membrane geometry and structural parameters as well as of the operating conditions is performed for reactors with catalytically coated membranes. Hereby, some aspects of the numerical solution of the corresponding differential equations are discussed from a mathematical point of view.

Chapter 5 describes results of a theoretical study of a packed-bed membrane reactor using models of different levels of complexity. Hereby, membrane reactor tubes are filled with particles of the solid catalyst. The chapter further summarizes the results of detailed experimental investigations of single- and multi-stage packed-bed membrane reactors as well as conventional packed-bed reactors carried out in laboratory- and pilot-scale reactor set-ups. Besides the ODH of ethane also the ODH of propane was considered.

Chapter 6 summarizes the results of an extensive experimental and theoretical investigation devoted to evaluating the potential of porous membranes which are dipped into a bed of fluidized catalyst particles and which serve as an oxygen distributor. This chapter also discusses benefits and limitations of this reactor concept compared to the classic co-feed reactant dosing applied in a conventional fluidized bed reactor.

Chapter 7 introduces electrochemical membrane reactors equipped with ion-conducting membranes, which are ideally impermeable for non-charged species. These reactors operate as electrochemical cells in which the oxidation and reduction reactions are carried out separately on catalyst/electrode layers located on opposite sides of the electrolyte. The working principle of a solid electrolyte membrane reactor is introduced. Further, material aspects and the modeling of solid electrolyte membrane reactors are discussed. The synthesis of maleic anhydride and again the oxidative dehydrogenation of ethane are considered as experimental examples.

Chapter 8 provides insight on nonlinear phenomena that may occur in membrane reactors. It is demonstrated that such reactors can become instable, when membranes are used for side-injection of reactants in order to enhance yields. Under certain conditions, spatially homogeneous solutions can give rise to concentration and temperature patterns which may be quite complex.

In the final Chapter 9 the results achieved for the different reactor configurations are compared and a short summary is given to evaluate the current state of membrane reactors of the distributor type.

This book is seen by the authors as a contribution to the actively investigated wider field of integrated chemical processes (Sundmacher *et al.*) and more specifically to membrane reactors combining reaction and separation steps (Sanchez Marcano and Tsotsis, 2002).

The editor is very grateful to all colleagues in Magdeburg who contributed to this book for their inspiring, fruitful and pleasant cooperation during the period of this project.

All authors thank the German Research Foundation for generous financial support, in particular Dr. Bernd Giernoth for his help in all organizational issues, and Hermsdorfer Institut für Technische Keramik for providing membrane samples.

We also would like to express our gratitude to Wiley-VCH, in particular Drs. Rainer Münz and Martin Graf, for their professional support and patience during the preparation of this manuscript.

Finally, the editor wants to express his personal thanks to Marion Hesse, Nancy Cassel and Dr. Christof Hamel for their immense support during the course of administrating the research project and editing this book.

Magdeburg, August 2009 *Andreas Seidel-Morgenstern*

References

Coronas, J., Menendez, M., and Santamaria, J. (1994) Methane oxidative coupling using porous ceramic membrane reactors. *Chem. Eng. Sci.*, **49**, 2015–2025.

Dittmeier, R., and Caro, J. (2008) Catalytic membrane reactors, in *Handbook of Heterogeneous Catalysis* (eds G. Ertl, H. Knözinger, F. Schüth, and J. Weitkamp), Wiley-VCH Verlag GmbH, Weinheim, pp. 2198–2248.

Ertl, G., Knözinger, H., Schüth, F., and Weitkamp, J. (eds) (2008) *Handbook of Heterogeneous Catalysis*, vol. 8, Wiley-VCH Verlag GmbH, Weinheim.

Hodnett, B.K. (2000) *Heterogeneous Catalytic Oxidation: Fundamental and Technological Aspects of the Selective and Total Oxidation of Organic Compounds*, John Wiley & Sons, Ltd.

Levenspiel, O. (1999) *Chemical Reaction Engineering*, 3rd edn, John Wiley & Sons, Inc., New York.

Lu, Y.L., Dixon, A.G., Moder, W.R., and Ma, Y.H. (1997) Analysis and optimization of cross-flow reactors with distributed reactant feed and product removal. *Catal. Today*, **35**, 443–450.

Moulijn, J.A., Makkee, M., and van Diepen, A.E. (2001) *Chemical Process Technology*, John Wiley & Sons, Ltd, Chichester.

Sanchez Marcano, J.G., and Tsotsis, T.T. (2002) *Catalytic Membranes and Membrane Reactors*, Wiley-VCH Verlag GmbH, Weinheim.

Seidel-Morgenstern, A. (2005) Analysis and experimental investigation of catalytic membrane reactors, in *Integrated Chemical Processes* (eds K. Sundmacher, A. Kienle, and A. Seidel-Morgenstern), Wiley-VCH Verlag GmbH, Weinheim, pp. 359–389.

Sundmacher, K., Kienle, A., and Seidel-Morgenstern, A. (eds) (2005) *Integrated Chemical Processes*, Wiley-VCH Verlag GmbH, Weinheim.

List of Contributors

Lyubomir Chalakov
Max-Planck-Institut für Dynamik komplexer technischer Systeme
Sandtorstr. 1
39106 Magdeburg
Germany

now

Güntner AG & Co. KG
Hans-Güntner-Str. 2–6
82256 Fürstenfeldbruck
Germany

Velislava Edreva
Otto-von-Guericke-Universität Magdeburg
Institut für Verfahrenstechnik
Lehrstuhl für Thermische Verfahrenstechnik
Universitätsplatz 2
39106 Magdeburg
Germany

Katya Georgieva-Angelova
Otto-von-Guericke-Universität Magdeburg
Institut für Strömungstechnik und Thermodynamik
Universitätsplatz 2
39106 Magdeburg
Germany

now

ANSYS Germany
Staudenfeldweg 12
83624 Otterfing
Germany

Henning Haida
Otto-von-Guericke-Universität Magdeburg
Institut für Verfahrenstechnik
Lehrstuhl für Chemische Verfahrenstechnik
Universitätsplatz 2
39106 Magdeburg
Germany

Christof Hamel
Otto-von-Guericke-Universität Magdeburg
Institut für Verfahrenstechnik
Lehrstuhl für Chemische Verfahrenstechnik
Universitätsplatz 2
39106 Magdeburg
Germany

Stefan Heinrich
Technische Universität Hamburg-Harburg
Institut für Feststoffverfahrenstechnik und Partikeltechnologie
Denickestr. 15
21073 Hamburg
Germany

Membrane Reactors: Distributing Reactants to Improve Selectivity and Yield.
Edited by Andreas Seidel-Morgenstern

ISBN: 978-3-527-32039-4

Arshad Hussain
Otto-von-Guericke-Universität
Magdeburg
Institut für Verfahrenstechnik
Lehrstuhl für Thermische
Verfahrenstechnik
Universitätsplatz 2
39106 Magdeburg
Germany

now

NTNU
Department of Chemical Engineering
7491 Trondheim
Norway

Milind Joshi
Otto-von-Guericke-Universität
Magdeburg
Institut für Verfahrenstechnik
Lehrstuhl für Chemische
Verfahrenstechnik
Universitätsplatz 2
39106 Magdeburg
Germany

now

Reaction Engineering
BASF SE
GCE/R–M311
67056 Ludwigshafen
Germany

Malte Kaspereit
Max-Planck-Institut für Dynamik
komplexer technischer Systeme
Sandtorstr. 1
39106 Magdeburg
Germany

Achim Kienle
Max-Planck-Institut für Dynamik
komplexer technischer Systeme
Sandtorstr. 1
39106 Magdeburg
Germany

and

Otto-von-Guericke-Universität
Magdeburg
Institut für Automatisierungstechnik
Universitätsplatz 2
39106 Magdeburg
Germany

Frank Klose
Max-Planck-Institut für Dynamik
komplexer technischer Systeme
Sandtorstr. 1
39106 Magdeburg
Germany

now

Südchemie AG
Waldheimerstr. 13
83052 Bruckmühl/ Heufeld

Michael Mangold
Max-Planck-Institut für Dynamik
komplexer technischer Systeme
Sandtorstr. 1
39106 Magdeburg
Germany

Lothar Mörl
Otto-von-Guericke-Universität
Magdeburg
Institut für Apparate und
Umwelttechnik
Universitätsplatz 2
39106 Magdeburg
Germany

Barbara Munder
Max-Planck-Institut für Dynamik
komplexer technischer Systeme
Sandtorstr. 1
39106 Magdeburg
Germany

Liisa Rihko-Struckmann
Max-Planck-Institut für Dynamik
komplexer technischer Systeme
Sandtorstr. 1
39106 Magdeburg
Germany

Jürgen Schmidt
Otto-von-Guericke-Universität
Magdeburg
Institut für Strömungstechnik und
Thermodynamik
Universitätsplatz 2
39106 Magdeburg
Germany

and

Max-Planck-Institut für Dynamik
komplexer technischer Systeme
Sandtorstr. 1
39106 Magdeburg
Germany

Andreas Seidel-Morgenstern
Otto-von-Guericke-Universität
Magdeburg
Institut für Verfahrenstechnik
Lehrstuhl für Chemische
Verfahrenstechnik
Universitätsplatz 2
39106 Magdeburg
Germany

Piotr Skrzypacz
Institut für Analysis und Numerik
Universitätsplatz 2
39106 Magdeburg
Germany

Yuri Suchorski
Institute of Materials Chemistry
Vienna University of Technology
Veterinärplatz 1
1210 Vienna
Austria

Kai Sundmacher
Max-Planck-Institut für Dynamik
komplexer technischer Systeme
Sandtorstr. 1
39106 Magdeburg
Germany

and

Otto-von-Guericke-Universität
Magdeburg
Institut für Verfahrenstechnik
Lehrstuhl für Systemverfahrenstechnik
Universitätsplatz 2
39106 Magdeburg
Germany

Sascha Thomas
Fraunhofer-Institut für Fabrikbetrieb
und -automatisierung
Sandtorstrasse 22
39106 Magdeburg
Germany

Lutz Tobiska
Institut für Analysis und Numerik
Universitätsplatz 2
39106 Magdeburg
Germany

Ákos Tóta
Otto-von-Guericke-Universität Magdeburg
Institut für Verfahrenstechnik
Lehrstuhl für Chemische Verfahrenstechnik
Universitätsplatz 2
39106 Magdeburg
Germany

now

Linde AG
Linde Engineering Division
Dr.-Carl-von-Linde-Str. 6–14
85049 Pullach
Germany

Desislava Tóta
Otto-von-Guericke-Universität Magdeburg
Institut für Apparate und Umwelttechnik
Universitätsplatz 2
39106 Magdeburg
Germany

now

Linde AG
Linde Engineering Division
Dr.-Carl-von-Linde-Str. 6–14
85049 Pullach
Germany

Evangelos Tsotsas
Otto-von-Guericke-Universität Magdeburg
Institut für Verfahrenstechnik
Lehrstuhl für Thermische Verfahrenstechnik
Universitätsplatz 2
39106 Magdeburg
Germany

Helmut Weiß
Otto-von-Guericke-Universität Magdeburg
Institut für Chemie
Universitätsplatz 2
39106 Magdeburg
Germany

Tanya Wolff
Max-Planck-Institut für Dynamik komplexer technischer Systeme
Sandtorstr. 1
39106 Magdeburg
Germany

Fan Zhang
Max-Planck-Institut für Dynamik komplexer technischer Systeme
Sandtorstr. 1
39106 Magdeburg
Germany

1
Basic Problems of Chemical Reaction Engineering and Potential of Membrane Reactors

Sascha Thomas, Christof Hamel, and Andreas Seidel-Morgenstern

1.1
Challenges in Chemical Reaction Engineering

Currently there are more then 30 000 specialty chemicals produced industrially from approximately 300 intermediate chemicals (Moulijn, Makkee, and Diepen, 2001). The vast majority of these intermediates are produced from a very limited number of approximately 20 simple base chemicals for example, ethylene, propylene, butane, ammonia, methanol, sulfuric acid and chlorine. To perform efficiently the large spectrum of chemical reactions of interest an arsenal of specific reactor types and dedicated operating regimes has been developed and is applied in various industries. The design of efficient and reliable reaction processes is the core subject of Chemical Reaction Engineering, a discipline which can be considered nowadays as rather mature. The progress achieved and important concepts developed are summarized in several excellent monographs (e.g., Froment and Bischoff, 1979; Schmidt, 1997; Levenspiel, 1999; Missen, Mims, and Saville, 1999; Fogler, 1999).

The main starting point of an analysis of reacting systems is typically an evaluation and quantification of the rates of the reactions of interest. Hereby, based on the specific physical an chemical properties of the reactants and products a wider range of temperature and pressure conditions has to be considered during the early development phases. The spectrum of reactor types available and operating principles applicable is very broad.

Reactions and reactors are often classified according to the phases present (Levenspiel, 1999). There are reactions that can be carried out in a single phase. However, in a reaction system often more phases are present requiring more sophisticated configurations and operation modes.

Another useful classification is based on the character of the process and distinguishes between continuous and discontinuous (batch) operations. Between these exist semi-batch processes which are often applied to carry out highly exothermal reactions exploiting adjusted dosing concepts (Levenspiel, 1999; Fogler, 1999).

To accelerate the desired reactions and/or to influence the selectivity in reaction networks with respect to the target products, frequently specific catalysts are

Membrane Reactors: Distributing Reactants to Improve Selectivity and Yield.
Edited by Andreas Seidel-Morgenstern

ISBN: 978-3-527-32039-4

applied. These catalysts might be present in the same phase as the reactants (homogeneous catalysis). To fix these often expensive materials in continuously operated reactors, catalysts are often deposited (immobilized) on the surface of solid porous supports (heterogeneous catalysis).

Despite the large efforts devoted to further develop the field of Chemical Reaction Engineering, the performance of how chemical reactions are carried out indutrially still suffers from several severe limitations. Very important and not sufficiently solved problems are:

Problem 1: The rates of chemical reactions leading to desired products are often too low to establish economically attractive processes.

Problem 2: The conversion of many reactions of interest is thermodynamically limited, that is, the reactions proceed also in the opposite direction and convert products back (reversible reactions).

Problem 3: The energy efficiency of endothermal and exothermal reactions performed industrially is often not satisfactory.

Problem 4: In reaction networks the selectivities and yields with respect to a certain target product are limited.

In recent decades several promising new approaches and innovative reactor concepts have been developed to tackle the mentioned problems.

Enhancing the rates of desired reactions (Problem 1) is the main field of catalysis. Significant progress has been achieved in recent years, both in homogeneous catalysis (e.g., Bhaduri and Mukesh, 2000) and heterogeneous catalysis (e.g., Ertl *et al.*, 2008).

To overcome equilibrium limitations (Problem 2) new reactor concepts have been suggested and developed. One of the most successful concepts in this area is reactive distillation which is based on separating certain reactants from each other directly in the reactor (column) by distillation. Thus, undesired backward reactions can be suppressed (Sundmacher and Kienle, 2003). The subject of integrating also other separation processes into chemical reactors is discussed, for example, in a review (Krishna, 2002) and a more recently published book (Sundmacher, Kienle, and Seidel-Morgenstern, 2005).

There has long been interest in applying reactor principles which allow for an efficient use of energy (Problem 3) when developing new reaction processes. Recently developed elegant autothermal reactor concepts exploit dedicated heat transfer processes and the dynamics of periodically operated reactors (Eigenberger, Kolios, and Nieken, 2007; Silveston, 1998). Examples of new reactor types include the reversed flow reactor (Matros and Busimovic, 1996) and the loop reactor (Sheintuch and Nekhamkina, 2005).

One of the most difficult problems in chemical reaction engineering is to navigate in a reaction network efficiently in order to optimize the production of the desired target component (Problem 4). In this field again catalysis is a main tool. In recent years many new and highly selective catalysts have been developed, allowing an increase in the selectivity and yield with which many base chemicals, intermediates and fine chemicals can be produced (Ertl *et al.*, 2008). Complementarily there are permanent activities devoted to identifying the most suitable reactor types and applying the most beneficial operating conditions in order to achieve high selectivities and yields. In this important field new reactor types can be expected for the future.

One promising option considered when tackling Problem 4 and the subject of this book is to apply optimized dosing strategies using specific membrane reactors. Before introducing the basic principle of these reactors, the broader field of membrane reactors is briefly introduced in the next section.

1.2 Concepts of Membrane Reactors

The application of membranes which divide two specific parts of a reactor possesses the potential to improve in various ways the performance of chemical reactors compared to conventional reactor concepts. For this reason membrane reactors have long been the focus of intensive research. The state of the art regarding this rather broad field has been described in several reviews (Zaspalis and Burggraaf, 1991; Saracco *et al.*, 1999; Dittmeyer, Höllein, and Daub, 2001; Dixon, 2003; Seidel-Morgenstern, 2005). Comprehensive summaries were recently given by (Sanchez Marcano and Tsotsis, 2002; Dittmeyer and Caro, 2008). Modern developments were reported on a regular basis during the "*International Congresses on Catalysis in Membrane Reactors*" (ICCMR, 1994–2009).

Due to the availability of the mentioned extensive overviews and conference proceedings it is not the goal of this chapter to review the field again. To introduce the main principles suggested and partly already applied, just a short overview is given below.

Figure 1.1 illustrates schematically six membrane reactor concepts (I–VI) related to different problems which should be tackled using membranes within the reactor. For illustration, and because it is frequently the competing principle, at the top of the figure the classic tubular reactor is shown. This reactor possesses closed walls. Thus, the reactants are typically introduced together at the reactor inlet (co-feed mode). Often tubular reactors are filled with solid catalyst particles in order to increase the rates and selectivities. This classic fixed-bed or packed-bed reactor (PBR) is intensively studied and used widely (Eigenberger, 1997). It serves as a reference in several sections of this book.

Concept I: Retainment of homogeneous catalysts

The first membrane reactor concept shown in Figure 1.1 exploits the membrane to retain in the reactor soluble (homogeneous) catalysts. Thus, it allows for continuous operation without the need to separate and recycle the typically valuable catalysts. An introduction into the concept is given, for example, by (Cheyran and Mehaaia, 1986; Sanchez Marcano and Tsotsis, 2002). Successful application for various synthesis reactions are described, for example, by (Kragl and Dreisbach, 2002).

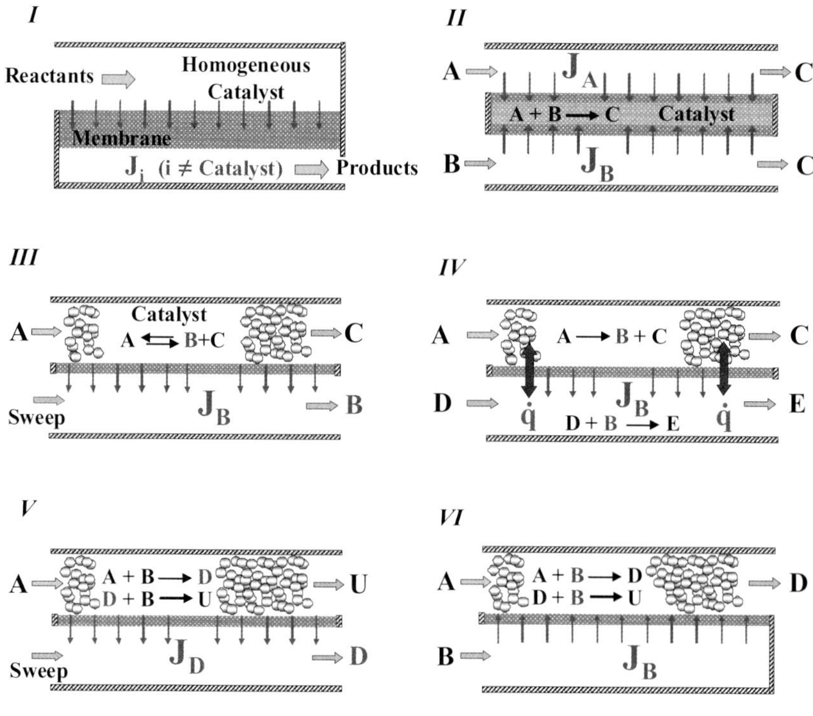

Figure 1.1 Illustration of the conventional packed bed reactor (PBR) and six membrane reactor concepts (I–VI). Concept I: catalyst retainment. Concept II: membrane as "contactor". Concept III: membrane as "extractor" (shift of equilibria). Concept IV: coupling of reactions. Concept V: membrane as "extractor" (removal of intermediates). Concept VI: "distributor" (reactant dosing).

Concept II: Contactor

Another interesting and promising membrane reactor principle is based on applying the membrane as an active "Contactor". The reactants are fed into the reactor from different sides and react within the membrane (Miachon *et al.*, 2003; Dittmeyer and Caro, 2008). There are significant efforts in order to exploit this principle for heterogeneously catalyzed gas/liquid reactions (three-phase membrane reactors) (Dittmeyer and Reif, 2003; Vospernik *et al.*, 2003).

Concept III: Extractor

A widely studied and rather well understood type of membrane reactors is the so-called "Extractor" which removes selectively from the reaction zone certain products via a membrane. As already recognized early (Pfefferie, 1966), this concept possesses the potential to enhance the conversion if the reactions are reversible. To remove the permeated products and to increase the driving force for the transport, additional sweep gases or solvents are needed to apply the "Extractor" principle. Several systematic studies were carried out (e.g., Itoh *et al.*, 1988; Ziaka *et al.*, 1993; Kikuchi, 1997; Schramm and Seidel-Morgenstern, 1999; Schäfer *et al.*, 2003). An evaluation of the potential considering also the additional sweep gas is given e.g. by (Seidel-Morgenstern, 2005).

Concept IV: Energetic coupling

Membranes can be also used to separate two reactor segments in which different reactions take place (Gryaznov, Smirnov, and Mischenko, 1974). The courses of these reactions are influenced when there is a selective transport of certain components which participate in both reactions (e.g., component B in the Figure 1.1). Reactive sweep gases might further improve the performance of the "Extractor" concept described above. If the two reactions are endothermal and exothermal an attractive thermal coupling can be realized (e.g., Gobina, Hou, and Hughes, 1995). In this case an additional heat flux over the membrane takes place which offers interesting degrees of freedom to optimize the reactor from an energetic point of view (Eigenberger, Kolios, and Nieken, 2007).

Concept V: Selectivity enhancement through withdrawal of a product

This concept resembles concept III. However, the component of interest that should be removed ("Extraction") via the membrane is an intermediate component generated in a network of reactions. This removal leads to the reduction or complete avoidance of undesired consecutive reactions and, thus, to enhanced selectivities with respect to this target component (Kölsch *et al.*, 2002; Dittmeyer and Caro, 2008). Unfortunately, the application of this elegant principle requires very selective membranes which are often not available for industrially relevant problems.

Concept VI: Selectivity enhancement through optimized reactant dosing (distributor)

The main focus of this book is to contribute to achieve higher selectivities and yields and thus tackling Problem 4 mentioned above. Hereby, an interesting and attractive approach is based on using membranes to dose (distribute) certain reactants into the reactor. Compared to conventional PBR operation different local concentrations and residence time characteristics can be established and exploited to enhance selectivities. Although the general idea has long been known and significant efforts have been undertaken to exploit the potential of the concept (e.g., Mallada, Menendez, and Santamaria, 2000; Al-Juaied, Lafarga, and Varma, 2001; references in Section 1.8), no industrial applications of such a "Distributor" type of membrane reactor have been reported. In implementing the concept, several degrees of freedom can be exploited. Some important questions considered in more or less detail in this book are:

- Which component should be dosed via the membrane and which should be introduced at the reactor inlet?
- Which kind of membrane and which separation mechanism should be exploited?
- To what extent does multi-stage dosing improve the performance compared to the application of a simple uniform dosing profile?
- Is a particulate catalyst (as used in the packed-bed membrane reactor, PBMR) more suitable than a thin catalytic layer on the membrane surface (as used in a catalytic membrane reactor, CMR)?
- What is the dynamic behavior of such a configuration?

Before discussing in more detail some reaction engineering aspects related to the selectivity problem, a short overview is given concerning the field of membrane materials and types.

1.3 Available Membranes

During the past decades a broad spectrum of different membrane types was developed. Extensive overviews are available (e.g., Bhave, 1991; Ohlrogge and Ebert, 2006; Peinemann and Pereira Nunes, 2007).

The two most suitable classification categories are related to: (a) the membrane materials and (b) the membrane permeabilities and selectivities.

Concerning the materials a distinction can be made between organic and inorganic membranes. Organic polymeric membranes can be synthesized with very specific properties using well developed concepts of macromolecular chemistry. Hereby, a large flexibility exists and a broad spectrum of materials can be made with properties adjusted to the specific separation problem. A drawback of organic membranes is their limited thermal stability. At higher temperatures only inorganic membranes can be applied. Also in this area there is a broad spectrum of membranes available based, for example, on ceramics, perovskites, metals, metal alloys and composites of these materials (e.g., Julbe, Farrusseng, and Guizard, 2001; Verweij, 2003).

Another classification distinguishes between dense and porous membranes. Whereas dense membranes offer typically high selectivities for certain components, they suffer from limited permeabilities. Overviews are given, for example, by (Dittmeyer, Höllein, and Daub, 2001) for metal membranes and by (Bouwmeester, 2003) for ion- and electron-conducting materials. The transport behavior is opposite when porous membranes are applied, allowing for higher fluxes but providing limited selectivities. Porous membranes are typically classified according to their pore size, defining the various types af membrane separation processes as, for example, microfiltration, ultrafiltration and nanofiltration (Li, 2008).

Besides pore size the chemistry of the membrane surface also plays an important role. Traditionally membranes are used to carry out transport and separation processes. In such applications they are chemically inert. Membranes might also possess certain surface properties which catalyze chemical reactions. Such catalytically active membranes are of particular interest for the "Contactor" type of membrane reactors. However, they might be applicable also in some of the other membrane reactor concepts depicted in Figure 1.1.

The quantitative description and prediction of component specific transport rates through dense and porous membranes has been studied intensively. Introductions into the transport theories available are given, for example, by (Mason and Malinauskas, 1983; Sahimi, 1995; Wesselingh and Krishna, 2000; Bird, Stewart, and Lightfoot, 2002). Specific problems of quantifying accurately transport rates are often related to the composite structure of the membranes of interest (Thomas *et al.*, 2001). The accurate prediction of permeabilities and separation factors is still a difficult task and the subject of further intensive research.

In general, the identification, provision and quantitative description of materials suitable to tackle a specific separation problem is still not a routine task.

There is a particular aspect related to membrane reactors which is addressed in Section 1.7. A successful operation requires a sufficient kinetic compatibility of the rates of the transport through the membranes and the rates of the reactions of interest.

Evaluating the general potential of membrane technology Burggraaf and Cot predicted already in 1996 that membrane reactors possess a significant and growing potential in particular for high-temperature reactions using inorganic membranes (Figure 1.2; Burggraaf and Cot, 1996).

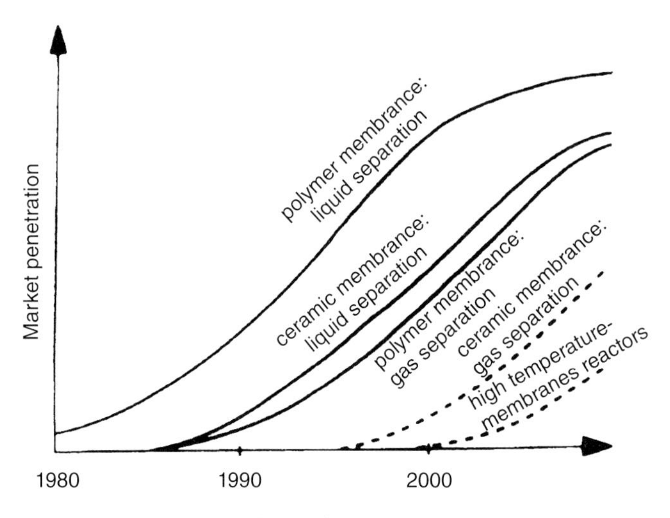

Figure 1.2 Quantitative scheme of expected market penetration as a function of time for different groups of membrane applications (reprinted from [Burggraaf and Cot, 1996], with permission).

1.4
Illustration of the Selectivity Problem

Impressive and frustrating examples characterizing the dilemma and importance of the selectivity problem introduced in Section 1.1 as Problem 4 were given for the industrially important class of partial oxidation reactions by (Haber, 1997; Hodnett, 2000). The latter author presented a large number of selectivity versus conversion plots for various hydrocarbon oxidations. An example is shown in Figure 1.3 for the partial oxidation of *n*-butane to maleic anhydride catalyzed by a vanadium phosphorus oxide (VPO) catalyst. In this plot the results of various studies reported in the literature are superimposed. Hereby, different reactor types and catalysts were applied. There is obviously a clear border which current technology cannot pass. For other reactions of this type which are applied in a large scale the "dream corner" (100% selectivity, 100% conversion) is even more remote. The problem described for oxidation reactions exists in a similar manner for the important class of selective hydrogenation reactions.

As mentioned above, selectivity improvements are the objective of intensive research in catalysis. Examples of successful new catalysts were summarized by (Ertl *et al.*, 2008). However, there are still many "dream reactions" for which satisfactory catalysts are not yet available. The alternative way to improve selectivities is to develop better reactors using currently available catalysts. In this case it is particularly important to understand the relation between local concentrations and temperatures and the selectivity–conversion behavior. To follow this second route is the focus of this book.

The next section summarizes a few basics of chemical reaction engineering which are important for understanding how membrane reactors of the distributor type can contribute to achieve improvements in selectivities and yields.

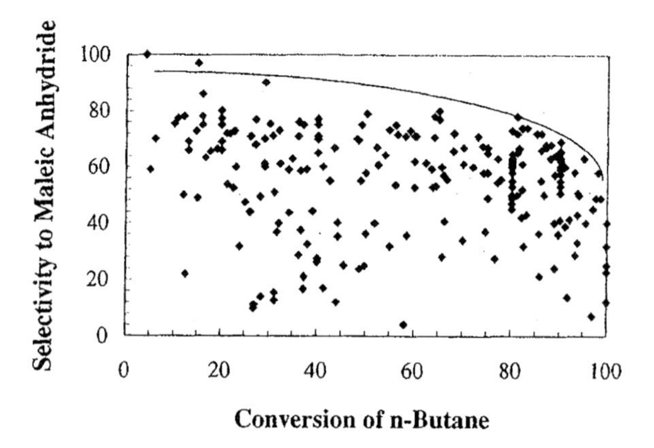

Figure 1.3 Multiple selectivity–conversion plot for *n*-butane selective oxidation to maleic anhydride over a range of catalysts and in a variety of reaction conditions (reprinted from Hodnett, 2000, with permission).

1.5 Reaction Rate, Conversion, Selectivity and Yield

In order to evaluate the potential of membrane reactors in general and "Distributors" in particular, the classic approaches of chemical reaction engineering are available. Basic aspects of analyzing and optimizing various types of chemical reactors have been discussed extensively in standard textbooks of the field (e.g., Levenspiel, 1999; Fogler, 1999). Below is given a selected summary introducing important quantities and performance criteria.

1.5.1 Reaction Rates

The reaction rates are the key information required to quantify chemical reactions and to describe the performance of chemical reactors.

The rate of a single reaction in which N components are involved is defined as:

$$r_{\text{Scale}} = \frac{1}{\text{Scale}} \frac{1}{\nu_i} \frac{dn_i}{dt}\bigg|_{\text{Reaction}} \qquad i = 1, N \tag{1.1}$$

The use of the stoichiometric coefficient ν_i guarantees that the reaction rate does not depend on the component i considered. There are several possibilities regarding the selection of an appropriate scale. For reactions taking place in a homogeneous phase, frequently the reaction volume V_R is used leading to a reaction rate which has the dimension [mol/m^3 s]. In heterogeneous catalysis often the mass or surface area of the catalyst, M_{Cat} or A_{Cat}, are more useful scaling quantities leading to reaction rates in [mol/kg s] or [mol/m^2 s]. Obviously, it is necessary to use r_{Scale} and the chosen scaling quantity consistently. If different scales are of relevance, for example, "a" and "b", it must hold:

$$\text{Scale}_a r_{\text{Scale}_a} = \text{Scale}_b r_{\text{Scale}_b} \tag{1.2}$$

To illustrate the relevance of the reaction rate, in this chapter the reactor volume is selected for scaling. For the sake of brevity no scale index is used. Please note that other chapters of this book also use mass-related reaction rates.

If the reactor volume V_R is assumed to be constant, the reaction rate r can be expressed as:

$$r = \frac{1}{\nu_i} \frac{d\tilde{c}_i}{dt} \qquad i = 1, N \tag{1.3}$$

where $\tilde{c}_i$ is the molar concentration of component i defined as:

$$\tilde{c}_i = \frac{n_i}{V_R} \qquad i = 1, N \tag{1.4}$$

or for open systems with the (also constant) volumetric flow rate $\dot{V}$ as:

$$\tilde{c}_i = \frac{\dot{n}_i}{\dot{V}} \quad i = 1, N \tag{1.5}$$

Reaction rates depend on temperature and the molar concentrations (Levenspiel, 1999), that is:

$$r = r(T, \tilde{c}_1, \tilde{c}_2, \ldots, \tilde{c}_N) \tag{1.6}$$

If only one reaction occurs, knowledge regarding the concentration change of a single key component is sufficient to describe all other concentration changes.

1.5.2 Conversion

If a reactant A is chosen as the key component, its conversion can be defined as:

$$X_A = \frac{n_A^0 - n_A}{n_A^0} \tag{1.7}$$

or for constant volumes:

$$X_A = \frac{\tilde{c}_A^0 - \tilde{c}_A}{\tilde{c}_A^0} \tag{1.8}$$

The mole numbers n_A^0 or concentrations $\tilde{c}_A^0$ stand here for the initial or inlet states. The conversion can be considered as a dimensionless concentration. Using this quantity Equation 1.6 can be reformulated:

$$r = r(T, \tilde{c}_A^0, X_A) \tag{1.9}$$

The temperature dependence of the reaction rate, $r(T)$, can be accurately described using the well known Arrhenius equation (Levenspiel, 1999).

Regarding the conversion (i.e., concentration) dependence, r can be split into a constant contribution r^0 (related to the initial or inlet state) and a conversion-dependent function $f(X_A)$ describing the rate law valid for the specific reaction considered:

$$r = r^0(\tilde{c}_A^0) f(X_A) \tag{1.10}$$

1.5.3 Mass Balance of a Plug Flow Tubular Reactor

One of the simplest models used to describe the performance of tubular reactors is the well known isothermal one-dimensional plug flow tubular reactor (PFTR) model. The mass balance of this model is for: (a) steady-state conditions, (b) a network of M simultaneously proceeding reactions and (c) a constant volumetric flow rate $\dot{V}$ (Froment and Bischoff, 1979; Levenspiel, 1999):

$$\frac{d\tilde{c}_i}{dz} = \frac{A_R}{\dot{V}} \sum_{j=1}^{M} \nu_{ij} r_j \quad i = 1, N \tag{1.11}$$

The ν_{ij} in Equation 1.11 are the elements of the stoichiometric matrix, A_R stands for the cross-sectional area of the tube and z is the axial coordinate.

With the residence time τ in a reactor section of length z:

$$\tau = \frac{A_R}{\dot{V}} z \tag{1.12}$$

the mass balance of the PFTR can be expressed also in the following manner:

$$\frac{d\tilde{c}_i}{d\tau} = \sum_{j=1}^{M} \nu_{ij} r_j \quad i = 1, N \tag{1.13}$$

The systems of ordinary differential equations (1.11) or (1.13) can be integrated numerically with the initial conditions $\tilde{c}_i^0 = \tilde{c}_i(z = 0 \; or \; \tau = 0)$ and the specific rate laws.

If only one reaction needs to be considered ($M = 1$) and the conversion of component A is chosen to be the state variable of interest, the mass balance of the PFTR can be also expressed as follows:

$$\frac{dX_A}{d\tau} = \frac{(-\nu_A) r^0 f(X_A)}{\tilde{c}_A^0} \tag{1.14}$$

Integration from 0 to the residence time corresponding to the reactor length L_R, that is, $\tau(L_R)$ and from 0 to $X_A(\tau)$ leads to the following dimensionless mass balance of the PFTR:

$$\mathrm{Da} = \int_0^{X_A(\tau)} \frac{dX_A}{f(X_A)} \tag{1.15}$$

In this equation Da is the Damköhler number (Levenspiel, 1999):

$$\mathrm{Da} = \frac{(-\nu_A) r^0}{\tilde{c}_A^0} \tau \tag{1.16}$$

which represents the ratio of the characteristic times for convection and reaction.

The dimensionless mass balance equation (1.15) can be solved analytically for various simple rate laws $f(X_A)$ providing instructive $X_A(\mathrm{Da})$ profiles.

If for example the rate of a reaction A $\rightarrow$ Products can be described by a simple first-order kinetic expression:

$$r = k\tilde{c}_A \tag{1.17}$$

the dimensionless balance provides with $r^0 = k\tilde{c}_A^0$ and $f(X_A) = 1 - X_A$:

$$X_A = 1 - e^{-\mathrm{Da}} \tag{1.18}$$

In contrast, for a second-order reaction with the rate expression:

$$r = k\tilde{c}_A^2 \tag{1.19}$$

holds:

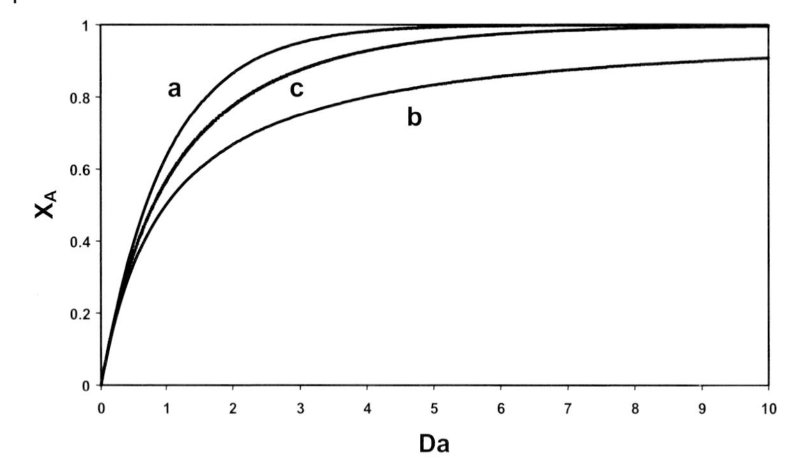

Figure 1.4 Dependence of conversion on Damköhler number for: (a) a first-order reaction (1.18), (b) a second-order reaction (1.20) and (c) a second-order reaction with two reactants and a non-stoichiometric feed composition (Equations 1.21 and 1.22, here for $\lambda = 0.5$).

$$X_A = \frac{\text{Da}}{1+\text{Da}} \tag{1.20}$$

In the case of a bimolecular reaction of the type $\nu_A A + \nu_B B \rightarrow$ Products the composition of the feed mixture is an important free parameter. This can be conveniently expressed using a stoichiometric feed ratio λ defined as follows:

$$\lambda = \frac{\nu_B \tilde{c}_A^0}{\nu_A \tilde{c}_B^0} \tag{1.21}$$

When component B is introduced in excess, that is, when $0 < \lambda < 1$, the solution of the mass balance provides:

$$X_A = \frac{e^{\text{Da}(1-\lambda)} - 1}{e^{\text{Da}(1-\lambda)} - \lambda} \tag{1.22}$$

The three different functions described by Equations 1.18, 1.20 and 1.22 are illustrated in Figure 1.4. The curves shown reveal the following two well known and important facts: (a) higher reaction orders require larger Da numbers (i.e., larger residence times) in order to reach the same conversion and (b) an excess of a reactant increases conversion of the other reactant.

1.5.4
Selectivity and Yield

In general, several reactions proceed simultaneously in a reactor. Thus, the selectivity and yield with respect to a certain desired target component D achievable in networks of parallel and series reactions are essential quantities.

The integral selectivity with respect to the desired component D, S_D, is related to the corresponding consumption of the reactant A. Considering the molar fluxes of the components at the inlet and outlet of a continuously operated reactor, S_D is defined as follows:

$$S_D = \frac{\dot{n}_D}{(\dot{n}_A^0 - \dot{n}_A)} \frac{(-\nu_A)}{\nu_D} \tag{1.23}$$

Of even more practical relevance is the yield of component D, Y_D, which is:

$$Y_D = \frac{\dot{n}_D}{\dot{n}_A^0} \frac{(-\nu_A)}{\nu_D} \tag{1.24}$$

Obviously, for the yield holds:

$$Y_D = S_D X_A \tag{1.25}$$

Let us consider a desired "reaction D" leading to the target product D:

$$A + B \rightarrow D \tag{1.26}$$

and an undesired consecutive "reaction U" leading to an undesired product U:

$$D + B \rightarrow U \tag{1.27}$$

The rates of these two reactions could be, for example, described by the following power law kinetics:

$$r_D = k_D \ \tilde{c}_A^{\alpha} \ \tilde{c}_B^{\beta_1} \tag{1.28}$$

$$r_U = k_U \ \tilde{c}_D^{\delta} \ \tilde{c}_B^{\beta_2} \tag{1.29}$$

The selectivity and the yield with respect to D depend strongly on the values of the two reaction rate constants k_D und k_U and on the reaction orders α, β_1, β_2, δ. Illustrative results assuming that all reaction orders are unity were obtained solving numerically the mass balance equations of the PFTR model (1.13) for three different ratios k_D/k_U. The courses of the $S_D(X_A)$ and $Y_D(X_A)$ curves shown in Figure 1.5 reveal the strong impact of the reaction rates. Obviously, it is very desirable to operate in the upper right ("dream") corners of these plots where all performance criteria (conversion, selectivity and yield) are unity. Obviously, this corner is closer when k_D/k_U is large.

1.6
Distributed Dosing in Packed-Bed and Membrane Reactors

In networks of parallel-series reactions optimal local reactant concentrations are essential for a high selectivity towards a certain target product. It is well known that it is advantageous to avoid back-mixing when undesired consecutive reactions can occur (e.g., Levenspiel, 1999; Fogler, 1999). This is one of the main reasons why partial hydrogenations or oxidations are performed preferentially in tubular

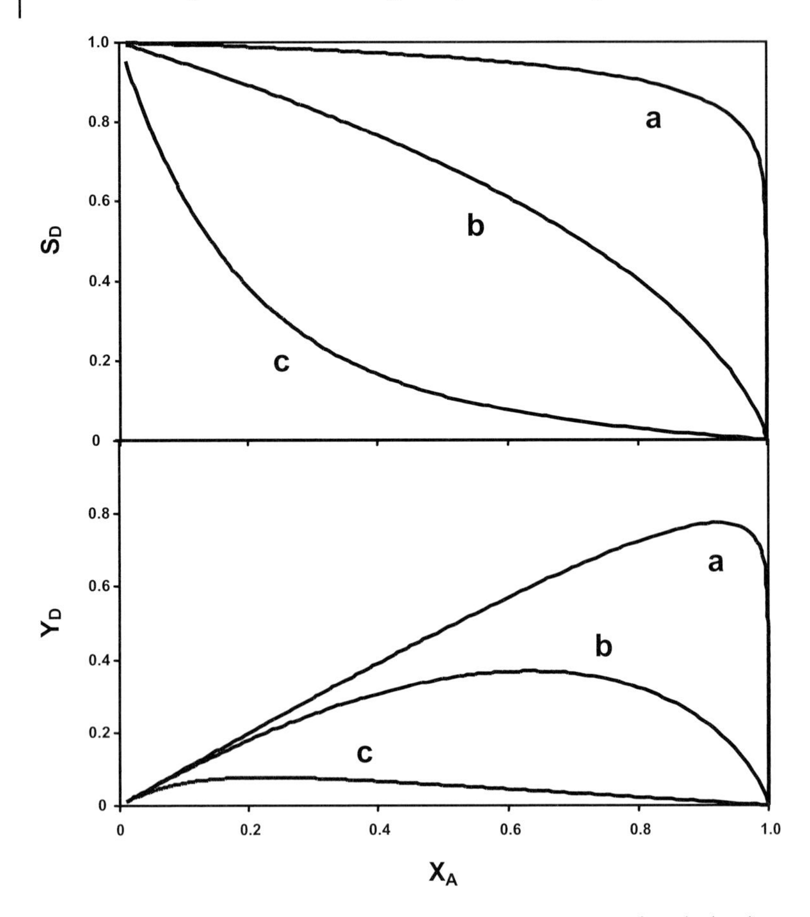

Figure 1.5 Selectivity S_D and yield Y_D as a function of conversion X_A for the two consecutive reactions A + B → D and D + B → U calculated with the PFTR model (1.11), fixing the residence time and varying the feed composition in a wide range. The reaction rates were described with Equations 1.28 and 1.29 assuming that all reaction orders are unity. Three different ratios of the rate constants of the desired and the undesired reaction were considered: (a) $k_D/k_U = 10$, (b) $k_D/k_U = 1$, (c) $k_D/k_U = 0.1$.

reactors. All reactants enter typically such tubular reactors together at the reactor inlet (co-feed mode, Mode 1 in Figure 1.6). In order to influence the reaction rates along the reactor length, essentially the temperature remains as the parameter that could be influenced. However, the realization of a defined temperature modulation in a tubular reactor is not trivial (Edgar and Himmelblau, 1989). An alternative and attractive possibility, also capable to influence the course of complex reactions in tubular reactors, is to abandon the co-feed mode and to install more complex dosing regimes. It is relatively simple to add one or several of the reactants to tubular reactors in a locally distributed manner. This approach obviously offers a large variety of options differing mainly in the positions at which the components

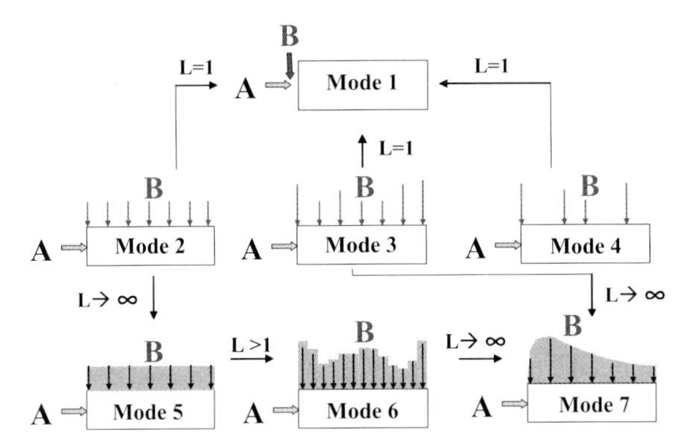

Figure 1.6 Illustration of possible dosing concepts for tubular reactors: conventional reactor (co-feed, Mode 1), possibilities of discrete dosing (Modes 2, 3, 4), possibilities of continuous dosing (Modes 5, 6, 7). L is the number of segments or stages.

are dosed. Figure 1.6 illustrates several possible scenarios differing from the conventional feeding strategy (Mode 1).

In the top row are depicted three scenarios of discrete dosing (Modes 2, 3, 4) differing in the positions and amounts of introducing a reactant B along the reactor length. If the number of discrete dosing points L is reduced to one the conventional reactor principle results. In the lower row of Figure 1.6 are illustrated schematically three concepts of dosing continuously over the reactor wall. These concepts are obviously connected with the membrane reactor Concept VI shown in Figure 1.1. Uniform dosing over one reactor segment (Mode 5), stage-wise segmented dosing (Mode 6) and the implementation of a fully continuous dosing profile (Mode 7) are possible options. The larger the number of segments L the more the concepts converge into the continuous dosing profile case (Mode 7).

1.6.1
Adjusting Local Concentrations to Enhance Selectivities

The different dosing concepts illustrated in Figure 1.6 provide different outlet compositions and, thus, performance criteria. To quantify and compare them it is instructive to introduce differential local selectivity with respect to a specific desired product D, $S_{\mathrm{D}}^{\mathrm{diff}}$, which depends on the local concentrations. Assuming that a valuable reactant A is converted the local selectivity $S_{\mathrm{D}}^{\mathrm{diff}}$ can be expressed as a function of the corresponding reaction rates of all M reactions occurring in the reaction network as follows:

$$S_{\mathrm{D}}^{\mathrm{diff}} = \frac{\sum_{j=1}^{M} \nu_{\mathrm{D},j} \cdot r_j}{\sum_{j=1}^{M} (-\nu_{\mathrm{A},j}) \cdot r_j} \tag{1.30}$$

To evaluate typical trends it is again instructive to consider the simple reaction scheme of two consecutive reactions introduced above (1.26 and 1.27) and the power law rate expressions given by Equations 1.28 and 1.29. In such a case for the differential selectivity with respect to D follows:

$$S_{\mathrm{D}}^{\mathrm{diff}} = \frac{r_{\mathrm{D}} - r_{\mathrm{U}}}{r_{\mathrm{D}}} = 1 - \frac{k_{\mathrm{U}}(T)\tilde{c}_{\mathrm{D}}^{\delta}}{k_{\mathrm{D}}(T)\tilde{c}_{\mathrm{A}}^{\alpha}} \tilde{c}_{\mathrm{B}}^{\beta_2 - \beta_1} \qquad (1.31)$$

Equation 1.31 clearly reveals that for this reaction scheme and the kinetics assumed an improved differential selectivity $S_{\mathrm{D}}^{\mathrm{diff}}$ can be achieved when:

- The reactions take please at a temperature which minimizes $k_{\mathrm{U}}/k_{\mathrm{D}}$,
- The concentration D is kept low (favoring a removal of D, for example, with a membrane reactor of the extractor type shown in Figure 1.1 as Concept V),
- The concentration of A is kept high (i.e., conversion should be restricted and back-mixing avoided; the former fact leading to concepts with a recycle of A, the latter fact favoring tubular reactors compared to stirred tanks),
- For $\beta_2 < \beta_1$ a high concentration of B is advantageous, which favors a concentrated feeding of this reactant at the reactor inlet,
- For $\beta_2 > \beta_1$ a low concentration of B is advantageous, which can be realized by a distributed feeding of this reactant.

Since these trends are specific for the reaction scheme and the rate equations considered no generalization is possible. However, a detailed inspection of the specific differential selectivities allows drawing similar conclusions for other cases.

Of special relevance for the example of partial oxidation reactions of hydrocarbons and for the chapters of this book is the following fact. Typically desired oxidation reactions leading to the intermediate products of interest possess lower reaction orders with respect to oxygen compared to the undesired total oxidation reactions leading to carbon dioxide and water (Mezaki and Inoue, 1991). Considering Equation 1.31 and the final conclusion listed above leads to the hypothesis that a low oxygen concentration achievable by implementing a spatial distribution can be beneficial for the selectivity with respect to a target component. Such a regime can be realized by the distributor type of membrane reactor shown in Figure 1.1 as Concept VI.

1.6.2
Optimization of Dosing Profiles

Knowing the structure of the reaction network of interest and the concrete concentration dependences of the reaction rates allows determining specific dosing profiles which are optimal for a certain reactor configuration. Only for a limited number of cases characterized by a small number of rate expressions and simple reactor models can analytical results be generated (Hamel *et al.*, 2003; Thomas, Pushpavanam, and Seidel-Morgenstern, 2004).

Using a modified PFTR model the different dosing modes illustrated in Figure 1.6 were analyzed. Mode 1 required just the direct application of Equation 1.11 with the boundary condition describing the co-feed mode of components A and B. Modes 2, 3 and 4 can be simulated using Equation 1.11 in a stage-wise manner applying for each segment the boundary conditions corresponding to the specific discrete dosing approach and the series connection of the segments. To describe dosing over the reactor walls (Modes 5, 6, 7) the mass balance equation of the PFTR has to be extended by an additional transport term as follows:

$$\frac{d\tilde{c}_i}{dz} = \frac{A_R}{\dot{V}} \sum_{j=1}^{M} \nu_{ij} r_j + \frac{P_R}{\dot{V}} J_i \quad i = 1, N \tag{1.32}$$

In this equation P_R is the perimeter of the tube and the J_i are the molar flux densities of the transport of component i through the reactor wall. Hereby, below-specific uniform dosing profiles (constant J_i) were assumed for each reactor segment.

To illustrate the principle and potential of distributed dosing selected results of a case study are summarized. The following three reactions of a parallel-consecutive reaction scheme were considered, converting the two reactants A and B into a desired product D and an undesired product U:

$$A + B \rightarrow D \tag{1.33}$$

$$A + 2B \rightarrow U \tag{1.34}$$

$$D + B \rightarrow U \tag{1.35}$$

The rates of these three reactions were described by the following power law kinetics:

$$r_D = k_D \ \tilde{c}_A^{\alpha_D} \ \tilde{c}_B^{\beta_D} \tag{1.36}$$

$$r_{U1} = k_{U1} \ \tilde{c}_A^{\alpha_{U1}} \ \tilde{c}_B^{\beta_{U1}} \tag{1.37}$$

$$r_{U2} = k_{U2} \ \tilde{c}_D^{\alpha_{U2}} \ \tilde{c}_B^{\beta_{U2}} \tag{1.38}$$

The reactor model given with Equation 1.32 was solved numerically for selected parameters of these rate expressions. With a sequential quadratic programming (SQP) optimization algorithm (Press *et al.*, 1992) the optimal amounts of component B that should be dosed were determined with the selected objective of maximizing the molar fraction of component D at the reactor outlet. The discrete dosing Mode 3 assuming equidistant feeding positions and the continuous dosing Mode 6 were compared assuming segments of identical length. In a series of optimizations in both cases the numbers of segments L were fixed to the following three values: 1, 3 or 10. Hereby the results of Mode 3 for $L = 1$ correspond to the conventional co-feed fixed-bed reactor (Mode 1). The specific degrees of freedom specified were the L molar flows of component B dosed, $\dot{n}_D^{dosed}$, at the inlet of each segment (Mode 3) or over the segment wall (Mode 6, calculated from the optimal J_D and the wall area).

Below for the purpose of illustration, selected results are presented in Figures 1.7 and 1.8. The calculations were done assuming that the reaction orders are $\alpha_D = \alpha_{U1} = \alpha_{U2} = \beta_D = 1$ and $\beta_{U1} = \beta_{U2} = 2$. This implies that in the undesired reactions (1.37) and (1.38) the order with respect to D is higher then in the desired reaction (1.36). The rate constants k_j were assumed to be identical. A stream of 1 mol/s of pure A was introduced at the inlet of the first segment of a reactor possessing an overall volume $V_R = 0.01\,m^3$. The figures show over the reduced reactor length the dosed amounts, the total flow rates, the local molar fractions of the dosed component B and those of the other three components, including the optimized molar fraction of D, x_D ($x_i = \dot{n}_i/\dot{n}_{tot}$).

The results obtained for this specific case provide the following conclusions:

- For both modes decreasing dosing profiles are found to be optimal, that is, the largest amounts are dosed in the segments close to the reactor inlet and lower amounts are dosed into the following segments. Some B is also dosed into the last segment.
- There is for both modes an increasing amount of D found at the reactor outlet for increasing segment numbers L.
- The diluted dosing of B leads in comparison to the conventional co-feed mode (discrete dosing, $L = 1$, Mode 1) to larger molar amounts of D.
- The continuous dosing (Mode 6, e.g., applied in a membrane reactor) outperforms for the same segment numbers the discrete dosing (Mode 3) as indicated by larger x_D at the reactor outlet.
- For $L = 1$ there is a significant performance increase of Mode 6 compared to the conventional co-feed operation.
- Already for $L = 3$ the potential of Mode 6 seems to be reached. Further segmentation does not lead to significant further enhancement in x_D.

The results of more systematic theoretical studies explaining in more detail the significance of the reaction orders regarding the selection of the component that should be dosed and regarding the shapes of suitable dosing profiles are available (Lu *et al.*, 1997a, 1997b, 1997c; Hamel *et al.*, 2003; Thomas, Pushpavanam, and Seidel-Morgenstern, 2004).

Examples for the application of the above theoretical considerations in concrete case studies are given in the next chapters of this book. As per (Kuerten *et al.*, 2004), limits of the above-used simplified one-dimensional isothermal membrane reactor model are also discussed.

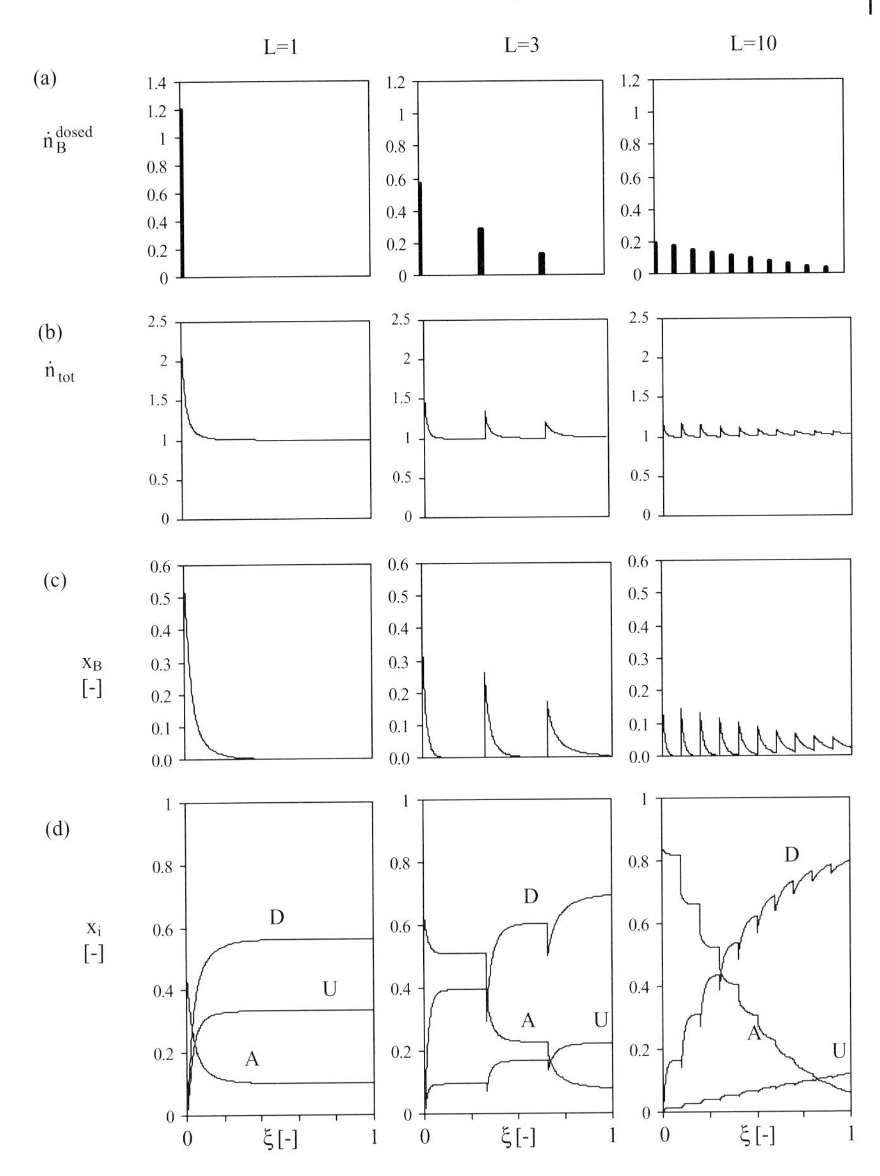

Figure 1.7 Optimized dosed amounts of B, local total molar fluxes and local molar fractions over the reduced reactor length for the discrete dosing Mode 3 and different segment numbers. Kinetic parameters: $\alpha_D = \alpha_{U1} = \alpha_{U2} = \beta_D = 1$, $\beta_{U1} = \beta_{U2} = 2$, $k_D = k_{U2} = k_{U1} = 10^4\,\mathrm{mol/(s{\cdot}m^3)}$.

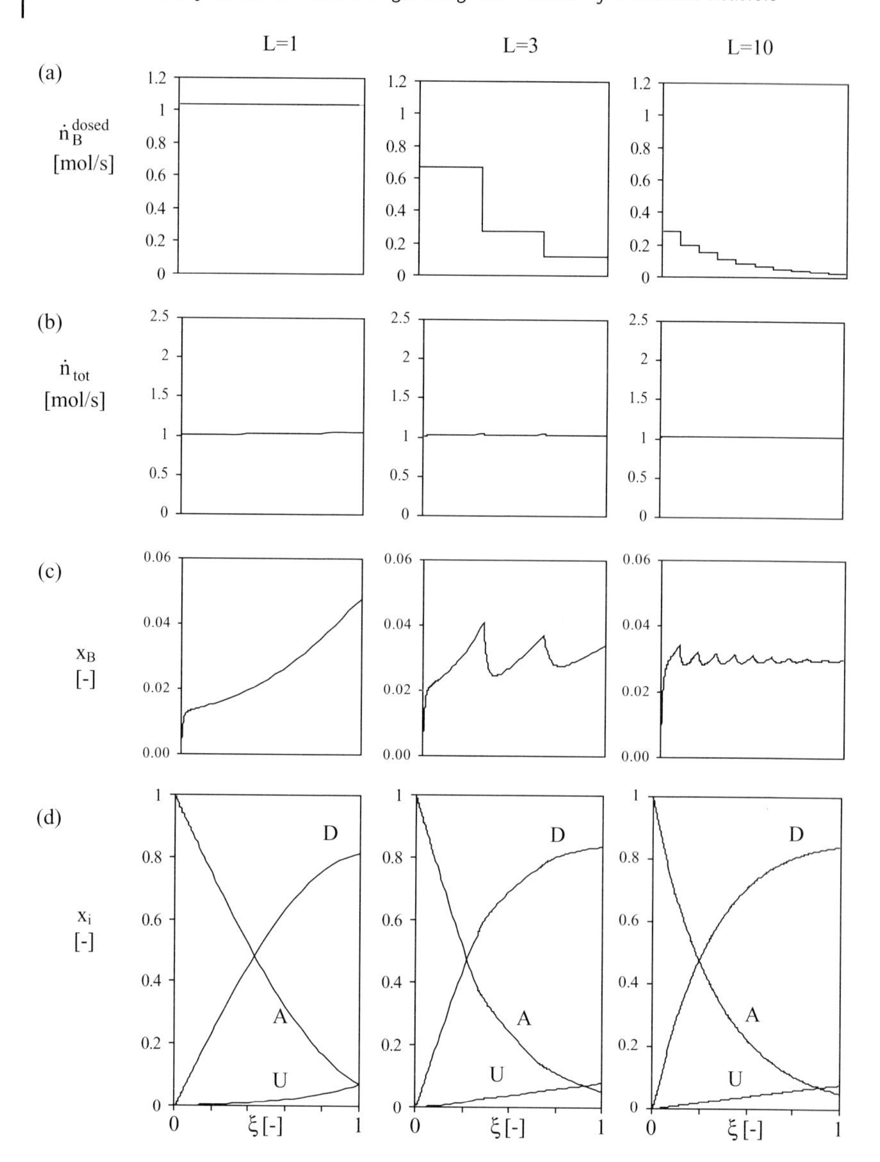

Figure 1.8 Optimized dosed amounts of B, local total molar fluxes and local molar fractions over the reduced reactor length for the continuous dosing Mode 6 and different segment numbers. Kinetic parameters: $\alpha_D = \alpha_{U1} = \alpha_{U2} = \beta_D = 1$, $\beta_{U1} = \beta_{U2} = 2$, $k_D = k_{U2} = k_{U1} = 10^4\,\text{mol}/(\text{s}\cdot\text{m}^3)$.

1.7 Kinetic Compatibility in Membrane Reactors

In order to achieve significant effects of membranes introduced into a reactor compared to conventional reactor operation, there should be certain compatibility between the fluxes that pass the membrane and the amounts consumed or produced during the chemical reactions. The specific amounts related to the simultaneous occurrence of M chemical reactions can be expressed based on Equation 1.1 as follows:

$$\left.\frac{dn_i}{dt}\right|_{\text{Reaction}} = \text{Scale}\sum_{j=1}^{M} \nu_{ij} r_{\text{Scale},j} \quad i = 1, N \tag{1.39}$$

As mentioned before different scales might be appropriate to quantify the reaction rates. Regarding the transport through membranes usually the membrane area A_M is the appropriate scaling parameter. The molar flux of a component i through a membrane can be expressed as:

$$\dot{n}_i|_{\text{Membrane}} = A_M\ J_i \quad i = 1, N \tag{1.40}$$

In the above J_i designates, as in Equation 1.32, the molar flux density of component i.

Provided there is information available regarding the amounts transformed by the reactions and the amounts that could be transported through the membranes (based on Equations 1.39 and 1.40, respectively), several important questions could be answered in early development stages as, for example,: "how much membrane area must be provided per scale of the reaction zone?" and "is a more detailed investigation of coupling reaction and mass transfer through a specific membrane justified?".

Following this approach recently a useful estimation was given by (van de Graaf *et al.*, 1999). Regarding the productivity of reactions, achievable space time yields (STY) of currently operated catalytic reactors were considered. Concerning this quantity currently the following "window of reality" holds:

$$\text{STY} = \frac{\dot{n}_{\text{Prod}}}{V_R} \approx 1 - 10 \quad \frac{\text{mol}}{\text{m}^3\text{s}} \tag{1.41}$$

The achievable fluxes through membranes, J, were designated by (van de Graaf *et al.*, 1999) as area time yields (ATY, in mol/m^{2}s). Figure 1.9 provides an estimation of the current state regarding the possibility of matching the two processes. For the wide range of considered membranes, the required ratios of membrane areas to reactor volumes (A_M/V_R) are between 10 and 100 m^{-1}. These values allow estimating that the diameter of applicable cylindrical tubular reactors should be in a range between 0.04 and 0.4 m. This appears to be a reasonable range for industrial applications indicating that a matching of the two processes under consideration is achievable with currently available membranes.

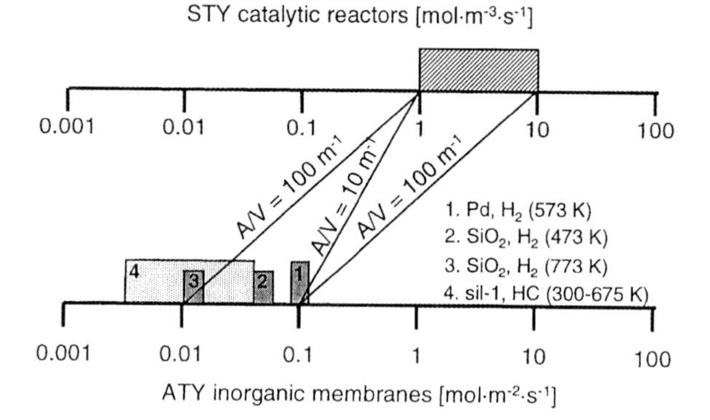

Figure 1.9 Comparison of the space time yield (STY) of catalytic reactors with the area time yield (ATY) of several inorganic membranes (reprinted from [van de Graaf *et al.*, 1999], with permission).

1.8 Current Status of Membrane Reactors of the Distributor Type

There are several profound theoretical and experimental studies at the laboratory scale available which focus on the application of various configurations of membrane reactors as a reactant distributor in order to improve selectivity–conversion performances.

In particular several industrially relevant partial oxidations were investigated. Examples include the oxidative coupling of methane (Coronas, Menedez, and Santamaria, 1994), the oxidative dehydrogenation of propane (Alonso *et al.*, 1999), butane (Tellez, Menedez, and Santamaria, 1997) and methanol (Diakov and Varma, 2003, 2004), the epoxidation of ethylene (Al-Juaied, Lafarga, and Varma, 2001) and the oxidation of butane to maleic anhydride (Mallada, Menedez, and Santamaria, 2000). Specific aspects of membrane reactors related to carrying out the oxidative dehydrogenation of ethane to ethylene, which are described and studied in detail in this book, were investigated by (Coronas, Menedez, and Santamaria, 1995; Tonkovich *et al.*, 1996).

There appears to be potential in using membrane reactors of the distributor type also for other types of reaction networks. Another promising field can be for example selective hydrogenations. The hydrogenation of acrolein to allyl alcohol was studied by (Hamel *et al.*, 2005).

All the studies mentioned were done exclusively at the laboratory or pilot scale. They focused on high-temperature reactions and applied different types of ceramic membranes. Currently there are no industrial applications applying a membrane reactor of the distributor type on a larger scale.

To further promote the promising concept systematic studies are required quantifying both the reaction and transport processes and describing in more detail the processes occurring in such membrane reactors. Hereby, various options applicable with respect to the types of membranes and the reactor principles should be considered and compared. This book contains various contributions to the mentioned problems.

The following chapter summarizes theoretical concepts required to model membrane reactors.

Notation used in this Chapter

A_{Cat}	m^2	catalyst surface area
A_M	m^2	membrane surface area
A_R	m^2	cross section area of a tubular reactor
Y_i	%	yield with respect to component i
J_i	$mol \cdot s^{-1} \cdot m^{-2}$	molar flux density of component i
k_j	$mol \cdot s^{-1} \cdot m^{-3}$	reaction rate constant
L		number of reactor segments
L_R	m	length of a tubular reactor or reactor segment
M	–	number of reactions
M_{Cat}	kg	catalyst mass
$\dot{n}$	mol/s	molar flux
P_R	m	perimeter of a tubular reactor
r	$mol \cdot s^{-1} \cdot m^{-3}$	rate of reaction
S_i	%	integral selectivity with respect to component i
S_i^{diff}	%	differential selectivity with respect to component i
T	K	temperature
V_R	m^3	reactor volume
$\dot{V}$	mol/s	volumetric flowrate
X_i	%	conversion of reactant i
x_i	–	molar fraction of component i
z	m	axial coordinate of tube

Greek Symbols

α	–	reaction order
β	–	reaction order
δ	–	reaction order
λ	–	stoichiometric feed ratio
ν	–	stoichiometric coefficient
ξ	–	non-dimensional axial reactor length, $\xi = z/L_R$
τ	s	residence time

Superscripts and Subscripts

i	component
j	reaction

tot	total
0	initial or inlet state

Abbreviations

CMR	Catalytic membrane reactor
PBR	Packed bed reactor
PBMR	Packed bed membrane reactor

References

Alonso, M.J., Julbe, A., Farrusseng, D., Menendez, M., and Santamaria, J. (1999) Oxidative dehydrogenation of propane on V/Al_2O_3 catalytic membranes. Effect of the type of membrane and reactant feed configuration. *Chem. Eng. Sci.*, **54**, 1265–1272.

Al-Juaied, M.A., Lafarga, D., and Varma, A. (2001) Ethylene epoxidation in a catalytic packed-bed membrane reactor: experiments and model. *Chem. Eng. Sci.*, **56**, 395–402.

Bhaduri, S., and Mukesh, D. (2000) *Homogenoeus Catalysis*, Wiley-VCH Verlag GmbH, Weinheim.

Bhave, R.R. (ed.) (1991) *Inorganic Membranes: Synthesis, Characteristics and Applications*, Reinhold, New York.

Bird, B.B., Stewart, W.E., and Lightfoot, E.N. (2002) *Transport Phenomena*, John Wiley & Sons, Inc., New York.

Bouwmeester, H.J.M. (2003) Dense ceramic membranes for methane conversion. *Catal. Today*, **82**, 141.

Burggraaf, A.J., and Cot, L. (eds) (1996) *Fundamentals of Inorganic Membrane Science and Technology*, Elsevier.

Cheyran, M., and Mehaaia, A. (1986) Membrane, and bioreactors, in *Membrane Separation in Biotechnology* (ed. W.C. McGregor), Marcel Dekker, New York, p. 255.

Coronas, J., Menendez, M., and Santamaria, J. (1994) Methane oxidative coupling using porous ceramic membrane reactors – II. Reaction studies. *Chem. Eng. Sci.*, **49**, 2015–2025.

Coronas, J., Menendez, M., and Santamaria, J. (1995) Use of a ceramic membrane reactor for the oxidative dehydrogenation of ethane to ethylene and higher hydrocarbons. *Ind. Eng. Chem. Res.*, **34**, 4229–4234.

Diakov, V., and Varma, A. (2003) Methanol oxidative dehydrogenation in a packed-bed membrane reactor: yield optimization experiments and model. *Chem. Eng. Sci.*, **58**, 801–807.

Diakov, V., and Varma, A. (2004) Optimal feed distribution in a packed-bed membrane reactor: the case of methanol oxidative dehydrogenation. *Ind. Eng. Chem. Res.*, **43**, 309–314.

Dittmeyer, R., and Caro, J. (2008) Catalytic membrane reactors, in *Handbook of Heterogeneous Catalysis* (eds G. Ertl, H. Knözinger, F. Schüth, and J., Weitkamp), Wiley-VCH Verlag GmbH, Weinheim, pp. 2198–2248.

Dittmeyer, R., and Reif, M. (2003) Porous, catalytically active ceramic membranes for gas–liquid reactions: a comparison between catalytic diffuser and forced through flow concept. *Catal. Today*, **82**, 3–14.

Dittmeyer, R., Höllein, V., and Daub, K. (2001) Membrane reactors for hydrogenation and dehydrogenation processes based on supported palladium. *J. Mol. Catal. A: Chem.*, **173**, 135–184.

Dixon, A.G. (2003) Recent research in catalytic inorganic membrane reactors. *Int. J. Chem. React. Eng.*, **1**, R6.

Edgar, T.F., and Himmelblau, D.M. (1989) *Optimization of Chemical Processes*, Mc Graw-Hill.

Eigenberger, G. (1997) Catalytic fixed-bed reactors, in *Handbook of Heterogeneous*

Catalysis, vol. 3 (eds G. Ertl, H. Knözinger, and J. Weitkamp), Wiley-VCH Verlag GmbH, Weinheim, pp. 1424–1487.

Eigenberger, G., Kolios, G., and Nieken, U. (2007) Thermal pattern formation and process intensification in chemical reaction engineering. *Chem. Eng. Sci.*, **62**, 4825–4841.

Ertl, G., Knözinger, H., Schüth, F., Weitkamp, J. (eds) (2008) *Handbook of Heterogeneous Catalysis*, vol. 8, Wiley-VCH Verlag GmbH, Weinheim.

Fogler, H.S. (1999) *Elements of Chemical Reaction Engineering*, 3rd edn, Prentice Hall, Upper Saddle River, New Jersey.

Froment, G., and Bischoff, K.B. (1979) *Chemical Reactor Analysis and Design*, John Wiley & Sons, Inc., New York.

Gobina, E., Hou, K., and Hughes, R. (1995) Ethane dehydrogenation in a catalytic reactor coupled with a retive sweep gas. *Chem. Eng. Sci.*, **50**, 2311–2319.

Gryaznov, V.M., Smirnov, V.S., and Mischenko, A.P. (1974) Catalytic reactor for coupled chemical reactions, GB patent 1342869.

Haber, J. (1997) Oxidation of hydrocarbons, in *Handbook of Heterogeneous Catalysis*, vol. 5 (eds G. Ertl, H. Knözinger, and J. Weitkamp), Wiley-VCH Verlag GmbH, Weinheim, pp. 2253–2274.

Hamel, C., Thomas, S., Schädlich, K., and Seidel-Morgenstern, A. (2003) Theoretical analysis of reactant dosing concepts to perform parallel-series reactions. *Chem. Eng. Sci.*, **58**, 4483–4492.

Hamel, C., Bron, M., Claus, P., and Seidel-Morgenstern, A. (2005) Experimental and model based study of the hydrogenation of acrolein to allyl alcohol. *Int. J. Chem. React. Eng.*, **3**, A10.

Hodnett, B.K. (2000) *Heterogeneous Catalytic Oxidation: Fundamental and Technological Aspects of the Selective and Total Oxidation of Organic Compounds*, John Wiley & Sons (Asia) Pte Ltd.

Itoh, N., Shindo, Y., Haraya, K., and Hakuta, T. (1988) A membrane reactor using microporous glass for shifting equilibrium of cyclohexane dehydrogenation. *J. Chem. Eng. Japan*, **21**, 399–404.

Julbe, A., Farrusseng, D., and Guizard, C. (2001) Porous ceramic membranes for catalytic reactors – overview and new ideas. *J. Membr. Sci.*, **181**, 3–20.

Kikuchi, E. (1997) Hydrogen-permselective membrane reactors. *CATTECH*, **1**, 67.

Kölsch, P., Smekal, Q., Noack, M., Schäfer, R., and Caro, J. (2002) Partial oxidation of propane to acrolein in a membrane reactor-experimental data and computer simulation. *Chem. Comm.*, **3**, 465–470.

Kragl, U., and Dreisbach, C. (2002) Membrane reactors in homogeneous catalysis, in *Applied Homogeneous Catalysis with Organometallic Compounds*, 2nd edn (eds B. Cornils and W.A. Herrmann), Wiley-VCH Verlag GmbH, Weinheim, p. 941.

Krishna, R. (2002) Reactive separations: more ways to skin a cat. *Chem. Eng. Sci.*, **57**, 1491–1504.

Kuerten, U., van Sint Annaland, M., and Kuipers, J.A.M. (2004) Oxygen distribution in packed bed membrane reactors for partial oxidation systems and its effect on product selectivity. *Int. J. Chem. React. Eng.*, **2**, A24.

Levenspiel, O. (1999) *Chemical Reaction Engineering*, 3rd edn, John Wiley & Sons, Inc., New York.

Li, N.N. (2008) *Advanced Membrane Technology and Applications*, Wiley-VCH Verlag GmbH.

Lu, Y.L., Dixon, A.G., Moder, W.R., and Ma, Y.H. (1997a) Analysis and optimization of cross-flow reactors with staged feed policies – isothermal operation with parallel-series, irreversible reaction systems. *Chem. Eng. Sci.*, **52**, 1349–1363.

Lu, Y.L., Dixon, A.G., Moder, W.R., and Ma, Y.H. (1997b) Analysis and optimization of cross-flow reactors with distributed reactant feed and product removal. *Catal. Today*, **35**, 443–450.

Lu, Y.L., Dixon, A.G., Moder, W.R., and Ma, Y.H. (1997c) Analysis and optimization of cross-flow reactors for oxidative coupling of methane. *Ind. Eng. Chem. Res.*, **36**, 559–567.

Mallada, R., Menendez, M., and Santamaria, J. (2000) Use of membrane reactors for the oxidation of butane to maleic anhydride under high butane concentrations. *Catal. Today*, **56**, 191–197.

Mason, E.A., and Malinauskas, A.P. (1983) *Gas Transport in Porous Media: The Dusty Gas Model*, Elsevier, Amsterdam.

Matros, Y.S., and Busimovic, G.A. (1996) Catalytic processes under unsteady state conditions, *Catal. Rev. Sci. Eng.*, **38**, 1–68.

Mezaki, R., and Inoue, H. (1991) *Rate Equations of Solid-Catalyzed Reactions*, University of Tokyo Press.

Miachon, S., Perz, V., Crehan, G., Torp, E., Raeder, H., Bredesen, R., and Dalmon, J.-A. (2003) Comparison of a contactor catalytic membrane reactor with a conventional reactor: example of wet air oxidation. *Catal. Today*, **82**, 75–81.

Missen, R.W., Mims, C.A., and Saville, B.A. (1999) *Introduction to Chemical Reaction Engineering and Kinetics*, John Wiley & Sons, Inc., New York.

Moulijn, J.A., Makkee, M., and van Diepen, A.E. (2001) *Chemical Process Technology*, John Wiley & Sons, Ltd, Chichester.

Ohlrogge, K., and Ebert, K. (2006) *Membranen: Grundlagen, Verfahren Und Industrielle Anwendungen*, Wiley-VCH Verlag GmbH, Weinheim. ISBN: 3-527-30979-9.

Peinemann, K.-V., and Pereira Nunes, S. (2007) *Membrane Technology*, Wiley-VCH Verlag GmbH.

Pfefferie, W.C. (1966) U.S. Patent App. 3290406.

Press, W., Flannery, B., Teukolsky, S., and Vetterling, W.T. (1992) *Numerical Recipes*, Cambridge University Press.

Proceedings of the International Congresses on Catalysis in Membrane Reactors. a) Villeurbanne (1994), b) Moscow (1996), c) Copenhagen (1998), d) Zaragoza (Catalysis Today, 2000, 56), e) Dalian (Catalysis Today, 2003, 82), f) Lahnstein (Catalysis Today, 2005,104), g) Cetraro (11–14 September, 2005), h) Kolkata (18–21 December, 2007), i) Lyon (28 June–2 July, 2009).

Sahimi, F. (1995) *Flow and Transport in Porous Media and Fractured Rock: from Classical Methods to Modern Approaches*, Wiley-VCH Verlag GmbH, Weinheim.

Sanchez Marcano, J.G., and Tsotsis, T.T. (2002) *Catalytic Membranes and Membrane Reactor*, Wiley-VCH Verlag GmbH, Weinheim.

Saracco, G., Neomagus, H.W.J.P., Versteeg, G.F., and van Swaaij, W.P.M. (1999) High-temperature membrane reactors: potential and problems. *Chem. Eng. Sci.*, **54**, 1997–2017.

Schäfer, R., Noack, M., Kölsch, P., Stöhr, M., and Caro, J. (2003) Comparison of different catalysts in the membrane-supported dehydrogenation of propane. *Catal. Today*, **82**, 15–23.

Schmidt, L. (1997) *The Engineering of Chemical Reactions*, Oxford University Press, Oxford.

Schramm, O., and Seidel-Morgenstern, A. (1999) Comparing porous and dense membranes for the application in membrane reactors. *Chem. Eng. Sci.*, **54**, 1447–1453.

Sheintuch, M., and Nekhamkina, O. (2005) The asymptotes of loop rectors. *AIChE J.*, **52**, 224–234.

Silveston, P.L. (1998) *Composition Modulation of Catatlytic Reactors*, Gordon and Breach, Amsterdam.

Sundmacher, K., and Kienle, A. (eds) (2003) *Reactive Distillation*, Wiley-VCH Verlag GmbH.

Sundmacher, K., Kienle, A., and Seidel-Morgenstern, A. (eds) (2005) *Integrated Chemical Processes*, Wiley-VCH Verlag GmbH, Weinheim.

Seidel-Morgenstern, A. (2005) Analysis and experimetnal investigation of catalytic membrane reactors, in *Integrated Chemical Processes* (eds K. Sundmacher, A. Kienle, and A. Seidel-Morgenstern), Wiley-VCH Verlag GmbH, Weinheim, pp. 359–390.

Tellez, C., Menendez, M., and Santamaria, J. (1997) Oxidative dehydrogenation of butane using membrane reactors. *AIChE J.*, **43**, 777–784.

Thomas, S., Schäfer, R., Caro, J., and Seidel-Morgenstern, A. (2001) Investigation of mass transfer through inorganic membranes with several layers. *Catal. Today*, **67**, 205–216.

Thomas, S., Pushpavanam, S., and Seidel-Morgenstern, A. (2004) Performance improvements of parallel–series reactions in tubular reactors using reactant dosing concepts. *Ind. Eng. Chem. Res.*, **43**, 969–979.

Tonkovich, A.L.Y., Zilka, J.L., Jimenez, D.M., Roberts, G.L., and Cox, J.L. (1996) Experimental investigations of inorganic membrane reactors: a distributed feed approach for partial oxidation reactions. *Chem. Eng. Sci.*, **51**, 789–806.

van de Graaf, J.M., Zwiep, M., Kapteijn, F., and Moulijn, J.A. (1999) Application of a silicalite-1 membrane reactor in metathesis reactions. *Appl. Catal. A: Gen.*, **178**, 225–241.

Verweij, H. (2003) Ceramic membranes: morphology and transport. *J. Mater. Sci.*, **38**, 4677–4695.

Vospernik, M., Pintar, A., Bercic, G., and Levec, J. (2003) Experimental verification of ceramic membrane potentials for supporting three-phase catalytic reactions. *J. Membr. Sci.*, **223**, 157–169.

Wesselingh, J.A., and Krishna, R. (2000) *Mass Transfer in Multicomponent Mixtures*, Delft University Press.

Zaspalis, V.T., and Burggraaf, A.J. (1991) *Inorganic Membranes: Synthesis, Characteristics and Applications* (ed. R.R. Bhave), Reinhold, New York.

Ziaka, Z.D., Minet, R.G., and Tsotsis, T.T. (1993) A high temperature catalytic membrane reactor for propane dehydrogenation. *J. Membr. Sci.*, **77**, 221–232.

2
Modeling of Membrane Reactors

Michael Mangold, Jürgen Schmidt, Lutz Tobiska, and Evangelos Tsotsas

2.1
Introduction

As indicated in Chapter 1, different configurations of membrane reactors can be obtained depending on:

- Placement of the catalyst (tube side, shell side, on the membrane, in a separate layer),
- Motion, or not, of particulate catalyst (packed beds, fluidized beds),
- Combination, or not, with electrochemical phenomena.

The present chapter outlines a frame for modeling of mass and heat transfer that is–in principle–common to all these configurations and accounts for both gas-filled and porous domains (Section 2.2). After an introduction to key kinetic phenomena (Section 2.3), opportunities for reducing the number of spatial coordinates in the geometrical regions involved are discussed (Section 2.4). Since the systems of model equations are, even after appropriate reduction, not easy to handle, solvability, discretization techniques and fast solvers for large systems of algebraic equations are treated in Section 2.5. Finally, specific tools used for solving the model equations and, in some cases, also for composing models from fundamental elements and interactions are presented in Section 2.6.

2.2
Momentum, Mass and Heat Balances

Both continuum and discrete (cell, pore-network, etc.) models are common in reaction engineering (Elnashaie and Elshishini, 1993; Tsotsas, 1991). From those two model families, only continuum models are used here in order to describe heat and mass transfer in membrane reactors. Continuum models are universally applicable down to the smallest scale which are significant for packed beds and for the operation of micro-reactors.

Membrane Reactors: Distributing Reactants to Improve Selectivity and Yield.
Edited by Andreas Seidel-Morgenstern

ISBN: 978-3-527-32039-4

Since fluid phases and porous regions (membranes, catalyst layers) must be treated simultaneously, multiphase approaches are necessary. In the porous regions themselves, it is possible to distinguish between the fluid and the particulate or solid phases. Several authors have presented calculations with full local resolution of field quantities for both ordered and random packed beds (Dixon and Nijemeisland, 2001; Dixon, Nijemeisland, and Stitt, 2003; Manjhi *et al.*, 2006; Tota *et al.*, 2007). However, such solutions are computationally expensive and, hence, seriously limited in regard of the number of particles that can be considered. Therefore, we decided to treat the packed-bed parts of packed-bed reactors (PBR) and packed-bed membrane reactors (PBMR) in a quasi-homogeneous way. Consequently, particles and the fluid are summarized to one equivalent phase in every porous region, which is then characterized by porosity (at small tube-to-particle diameter ratios: local porosity) and effective transport coefficients.

Irrespective of this approximation, the complete equations for the conservation of momentum, total mass, component mass and energy must be considered in every real (fluid) or assumed (porous) phase. This gives rise to a system of coupled nonlinear partial differential equations of second order, which consists of:

- The continuity equation,
- Three equations for momentum transport, for example, the Navier-Stokes equations,
- The energy balance,
- $(n-1)$ species balance equations.

Solution of this system provides the fields of flow velocity (with the components v_r, v_φ, v_z, in the case of cylindrical coordinates), pressure p, temperature T, and $(n-1)$ mass fractions y_i for n species components. According to (Bird, Stewart, and Lightfoot, 2002) the differential equations for fluid-filled regions of the membrane reactor are summarized in tensor form in Table 2.1.

The following should be noticed in connection with Table 2.1:

Density ρ and enthalpy h can, in general, be obtained from state equations. Here, ideal gas behavior is assumed for all fluid phases. In ideal gases, partial molar volume $\tilde{v}_i^*$ and molar volume $\tilde{v}_i$ of every species are equal to the molar volume of the mixture. It is:

$$\tilde{v}_i^* = \tilde{v}_i = \tilde{v} = \frac{1}{\tilde{c}}, \tag{2.1}$$

$$\rho = \sum_{i=1}^{n} \rho_i^* = \sum_{i=1}^{n} y_i \rho = \frac{p}{\tilde{R}T} \sum_{i=1}^{n} \tilde{M}_i \tilde{y}_i \tag{2.2}$$

Respective relationships for enthalpy are:

$$\tilde{h}_i^* = \tilde{h}_i = \tilde{M}_i h_i, \tag{2.3}$$

$$h = \sum h_i y_i = \sum y_i \int_{T_{ref}}^{T} c_p(T)\,dT,\ h_{i,ref}(T_{ref}) = 0. \tag{2.4}$$

Equations 2.7–2.10 are special cases of the general balance:

Table 2.1 Model equations for the fluid zones, independent from the choice of the coordinate system.

Conservation of total mass

$$\frac{\partial \rho}{\partial t} + \nabla \cdot (\rho \vec{v}) = 0 \tag{2.7}$$

Momentum conservation equations

$$\frac{\partial (\rho \vec{v})}{\partial t} + \nabla \cdot (\rho \vec{v} \vec{v}) + \nabla p - \nabla \cdot (\bar{\bar{\tau}}) - \rho \vec{g} = 0 \tag{2.8}$$

Transport equations for species

$$\frac{\partial (\rho y_i)}{\partial t} + \nabla \cdot (\rho \vec{v} y_i) + \nabla \cdot \vec{\dot{m}}_i^{\text{Diff}} = 0 \tag{2.9}$$

$$\rho \frac{\partial y_i}{\partial t} + \rho \vec{v} \nabla y_i + \nabla \cdot \vec{\dot{m}}_i^{\text{Diff}} = 0 \tag{2.9a}$$

Energy conservation equation

$$\frac{\partial (\rho e)}{\partial t} + \nabla \cdot [\vec{v}(\rho e + p)] + \nabla \cdot \left(\dot{q} + \sum_{i=1}^{n} h_i \vec{\dot{m}}_i^{\text{Diff}} \right) - \nabla (\bar{\bar{\tau}} \cdot \vec{v}) = 0 \tag{2.10}$$

$$c_p \rho \frac{\partial T}{\partial t} + c_p \rho \vec{v} \nabla T + \nabla \cdot \left(\dot{q} + \sum_{i=1}^{n} h_i \vec{\dot{m}}_i^{\text{Diff}} \right) - \frac{\partial p}{\partial t} - \vec{v} \nabla p - \bar{\bar{\tau}} \nabla \vec{v} = 0 \tag{2.10a}$$

With

$$e = h - \frac{p}{\rho} + \frac{v^2}{2}; h = \sum_{i=1}^{n} y_i h_i; h_i = \int_{T_{\text{ref}}}^{T} c_{p,i} dT \tag{2.11}$$

$$\frac{\partial (\rho \varphi)}{\partial t} + \nabla \cdot (\rho \varphi \vec{v}) + \nabla \vec{j}_{\varphi} = \dot{s}_{\varphi} \tag{2.5}$$

for an arbitrary field quantity per unit mass of mixture φ. At the left-hand side they consider temporal changes in the volume element dV in respect to a fixed coordinate system, convective transport and non-convective flux of the field quantity $\vec{j}_{\varphi}$. At the right-hand side a source term $\dot{s}_{\varphi}$ (rate of production of field quantity per unit volume) is considered.

Volume reactions in the fluid phase (homogeneous reactions) are excluded, so that the source term vanishes in the species balances of Table 2.1.

For the multi-component systems under consideration the mass average velocity $\vec{v}$ that corresponds to a mass flux defined as:

$$\vec{\dot{m}} = \sum_{i=1}^{n} \vec{\dot{m}}_i = \sum_{i=1}^{n} \rho_i^* \vec{u}_i = \rho \vec{v}, \vec{v} = \sum_{i=1}^{n} \frac{\rho_i^*}{\rho} \vec{u}_i = \sum_{i=1}^{n} y_i \vec{u}_i \tag{2.6}$$

is used as the mixture velocity. Here $\vec{u}_i$ denotes the velocity of component i. The sum of the balances for all species is the continuity equation. Consequently, when using the equation of continuity of total mass of the mixture only $(n - 1)$ species balances are independent from each other.

The mass average frame should always be preferred when having to solve not only mass transfer equations, but also the equations of motion. Only in this way is it possible to express the momentum balance for pure fluids and multi-component mixtures in a uniform way according to the principle of conservation of total momentum. The use of other reference velocities, such as the average volume velocity:

$$\vec{u} = \sum_{i=1}^{n} \tilde{c}_i \tilde{v}_i^* \vec{u}_i \tag{2.6a}$$

in the so-called Fickian reference system, in general yields significantly more complicated expressions for the momentum balance. For ideal gas mixtures it is $\tilde{v}_i^* = 1/\tilde{c}$, and the average volume velocity is equal to the average of species velocities:

$$\vec{u} = \sum \tilde{y}_i \vec{u}_i. \tag{2.6b}$$

Momentum transport by diffusive fluxes has been neglected in Equation 2.8. Without this simplification, it would be necessary to treat momentum transport in the same way as mass transport, writing separate balances for the momentum of every individual species (Baranowski, 1975; Kerkhof and Geboers, 2005). However, it is difficult to obtain experimental validation for any method of splitting the pressure and the viscosity tensor down to individual components. Therefore, such a split is reasonable only when the influence of external force fields, for example, electrostatic field, must be considered.

For a multi-component mixture, enthalpy transport by diffusive fluxes has to be accounted for in the energy balance (2.10). Dissipation can be neglected in presence of heat transport for the considered applications and flow velocities.

Equation 2.9a for the calculation of molar fraction fields and Equation 2.10a for the calculation of the temperature field are obtained from Equations 2.9 and 2.10 by inserting in them the continuity condition of Equation 2.7 and by inserting additionally a balance of kinetic energy in Equation 2.10.

Membrane, membrane-supported catalyst layer and packed bed are regarded – as already mentioned – as quasi-homogeneous phases with the porosity ε. For every such phase an additional system of equations is required, according to Table 2.2. Anisotropy is excluded. Especially for the packed bed, heat and mass transfer limitations between the bulk phase and catalyst particles are neglected.

Concerning the porous zones and Table 2.2, the following should be noticed.

Velocities denoted by v are superficial velocities that refer to the total cross-sectional area or volume of the bed (sum of fluid and particle volume, $V = V_f + V_p$). Volume flow rate and area yield the average of superficial velocity in any cross-section of the bed:

$$u = \dot{V}/A \text{ and } v = \frac{\dot{M}}{\bar{\rho}_f A}. \tag{2.17}$$

Interstitial (or fluid) velocities, which refer only to the volume V_f or area occupied by the fluid, may also be used in porous regions. They are then denoted by v_f. The local porosity:

Table 2.2 Model equations for the porous zones, independent from the choice of the coordinate system.

Conservation of total mass

$$\frac{\partial(\varepsilon\rho_f)}{\partial t}+\nabla\cdot(\varepsilon\rho_f\vec{v}_f)=0 \tag{2.12}$$

Momentum conservation equations

$$\frac{\partial(\varepsilon\rho_f\vec{v}_f)}{\partial t}+\nabla\cdot(\varepsilon\rho_f\vec{v}_f\vec{v}_f)+\nabla\varepsilon p-\nabla\cdot(\varepsilon\bar{\bar{\tau}})-\varepsilon\rho_f\vec{g}=-\varepsilon\vec{f} \tag{2.13}$$

Source terms in the momentum conservation equations

$$\vec{f}=f_1\vec{v}_f+f_2|v_f|\vec{v}_f \tag{2.14}$$

Transport equations for species

$$\frac{\partial}{\partial t}(\varepsilon\rho_f y_i)+\nabla\cdot(\varepsilon\rho_f\vec{v}_f y_i)+\nabla\cdot\vec{m}_i^{\mathrm{Diff}}=R_i \tag{2.15}$$

Energy conservation equation

$$\frac{\partial}{\partial t}(\varepsilon\rho_f e_f+(1-\varepsilon)\rho_s e_s)+\nabla\cdot[\varepsilon\vec{v}_f(\rho_f e_f+p)]+\nabla\cdot\left(\dot{q}+\sum_{i=1}^{n}h_i\vec{m}_i^{\mathrm{Diff}}\right)-\nabla(\varepsilon\bar{\bar{\tau}}\cdot\vec{v}_f)=\sum_{i=1}^{n}\frac{\tilde{h}_i^0R_i}{\tilde{M}_i} \tag{2.16}$$

$$\varepsilon=dV_f/dV=1-(dV_s/dV) \tag{2.18}$$

interrelates superficial and interstitial velocities in the case of isotropic porous media according to this equation:

$$\vec{v}=\varepsilon\vec{v}_f. \tag{2.19}$$

Friction and inertial forces caused by flow through pores lead to an additional loss of momentum – accounted for by the source term $\vec{f}$ in Equation 2.13. On contrary to the frequently used, simplified relationship after (Forchheimer, 1901):

$$\nabla p=-f_1\vec{v}-f_2|\vec{v}|\vec{v} \tag{2.20}$$

Equation 2.13 takes into consideration all kinds of momentum transport, especially viscous transport in the fluid. The limiting case of $f_2=0$, which is always applicable to the membrane and membrane-supported catalyst layers, corresponds to Darcy's law. For the packed bed, the coefficients f_1 and f_2 are calculated according to (Ergun, 1952):

$$f_1=150\frac{(1-\varepsilon)^2}{\varepsilon^3}\frac{\eta_f}{d_p^2},\; f_2=1.75\frac{(1-\varepsilon)^2}{\varepsilon^3}\frac{\rho_f}{d_p}. \tag{2.21}$$

This is done on the basis of local values of porosity after (Hunt and Tien, 1990):

$$\varepsilon(r)=\varepsilon_\infty+(1-\varepsilon_\infty)\exp\left(-6\frac{R-r}{d_p}\right) \tag{2.22}$$

Heterogeneous catalytic reactions appear now in the source term R_i of the species balances. This can be expressed as:

$$R_i = \tilde{M}_i \sum_{j=1}^{nr} \nu_{ij}(1-\varepsilon)\rho_{cat}^{*} r_j \text{ with } \rho_{cat}^{*} = \frac{M_{cat}}{V_s} \tag{2.23}$$

In the case of packed beds, the volume of the solid phase V_s is given by the volume of the particles. The molar rate of reaction j, r_j, refers here to the mass of catalyst used.

Component adsorption and desorption on the solid phase are neglected in relation to the accumulation term.

In the energy balance, the heat capacity of the solids, which fill $(1 - \varepsilon)$ of the space, is considered. Local thermal equilibrium is assumed between solids and the fluid. A source term accounts for reaction enthalpies, where $\tilde{h}_i^0$ is the molar enthalpy of formation of species i.

Since tubular reactors are used in most applications, axisymmetric conditions can be assumed. This saves much computational time in comparison to 3-D calculations. The 2-D equations resulting in cylindrical coordinates using the z-axis as a rotation axis are exemplarily summarized for porous domains in Table 2.3.

Table 2.3 Model equations for the porous zones in 2-D axisymmetric geometry.

Conservation of total mass

$$\frac{\partial(\varepsilon\rho_f)}{\partial t} + \frac{\partial(\varepsilon\rho_f v_{z,f})}{\partial z} + \frac{\partial(\varepsilon\rho_f v_{r,f})}{\partial r} + \frac{\varepsilon\rho_f v_{r,f}}{r} = 0 \tag{2.24}$$

Momentum conservation equations z-coordinate:

$$\frac{\partial(\varepsilon\rho_f v_{z,f})}{\partial t} + \frac{\partial(\varepsilon\rho_f v_{z,f} v_{z,f})}{\partial z} + \frac{1}{r}\frac{\partial(r\varepsilon\rho_f v_{r,f} v_{z,f})}{\partial r} + \frac{\partial(\varepsilon p)}{\partial z} - \frac{\partial}{\partial z}(\varepsilon\tau_{zz}) - \frac{1}{r}\frac{\partial}{\partial r}(r\tau_{rz}) - \rho_f g_z = -\varepsilon\left(f_1 v_{z,f} + f_2|v_f|v_{z,f}\right) \tag{2.25}$$

r-coordinate:

$$\frac{\partial(\varepsilon\rho_f v_{r,v})}{\partial t} + \frac{\partial(\varepsilon\rho_f v_{r,f} v_{z,f})}{\partial z} + \frac{1}{r}\frac{\partial(r\varepsilon\rho_f v_{r,f} v_{r,f})}{\partial r} + \frac{\partial(\varepsilon p)}{\partial r} - \frac{\partial}{\partial z}(\varepsilon\tau_{rz}) - \frac{1}{r}\frac{\partial}{\partial r}(\varepsilon r\tau_{rr}) - \varepsilon\rho_f g_r = -\varepsilon\left(f_1 v_{r,f} + f_2|v_f|v_{r,f}\right) \tag{2.26}$$

Transport equations for species

$$\frac{\partial}{\partial t}(\varepsilon\rho_f y_{i,f}) + \frac{1}{r}\frac{\partial(r\varepsilon\rho_f v_{r,f} y_i)}{\partial r} + \frac{\partial(\varepsilon\rho_f v_{z,f} y_i)}{\partial z} + \frac{1}{r}\frac{\partial}{\partial r}(r\dot{m}_{i,r}^{Diff}) + \frac{\partial \dot{m}_{i,z}^{Diff}}{\partial z} = R_i \tag{2.27}$$

Energy conservation equation

$$\frac{\partial}{\partial t}(\varepsilon\rho_f e_f + (1-\varepsilon)\rho_s e_s) + \frac{1}{r}\frac{[\partial r v_{r,f}(\varepsilon\rho e_f + p)]}{\partial r} + \frac{\partial[v_{z,f}(\varepsilon\rho e_f + p)]}{\partial z} + \left[\frac{1}{r}\frac{\partial}{\partial r}(r\dot{q}_r) + \frac{\partial \dot{q}_z}{\partial z}\right] + \left[\frac{1}{r}\frac{\partial}{\partial r}(r h_i \dot{m}_{i,r}^{Diff}) + \frac{\partial h_i \dot{m}_{i,z}^{Diff}}{\partial z}\right] + \left[\frac{1}{r}\frac{\partial}{\partial r}(r\varepsilon(\tau_{rr} v_{r,f} + \tau_{rz} v_{z,f})) + \frac{\partial(\varepsilon\tau_{zz} v_{z,f} + \varepsilon\tau_{rz} v_{r,f})}{\partial z}\right] = \sum_{i=1}^{n} \frac{h_i^0 R_i}{M_i} \tag{2.28}$$

Apart from the system of governing differential equations, it is necessary to have:

- A precise definition of control volume (solution domain),
- An adequate selection of boundary conditions.

These are outlined in Section 2.4. Before that, kinetic expressions needed in order to calculate the fluxes in the governing equations are discussed.

2.3 Transport Kinetics

2.3.1 Fluid-Filled Regions

Comprehensive treatments of diffusion in multi-component mixtures have been provided by, among others, (Haase, 1963; de Groot and Mazur, 1984; Taylor and Krishna, 1994; and Cussler, 1997). Appropriate simplified approaches are used in the present work. Cross-effects like thermal diffusion or interrelation between reaction and momentum transport and the influence of external force fields are neglected. It should be noticed that kinetic equations have a similar structure for all transport processes, correlating the flux of momentum, mass and energy with relevant driving forces, which are gradients of, respectively, velocity, chemical potential and temperature in the assumed continuum.

2.3.1.1 Molecular Transport of Momentum

The kinetics of molecular transport of momentum is described in a linear way by the generalized form of Newton's law of viscosity. Taking into account that the shear stresses are a symmetric combination of velocity gradients and that the fluid is isotropic, it follows after (Bird, Stewart, and Lightfoot, 2002):

$$\bar{\bar{\tau}} = \eta(\nabla\vec{v} + \nabla\vec{v}^{t}) - \left(\frac{2}{3}\eta - \kappa\right)(\nabla\cdot\vec{v})I \tag{2.29}$$

The quantity κ in Equation 2.29 is called the dilatational viscosity. It can be neglected for low density gases at low Mach number flows.

2.3.1.2 Heat Conduction

The law of heat conduction, also known as Fourier's law, states that the time rate of heat transfer through a material is proportional to the negative gradient in the temperature and to the cross-sectional area to that gradient. Defined per flow area, the heat flux can be calculated as:

$$\vec{q} = -\lambda\nabla T \tag{2.30}$$

Cross-effects described by the thermodynamics of irreversible processes are neglected. The thermal conductivity λ corresponds to the molecular quantity λ_g in the gas phases and in the porous media to the effective value λ_{eff}.

2.3.1.3 Molecular Diffusion

The mentioned assumption that the influence of external force fields and all cross-effects foreseen in general irreversible thermodynamics can be neglected leads to the Stefan–Maxwell equations (Taylor and Krishna, 1994):

$$\frac{\tilde{y}_i}{\tilde{R}T}(\nabla\mu_i)_{T,p} = -\sum_{j=1}^{n} \frac{\tilde{y}_i\tilde{y}_j(\vec{u}_i - \vec{u}_j)}{D_{ij}^M} \tag{2.31}$$

for describing diffusion in multi-component mixtures. The derivation of these equations is based on momentum transfer during collisions between different species. Equation 2.31 is independent from the average velocity of the mixture and uses gradients of chemical potential μ_i as the driving forces of mass transport. These gradients vanish at equilibrium.

By definition of molar diffusion fluxes in respect to the average component velocity (2.6b):

$$\vec{n}_i^{\mathrm{Diff}} = \vec{n}_i - \tilde{c}_i\vec{u} = \tilde{c}_i(\vec{u}_i - \vec{u}), \sum_{i=1}^{n}\vec{n}_i^{\mathrm{Diff}} = 0 \tag{2.32}$$

the dependence between fluxes and driving forces follows from Equation 2.32 to:

$$\frac{\tilde{c}_i}{\tilde{R}T}(\nabla\mu_i)_{T,p} = -\sum_{j=1}^{n} \frac{\tilde{y}_j\vec{n}_i^{\mathrm{Diff}} - \tilde{y}_i\vec{n}_j^{\mathrm{Diff}}}{D_{ij}^M} \tag{2.33}$$

Alternatively, the relationships:

$$\vec{m}_i^{\mathrm{Diff}} = \vec{m}_i - \rho_i\vec{v} = \rho_i^*(\vec{u}_i - \vec{v}), \sum_{i=1}^{n}\vec{m}^{\mathrm{Diff}} = 0 \tag{2.34}$$

$$\frac{\rho_i^*}{\tilde{R}T}(\nabla\mu_i)_{T,p} = -\sum_{j=1}^{n} \frac{\tilde{y}_j\vec{m}_i^{\mathrm{Diff}} - \dfrac{\tilde{M}_i}{\tilde{M}_j}\tilde{y}_i\vec{m}_j^{\mathrm{Diff}}}{D_{ij}^M} \tag{2.35}$$

are obtained by using mass-average velocity (2.6) and mass-related diffusion fluxes. It should be noticed that with n components $(n-1)$ Equations 2.33 and 2.35 are independent from each other.

A clear disadvantage when computing with the Stefan–Maxwell equations is that the diffusion fluxes can not be expressed explicitly. Matrix formulation according to (Taylor and Krishna, 1994) in the form:

$$(\vec{n}) = -\frac{\tilde{c}}{\tilde{R}T}[B^{-1}][\nabla\mu] = -\tilde{c}[B^{-1}][\Gamma](\nabla\tilde{y}) \tag{2.36}$$

with:

$$B_{ii} = \frac{\tilde{y}_i}{D_{in}^M} + \sum_{\substack{j=1 \\ j\neq i}}^{n}\frac{\tilde{y}_i}{D_{ij}^M}, \; B_{ij} = -\tilde{y}_i\left(\frac{1}{D_{ij}^M} - \frac{1}{D_{in}^M}\right) \tag{2.37}$$

does not remove this disadvantage. Here $[\Gamma]$ is a matrix that should be calculated from activities in order to–when necessary–account for real mixture behavior.

Friendlier for calculations are the generalized expressions according to Fick, which provide explicitly the diffusion fluxes as:

$$\vec{n}_i^{\text{Diff}} = -\tilde{c}\sum_{j=1}^{n-1} D_{ij}\nabla\tilde{y}_j \tag{2.38}$$

or:

$$\vec{m}_i^{\text{Diff}} = -\rho\sum_{j=1}^{n-1} D_{ij}^{*}\nabla y_j \tag{2.39}$$

The disadvantage of these expressions is a relatively strong dependence of the multi-component diffusion coefficients D_{ij} or D_{ij}^{*} upon the composition of the mixture.

A further simplification is to use pseudo-binary diffusion coefficients D_i^{pb} and relate the flux of every species i to the driving force $\nabla\tilde{y}_i$ of only this component:

$$\vec{n}_i^{\text{Diff}} = -\tilde{c}D_i^{\text{pb}}\nabla\tilde{y}_i \tag{2.40}$$

However, if one requires the independence of diffusion coefficients from driving forces, the relationship:

$$\sum_{i=1}^{n}\vec{n}_i^{\text{Diff}} = -\tilde{c}\sum_{i=1}^{n} D_i^{\text{pb}}\nabla\tilde{y}_i = -c\sum_{i=1}^{n}\left(D_i^{\text{pb}} - D_n^{\text{pb}}\right)\nabla\tilde{y}_i = 0 \tag{2.41}$$

is obtained. This leads immediately to:

$$D_i^{\text{pb}} = D_n^{\text{pb}} \quad i = 1, 2, \ldots, n-1. \tag{2.42}$$

Equation 2.42 means that all diffusion coefficients D_i^{pb} should have the same value, which is obviously not true in multi-component mixtures.

A comprehensive discussion of the complex interrelations between diffusion coefficients defined in different ways for multi-component mixtures is given by (Wesselingh and Krishna, 2000).

Fortunately, ideal gas behavior can usually be assumed. Additionally, for the membrane reactors under consideration it can be assumed that every species $i \neq n$ is present in strong dilution to the inert component n. Significant simplifications are possible on this basis. From the definition of the chemical potential of ideal gases it follows:

$$\frac{\tilde{y}_i}{\tilde{R}T}(\nabla\mu_i)_{T,p} = \sum_{j=1}^{n-1}\Gamma_{ij}\nabla\tilde{y}_j = \sum_{j=1}^{n-1}\nabla\tilde{y}_j \tag{2.43}$$

For diluted mixtures with (at the limit):

$$\tilde{y}_{i\neq n} = 0,\ \tilde{y}_n = 1 \tag{2.44}$$

Equation 2.39 simplifies (Taylor and Krishna, 1994), leading to multi-component diffusion coefficients of:

$$D_{ij} = 0 \text{ and } D_{ii} = D_{in}^{M} \quad i = 1, 2, \ldots, n-1 \tag{2.45}$$

in Equation 2.45. These diffusion coefficients can be calculated according to (Fuller, Schettler, and Giddings, 1966) – as recommended by (Reid, Prausnitz, and Sherwood, 1977) – by means of the equation:

$$D_{in} = 1.013 \cdot 10^{-7} T^{1.75} \frac{\left(\frac{\tilde{M}_i + \tilde{M}_n}{\tilde{M}_i \tilde{M}_n}\right)}{p\left[\left(\sum v\right)_i^{0.333} + \left(\sum v\right)_n^{0.333}\right]^2} \tag{2.46}$$

(T in K, p in bar, D_{in} in m^2/s).

The so-called diffusion volumes $(\Sigma v)_i$ are found for each component by summing of the atomic and structural increments and can be taken from (Poling, Prausnitz, and O'Connell, 2001) or (Lucas and Luckas, 2002).

Because of inaccuracies of modeling and computation, the sum of diffusion fluxes is never exactly equal to zero:

$$\sum_{i=1}^{n} \vec{\dot{m}}_i^{\mathrm{Diff}} = \vec{\dot{m}}_{\mathrm{cor}} \neq 0. \tag{2.47}$$

To still guarantee the conservation of mass and fulfill Equations 2.32, 2.34 either all fluxes must be adjusted by weighted application of some correction terms, or the main mixture component must be fitted. In our case it is reasonable to implement the correction in the excess component (inert species n):

$$\vec{\dot{m}}_{n,\mathrm{cor}}^{\mathrm{Diff}} = \vec{\dot{m}}_n^{\mathrm{Diff}} - \vec{\dot{m}}_{\mathrm{cor}}. \tag{2.48}$$

Mass-average velocity $\vec{v}$, the composition fields obtained from the $(n-1)$ independent material balances and the diffusion fluxes $\vec{\dot{m}}_{i,i\neq n}$ have to be treated correspondingly in the course of numerical calculations. The correction of diffusion fluxes is especially important in the case of creeping flow as well as for closure of the material balances over the entire computational domain.

2.3.2 Porous Domains

In literature it is usual to distinguish between macro-pores ($d_{\mathrm{pore}} > 50\,\mathrm{nm}$), meso-pores ($2\,\mathrm{nm} < d_{\mathrm{pore}} < 50\,\mathrm{nm}$) and micro-pores ($d_{\mathrm{pore}} < 2\,\mathrm{nm}$) and to classify porous materials correspondingly (Melin and Rautenbach, 2007). This classification does not say anything about the prevailing transport mechanisms, because it does not put the pore diameter d_{pore} in relation to the size of the transported species d_{mol} and to the free mean path of gas molecules l. By introducing these quantities and the Knudsen number $\mathrm{Kn} = l/d_{\mathrm{pore}}$ we can better recognize regions of:

- Molecular diffusion or viscous flow, when $l < d_{\mathrm{pore}}$ ($\mathrm{Kn} < 1$),
- Knudsen diffusion, when $d_{\mathrm{mol}} < d_{\mathrm{pore}} < l$ ($\mathrm{Kn} > 1$),
- Configurational diffusion (molar sieving effect), when $d_{\mathrm{mol}} \approx d_{\mathrm{pore}}$.

The mentioned transport mechanisms are illustrated in Figure 2.1 and are briefly explained in the following – with the exception of configurational diffusion, which is not of interest for the used gases and porous media.

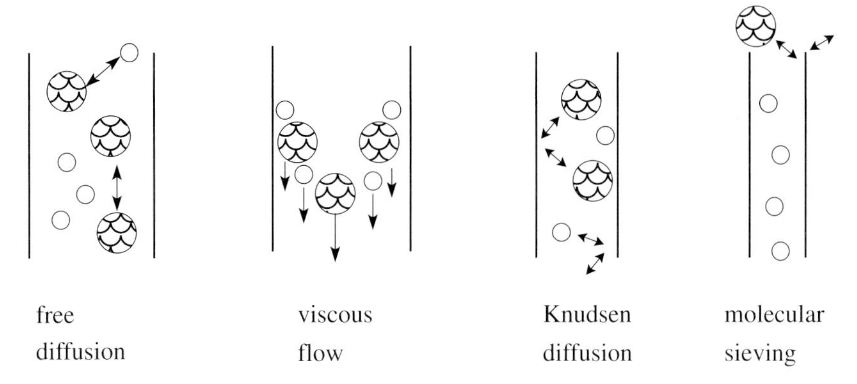

Figure 2.1 Transport mechanisms in porous media.

2.3.2.1 Molecular Diffusion

Molecular (or free) diffusion is dominated by the interactions between gas molecules and, thus, prevails when the pore diameter is considerably larger than the mean free path. If collisions with the wall can be completely neglected, the total momentum of the molecules remains constant. Mass transport kinetics can be modeled as previously explained. Diffusivities are, however, reduced in a porous medium by a factor of F_0, which is often expressed as the ratio between porosity ε and tortuosity τ. This gives rise to effective diffusion coefficients:

$$D_{ij}^{e} = F_0 D_{ij} = \frac{\varepsilon}{\tau} D_{ij} \tag{2.49}$$

2.3.2.2 Knudsen Diffusion

When the pore diameter is comparable to or smaller than the mean free path, momentum is transferred primarily by collisions of molecules with the wall. Gas–wall interactions dominate, and molecules of different species are transported independently of each other, corresponding to their mobility. By application of kinetic gas theory to a single, straight and cylindrical pore the coefficient of Knudsen diffusion can be derived to:

$$D_{K,i} = \frac{1}{3} d_{\text{pore}} \sqrt{\frac{8\tilde{R}T}{\pi \tilde{M}_i}} \tag{2.50}$$

For a porous medium, a more general form of this equation can be used, namely:

$$D_{K,i} = \frac{4}{3} K_0 \sqrt{\frac{8\tilde{R}T}{\pi \tilde{M}_i}}, \quad K_0 = \frac{\varepsilon}{\tau} \frac{d_{\text{pore}}}{4} \tag{2.51}$$

By analogy to molecular diffusion, the Knudsen diffusion flux can be calculated to:

$$\vec{\dot{n}}_i^K = \frac{\varepsilon}{\tau} \frac{D_{K,i}}{\tilde{R}T} \nabla p_i \tag{2.52}$$

The gradient of partial pressure is the driving force of the process.

2.3.2.3 Viscous Flow

A total pressure gradient gives rise to mass transport by convective, viscous flow. Both molecular diffusion and viscous flow gradually disappear when moving into the Knudsen region. Though viscous flow does not contribute to the separation of different species, its role may be very important in membrane reactors. For a single capillary, viscous flow can be calculated according to Hagen and Poiseuille (Jackson, 1977). Due to laminar conditions, average velocity is directly proportional to the pressure gradient. A modified version of the Hagen–Poiseuille law that goes back to Darcy can be used for porous media, leading to a flux of:

$$\dot{n}_i^{\nu} = -\frac{1}{\tilde{R}T}\frac{B_0}{\eta}p\nabla p \tag{2.53}$$

with:

$$B_0 = \frac{\varepsilon}{\tau}\frac{d_{\text{pore}}^2}{32}. \tag{2.54}$$

For the convective flow of a mixture it follows:

$$\dot{n}^{\nu} = \sum_{i=1}^{n}\dot{n}_i^{\nu}. \tag{2.55}$$

2.3.2.4 Models for Description of Gas Phase Transport in Porous Media

According to Jackson – and based on ideas that go back to Clerk Maxwell – porous media can be considered as either networks of interconnected capillaries or assemblies of stationary obstacles dispersed in the gas at a molecular scale for modeling mass transport. Models of the first type can be closely related to the real structure of the medium, but have some difficulties in implementing the Knudsen effect.

A prominent representative of models of the second type is the dusty gas model (DGM), which was rediscovered independently three times by (Deriagin and Bakanov, 1957; Evans, Watson, and Maso, 1962; Mason, Malinauskas, and Evans, 1967; and Mackey, 1971). The model considers the porous medium as composed of giant molecules that are fixed and uniformly distributed in space. These so-called dust particles are treated as one additional component of the gas mixture in the frame of a Stefan–Maxwell approach. By consequence, Knudsen diffusion is automatically accounted for as a result of molecule–dust interactions. In contrast, the porous medium is represented by formal parameters in the model, which are denoted by F_0, K_0, B_0 in the foregoing discussion. To connect these parameters with characteristic geometrical features of the medium (like the mentioned ratio of porosity and tortuosity, and the mean pore diameter) additional assumptions are required. Micro-structural characteristics get lost, in the same way as heterogeneities at the macro-scale.

Figure 2.2 illustrates the combination of molecular diffusion and Knudsen diffusion with viscous flow in the DGM by an electrical analog. According to it, the total flux is split into parallel paths of diffusion and viscous flow:

$$\dot{n}_i = \dot{n}_i^D + \tilde{y}_i\dot{n}^{\nu} \tag{2.56}$$

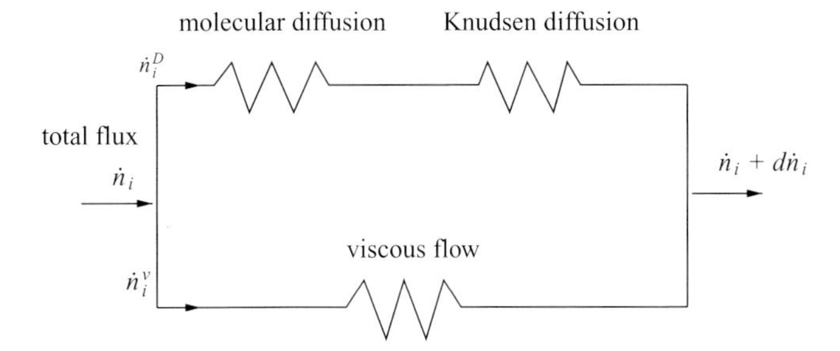

Figure 2.2 Electrical analog illustrating transport phenomena in porous media.

Molecular diffusion and Knudsen diffusion are combined in series. This corresponds to the law of (Bosanquet, 1944) for an overall diffusion coefficient of:

$$D_{i,\text{eff}} = \frac{1}{\frac{1-a\tilde{y}_i}{D_{ij}^e} + \frac{1}{D_{K,i}}} \tag{2.57}$$

The parameter α in Equation 2.57 takes into account the ratio of molar fluxes or, according to Graham's law (Evans, Watson, and Maso, 1962), the ratio of the square root of the molar masses of the considered components:

$$\alpha = 1 - \frac{\dot{n}_i}{\dot{n}_j} = 1 - \sqrt{\frac{\tilde{M}_j}{\tilde{M}_i}} \tag{2.58}$$

In the case of equimolar diffusion, the transported amounts of different species are equal to each other. Hence, $\alpha = 0$ and Equation 2.57 can be reduced to the form:

$$D_{i,\text{eff}} = \frac{1}{\frac{1}{D_{ij}^e} + \frac{1}{D_{Ki}}} \tag{2.59}$$

Based on the assumptions of spatially uniform dust concentration, motionless dust (such that $\dot{n}_{n+1} = 0$) and very massive dust molecule ($M_{n+1} \to \infty$) equations describing the dusty gas model can be obtained. A concise form of these equations is:

$$\frac{p}{\tilde{R}T}\frac{d\tilde{y}_i}{dr} + \frac{\tilde{y}_i}{\tilde{R}T}\left(1 + \frac{B_0 p}{\eta D_{K,i}}\right)\frac{dp}{dr} = \sum_{j=1, i\neq j}^{n} \frac{\tilde{y}_i \dot{n}_j - \tilde{y}_j \dot{n}_i}{D_{ij}^e} - \frac{\dot{n}_i}{D_{K,i}} \tag{2.60}$$

A matrix formulation that is more convenient for numerical calculations reads:

$$\dot{n} = -\frac{1}{RT} C^{-1} F,$$

$$C_{ii} = \frac{1}{D_{K,i}} + \sum_{i=1, j\neq 1}^{n} \frac{\tilde{y}_i}{D_{ij}^e}; \quad C_{ij} = \frac{\tilde{y}_i}{D_{ij}^e}, \tag{2.61}$$

$$F_i = p\nabla\tilde{y}_i + \tilde{y}_i\left(1 + \frac{B_0 p}{\eta D_{K,i}}\right)\nabla p$$

The dusty gas model can be extended to include thermal diffusion effects, non-ideal fluid behavior, external body forces, surface diffusion and selective viscous flow. Criticism to the DGM concerns especially the accounting of the viscous contribution (Kerkhof and Geboers, 2005; Krishna and van Baten, 2009).

2.4 Reduced Models

Depending on the number and kind of zones (domains), the models discussed in the foregoing sections can be adapted for simulating three different types of reactors, namely:

- Conventional packed-bed reactor (PBR),
- Packed-bed membrane reactor (PBMR),
- Catalytic membrane reactor (CMR).

A PBR consists only of the catalyst-filled tube side. In a PBMR, the membrane separates the tube side from the shell side. One of them is empty, and the other (usually the tube side) is filled with catalyst particles. In a CMR both sides are empty, because the membrane itself has a catalytically active layer.

For the description of different reactors and conditions, transformations of model equations are necessary. Models of different spatial dimensionality in the treatment of the considered domains have been used, such as:

- 1-D PBMR (membrane not modeled)
- 1-D + 1-D (membrane in r-direction, channels in z-direction)
- 2-D PBMR (membrane not modeled)
- 2-D CMR (membrane, tube side, shell side)

To completely specify the model used for a certain mode or operation, it is necessary to go through several steps:

- Identification of the domains that need to be modeled,
- Selection of the spatial dimensionality of equations for each domain,
- Introduction of appropriate kinetic expressions into the balance equations,
- Definition of all necessary boundary conditions.

For reasons of space, it is not possible to provide here a comprehensive set of equations for all operational modes that have been modeled. Every specific implementation is explained in the following chapters, as it comes to use.

However, the procedure should be illustrated by one example, specifically the example of CMR modeling.

The complete CMR model is based on the standard consideration of convective and diffusive transport taking into account the necessary source and sink terms. Because of the symmetry conditions in the tube geometries, it is possible to model each zone of the membrane reactor in a 2-D way without information loss. Because of the multi-dimensional modeling, the heat and mass transfer between the zones are integral parts of the model and no correlations for Nusselt or Sherwood numbers are needed (Georgieva *et al.*, 2005). The simulated axisymmetric geometry is presented in Table 2.4. The same table summarizes the steady-state balance equations for the catalytic active membrane, model assumptions and the boundary conditions for the entire computational domain.

The model description of the empty channels is not presented in Table 2.4, but it can be derived from the equations for the catalytic active zone by removing the source terms and by replacing the effective transport coefficients with molecular values. Because of the small geometrical dimensions, laminar flow conditions are assumed and plug flow profiles are defined at the reactor inlets. The influence of gravitation can also be neglected in gas systems.

The pressure drop through each membrane layer is considered by means of a sink term in the momentum equations. The parameters f_1 and f_2 are calculated using the coefficients determined by application of the DGM:

$$f_1 = \frac{\eta_{N_2}}{B_0 + \dfrac{D_{K,N_2}\eta_{N_2}}{p}}, \quad f_2 = 0, \quad B_0 = \frac{\varepsilon}{\tau}\frac{d_{pore}^2}{32}, \quad D_{K,N_2} = \frac{4}{3}K_0\sqrt{\frac{8\tilde{R}T}{\pi\tilde{M}_{N_2}}}, \quad K_0 = \frac{\varepsilon}{\tau}\frac{d_{pore}}{4} \tag{2.62}$$

These expressions are derived from the DGM for the transport of N_2 through the membrane and take into account the viscous slip at the pore walls. At very small Knudsen numbers, the laminar viscous flow of a single species is known as Poiseuille flow. In this particular case, the non-slip boundary condition can be used reliably. Viscous slip is observed in the region of Knudsen diffusion and in the transition region, that means for $Kn > 0.1$ (Young and Todd, 2005). Because the developed model for description of the porous media is pseudo-homogenous, the viscous slip on the pore walls can be expressed only by integral and apparent quantities. The parameter f_2 was set to zero due to the laminar character of the flow.

The investigated membranes have an asymmetric structure. The experimentally determined parameters of each membrane layer (see Chapter 4) are used for the definition of the sink terms in the momentum equations and for the calculation of transport coefficients.

In CMR, the reactions take place only in the catalyst layer. Therefore, the reaction terms are considered in the conservation equations for mass and energy only in this zone. To apply the experimentally determined reaction kinetics, the reaction constants have to be converted for the coated vanadium amount in the catalyst layer on the basis of the Turnover number (Masel, 2001). Because of the strong dilution of the educts, the material properties are based on correlations for nitrogen. Therefore, a quasi-binary gas mixture was assumed for the calculation of diffusion coefficients, in which each species is located only in nitrogen environment. The effective diffusion coefficients are calculated according to Equation 2.59. The determination of effective thermal conductivity are discussed in Chapter 4.

For a closed solution of the equation system, appropriate boundary conditions are needed. At the channel entrances, the total mass flow rates, the mass fractions of $n - 1$ components and the initial temperature have to be defined. Additionally, the operating pressure has to be specified and set as a boundary condition at the reactor outlet. On this basis, the pressure drops through the packed bed,

Table 2.4 Model equations for the catalytical membrane layer.

Model assumptions:
Steady state, laminar, pseudo-homogeneous, extended Fickian law, highly diluted reaction system, no homogeneous reaction, plug flow profiles at reactor inlets, no gravity influence
Conservation of total mass:

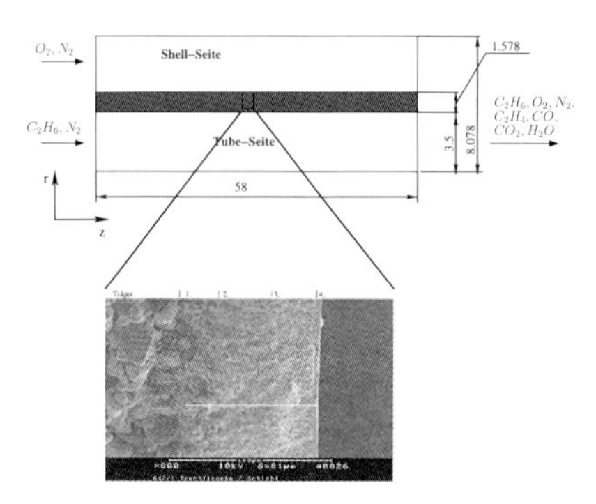

$$\frac{\partial(\rho_f v_z)}{\partial z} + \frac{\partial(\rho_f v_r)}{\partial r} + \frac{\rho_f v_r}{r} = 0$$

Momentum conservation equations
z-Coordinate:

$$\frac{\partial(\rho_f v_z v_z)}{\partial z} + \frac{1}{r}\frac{\partial(r\rho_f v_r v_z)}{\partial r} + \frac{\partial p}{\partial z} - \frac{\partial}{\partial z}\left[\eta\left(2\frac{\partial v_z}{\partial z} - \frac{2}{3}\left(\frac{\partial v_z}{\partial z} + \frac{\partial v_r}{\partial r} + \frac{v_r}{r}\right)\right)\right] - \frac{1}{r}\frac{\partial}{\partial r}\left[r\eta\left(\frac{\partial v_z}{\partial r} + \frac{\partial v_r}{\partial r}\right)\right] = -f_1 v_z$$

r-Coordinate:

$$\frac{\partial(\rho_f v_r v_z)}{\partial z} + \frac{1}{r}\frac{\partial(r\rho_f v_r v_r)}{\partial r} + \frac{\partial p}{\partial r} - \frac{\partial}{\partial z}\left[\eta\left(\frac{\partial v_r}{\partial z} + \frac{\partial v_z}{\partial z}\right)\right] - \frac{1}{r}\frac{\partial}{\partial r}\left[r\eta\left(2\frac{\partial v_r}{\partial r} - \frac{2}{3}\left(\frac{\partial v_z}{\partial z} + \frac{\partial v_r}{\partial r} + \frac{v_r}{r}\right)\right)\right] = -f_1 v_r$$

membrane or empty channels are calculated. The gradient of the axial component of the velocity vector is set to zero at the axis of the tube, where the radial velocity is also zero. Adiabatic boundary conditions are applied at the reactor walls, so that heat and mass streams are vanishing there.

Model development, reduction, implementation and solution can be organized and supported with the help of modern computational tools, as discussed in the following.

2.5
Solvability, Discretization Methods and Fast Solution

As seen in the previous sections the modeling of membrane reactors leads to a coupled system of nonlinear partial differential equations completed by appropriate boundary and initial conditions. The solvability of these types of initial boundary value problems is not at all a trivial problem and even in special situations

Table 2.4 *(Continued)*

Transport equations for species:

$$\frac{1}{r}\frac{\partial(r\rho_f v_r y_i)}{\partial r}+\frac{\partial(\rho_f v_z y_i)}{\partial z}+\frac{1}{r}\frac{\partial}{\partial r}\left(r\rho_f D_{i,\mathrm{eff}}\frac{\partial y_i}{\partial z}\right)+\frac{\partial}{\partial z}\left(\rho_f D_{i,\mathrm{eff}}\frac{\partial y_i}{\partial z}\right)=M_i\sum_{i=1}^{nr}v_{ij}(1-\varepsilon)\rho_{\mathrm{cat}}r_j$$

Energy conservation equation:

$$\frac{1}{r}\frac{\partial[rv_r(\rho e_f+p)]}{\partial r}+\frac{\partial[v_z(\rho e_f+p)]}{\partial z}+\left[\frac{1}{r}\frac{\partial}{\partial r}\left(r\lambda^e\frac{\partial T}{\partial r}\right)+\frac{\partial}{\partial z}\left(r\lambda^e\frac{\partial T}{\partial z}\right)\right]+$$

$$\left[\frac{1}{r}\frac{\partial}{\partial r}\left(rh_i\rho_f D_{i,\mathrm{eff}}\frac{\partial y_i}{\partial r}\right)+\frac{\partial}{\partial z}\left(h_i\rho_f D_{i,\mathrm{eff}}\frac{\partial y_i}{\partial z}\right)\right]=\sum_{i=1}^{n}h_i^0\sum_{i=1}^{nr}v_{ij}(1-\varepsilon)\rho_{\mathrm{cat}}r_j$$

Boundary conditions:

Geometric position	Momentum balance	Mass balance	Energy balance
Tube side, $z = 0$	$\dot{M}_{\mathrm{TS},0}$	$y_{i},_{\mathrm{TS},0}$	$T_{\mathrm{TS},0}$
Shell side, $z = 0$	$\dot{M}_{\mathrm{SS},0}$	$y_{i},_{\mathrm{SS},0}$	$T_{\mathrm{SS},0}$
Tube side, $z = L$	$p_{\mathrm{TS}} = p_0$		
Shell side, $z = L$	$v_{z,\mathrm{SS}} = 0,\ v_{r},_{\mathrm{SS}} = 0,$	$\frac{\partial y_{i,\mathrm{SS}}}{\partial r}=0, \frac{\partial y_{i,\mathrm{SS}}}{\partial z}=0$	$\frac{\partial T_{\mathrm{SS}}}{\partial r}=0, \frac{\partial T_{\mathrm{SS}}}{\partial z}=0$
Axis, $r = 0$	$\frac{\partial v_{z,\mathrm{TS}}}{\partial r}=0, v_{r,\mathrm{TS}}=0$	$\frac{\partial y_{i,\mathrm{TS}}}{\partial r}=0$	$\frac{\partial T_{\mathrm{TS}}}{\partial r}=0$
Interiors, $\forall r_j$	$p_j=p_{j+1}, \frac{\partial v_r^j}{\partial r}=\frac{\partial v_r^{j+1}}{\partial r}$	$y_j^i=y_i^{j+1}, \frac{\partial y_i^j}{\partial r}=\frac{\partial y_i^{j+1}}{\partial r}$	$T^j=T^{j+1}, \frac{\partial T_j}{\partial r}=\frac{\partial y^{j+1}}{\partial r}$
Walls	$v_z = 0,\ v_r = 0$	$\frac{\partial y_i}{\partial r}=0, \frac{\partial y_i}{\partial z}=0$	$\frac{\partial T}{\partial r}=0, \frac{\partial T}{\partial z}=0$

is the subject of current mathematical research, for example, see (Lions, 1996; Galdi, 1994; Girault and Raviart, 1986; Amann, 2001; Russo and Simader, 2006; Zajaczkowski, 2007). In general there are no classic solutions, that is, solutions which are continuously differentiable, and one has to reformulate the partial differential equations in a so-called weak form for which the existence of weak solutions can be established. Without going into the mathematical details let us summarize some existence and uniqueness results for a flow problem described by the incompressible Navier–Stokes equations in a bounded domain Ω:

$$\frac{\partial\vec{v}}{\partial t}+\nabla\cdot(\vec{v}\otimes\vec{v})+\nabla p-\frac{1}{\mathrm{Re}}\Delta\vec{v}=\vec{f}, \nabla\cdot\vec{v}=0, \text{ in } \Omega, \vec{v}=\vec{v}_b \text{ on } \partial\Omega, \vec{v}|_{t=0}=\vec{v}_0 \quad (2.63)$$

In the two-dimensional case, the Navier–Stokes equations with given velocity field $\vec{v}_b$ at the boundary $\partial\Omega$ of the domain have, on any time interval $[0, t_e]$, a unique solution that is also a classic solution provided that all data of the problem are smooth enough. But in the three-dimensional case, the existence of such solutions has been proved only for sufficiently small data $\vec{v}_0, \vec{v}_b, \vec{f}$ or on sufficiently short intervals of time (Lions, 1996). The existence of weak solutions for the stationary Navier–Stokes equations can be guaranteed in both the two- and three-dimensional cases, but uniqueness has not been proved for all data and all Reynolds numbers; in particular a smallness condition on the data is required which for fixed $\vec{f}$ and $\vec{v}_b$ results in an upper bound for the Reynolds numbers (Galdi, 1994). Another delicate question is the correct formulation of boundary conditions since a prescribed velocity profile – as in the mentioned solvability results above – seems to be questionable, at least at the outflow part of the boundary. A discussion of this issue and its numerical aspects can be found in (Sani and Gresho, 1994). A possible analytical approach to handle this situation can be found in (Kracmar and Nestupa, 2001). In the following, we assume that solutions of the underlying partial differential equations with certain smoothness properties do exist. Only in very special cases and simple domain geometries, like for example the channel flow of an incompressible fluid with no slip boundary condition at the walls and a prescribed parabolic in- and outflow boundary condition, are the Hagen–Poiseuille flow, analytical expressions for the solutions of the Navier–Stokes equations available. Therefore, analytical and/or numerical approximations of solutions play an essential role.

In general the solution of the partial differential equation depends on some dimensionless parameters like the Reynolds, Péclet, Damköhler, or some other numbers. If we are looking for solutions in cases where these parameters achieve very large or small values the method of matched asymptotic expansions can often be applied (Eckhaus, 1973; Varma and Morbidelli, 1997). It is based on the solution of the reduced problem by setting the corresponding parameter equal to the limit case and correcting the reduced solution in order to fulfill the remaining conditions which could not yet be satisfied. The correction solves a simplified local problem generated by a proper scaling and tending the parameter again to the limit. As a result of the matched asymptotic expansion method an analytical approximation of the solution is available where the accuracy depends on how close the parameter is to the limit. If the reduced and/or the local problem cannot be solved analytically, the subproblems can also be approximated numerically. An example of this approach is given in Chapter 4.

In most cases, however, it is hopeless to look for analytical approximations. Then, one has to discretize the partial differential equations. Among the numerical schemes finite difference, finite volume or finite element methods are most popular. In computational fluid dynamics (CFD), discretizations by finite volume (FVM) and finite elements (FEM) are common due to their flexibility to adapt to the geometry of the computational domain. In finite volume methods, the computational domain is subdivided into control volumina (cells) and an integral form of the balance equations is derived by integration over theses cells. Then, in each cell we look for values of the solution of the problem at the barycentre (cell-centered FVM) or at the vertices (cell-vertex FVM). An important problem is the discretization of convective terms which for the standard cell-centered FVM consists of:

$$\int_V \nabla\cdot(\rho\varphi\vec{v})dV = \int_A \rho\varphi\vec{v}\cdot\vec{n}dA \approx \{\rho uA\}_e\{\varphi\}_e - \{\rho uA\}_w\{\varphi\}_w + \{\rho vA\}_n\{\varphi\}_n - \{\rho vA\}_s\{\varphi\}_s + \{\rho wA\}_t\{\varphi\}_t - \{\rho wA\}_b\{\varphi\}_b \quad \text{with} \quad \vec{v} = (u, v, w) \tag{2.64}$$

where V denotes the control volume, A its boundary. The values of the quantities on the east, west, north, south, top and bottom parts of the boundary are expressed by linear interpolation from the neighboring cells. However, it turns out that the resulting difference scheme becomes unstable in the case of dominated convection, that is, at high Reynolds numbers. Therefore, instead of using linear interpolation to evaluate the value of φ on a boundary part of A, the value of φ at the next upwind cell is taken, which corresponds more strongly to the physics of the problem and leads to a stable discretization. When solving incompressible flow problems with a cell-centered FVM for velocity and pressure on the same grid of control volumina unphysical oscillations may appear. To overcome this type of instability staggered grid methods can be used, in which the control volumina for the velocity components and the pressure do not coincide (Noll, 1993). However, on general non-orthogonal grids the staggered grid approach becomes costly due to the need of using grid-oriented components of vectors and tensors (Ferziger and Peric, 1996). Improved pressure–velocity coupling algorithms based on special interpolation formulas on the cell boundaries have been developed to allow also non-staggered grids and to avoid unphysical oscillations of the pressure (Noll, 1993; Ferziger and Peric, 1996).

The discretization of the convective and diffusive terms of the transport equations by FVM is conservative which means that the sum of in and outgoing mass fluxes over a cell is equal to zero. This property is an essential advantage compared to finite differences. However, it is difficult to extend and analyze FVM to higher than first order.

Finite element methods are based directly on the weak formulation of the problem which is also used to investigate the existence and uniqueness of solutions. For them most mathematical tools are available, even the convergence properties of FVM are analyzed by tools developed for FEM (Chou and Ye, 2007). Using higher-order polynomials in FEM any approximation order can be achieved provided that the solutions of the partial differential equations are smooth enough. Another option for increasing the accuracy of finite element solutions is to take profit from superconvergence properties, that is, that the error to the interpolation is of a higher order than the error to the solution itself. A proper postprocessing for the computed finite element solutions allows recovery of this higher order (Matthies, Skrzypacz, and Tobiska, 2005). Examples of this technique are given in Chapter 4. Instabilities caused by dominated convection can be handled for low-order finite element approximations again by upwinding (Tabata, 1977). Unfortunately, as in FVM, upwinding has the tendency to smear out sharp layers of the solution. Therefore, starting in the 1980s new mathematical approaches have been developed which are able to suppress oscillations caused by dominated convection and can be applied to any order of finite elements (Brooks and Hughes, 1982). The streamline-upwind Petrov–Galerkin (SUPG) method consists in adding weighted residuals of the differential equation to the standard finite element method. Other approaches, like the Galerkin least square (GLS) and the residual free bubble (RFB) method have been suggested and analyzed. All these approaches fall into the class of stabilized finite element methods in which the recently

Figure 2.3 Examples of inf-sup stable finite element pairs.

developed stabilization by local projection (LPS) (Matthies, Skrzypacz, and Tobiska, 2007; Ganesan, Matthies, and Tobiska, 2008) seems to be most attractive since it allows to attain the same stability as the SUPG method (Knobloch and Tobiska, 2008) but with less computational costs and in an symmetric manner which is important for optimization. Note that, in particular for higher-order finite elements, stabilized methods do not remove oscillations completely; but in contrast to standard finite element methods they are strongly localized to the regions where there are sharp layers in the solution. Adaptive mesh-refinements in these regions or additional tools like shock capturing techniques can try to improve the remaining situation.

The mass conservation in FEM is in contrast to FVM not automatically guaranteed, however the use of discontinuous pressure approximations in incompressible flow computations can considerably improve the fulfillment of the incompressibility constraint. Moreover, for coupled flow–transport problems advanced finite element methods are available (Matthies and Tobiska, 2007). As in FVM (where staggered grids or special interpolations have been used to solve the Navier–Stokes equation) a compatibility condition between the finite element spaces approximating velocity and pressure is needed to avoid non-physical oscillations in the pressure field. This phenomenon is well understood and results in the use of inf-sup stable finite element[1] pairs for velocity and pressure (Girault and Raviart, 1986; Brezzi and Fortin, 1991). For example, continuous, piecewise quadratic functions for both the velocity space V_h and the pressure space Q_h on triangles or tetrahedrons do not satisfy the inf-sup or Babuska–Brezzi condition:

$$\inf_{q \in Q_h} \sup_{v \in V_h} \frac{\int_\Omega q \nabla \cdot v \, dV}{\sqrt{\int_\Omega q^2 dV} \sqrt{\int_\Omega |\nabla v|^2 dV}} \geq \beta > 0 \tag{2.65}$$

(they are unstable), whereas continuous, piecewise quadratic velocities and continuous, piecewise linear pressures are inf-sup stable. A popular inf-sup stable element pair on quadrilaterals consists of continuous, piecewise biquadratic velocities and discontinuous, piecewise linear pressures (see Figure 2.3). It can be directly extended to the 3-D case.

Summarizing the aspect of inf-sup stable finite element approximations we find that a large number of pairs are known leading to optimal error estimates (Girault and Raviart, 1986; Brezzi and Fortin, 1991; Matthies and Tobiska, 2002, Matthies and Tobiska, 2005). Moreover, since the mid 1980s the technique of adding

1) Inf-sup stable finite element pairs satisfy the Babuska–Brezzi condition.

weighted residuals of the differential equation has been also used to circumvent the inf-sup condition and to allow equal order interpolations for velocity and pressure. It is called the pressure-stabilized Petrov–Galerkin (PSPG) approach. Note that the combined SUPG/PSPG method is widely used in the CFD community due to its ability to handle instabilities caused by dominated convection and violating the inf-sup condition. The LPS method is even more attractive since the same stabilizing effect can be attained with less computational effort and by adding symmetric stabilizing terms.

Let us finally mention that each discretization method (FDM, FVM, FEM) leads to a large system of algebraic equations, the solution of which requires powerful computers. In particular for complex three-dimensional problems and higher order discretizations one quickly reaches the limit of memory and processor speed. Therefore, efficient methods for solving the large systems of algebraic equations are also needed. Fast iterative methods are preferable since they do not allocate additional memory during the solution process. Note that the discretization of the incompressible Navier–Stokes equations leads to a coupled system for velocity and pressure unknowns for which special iterative methods have been developed. The semi-implicit method for pressure-linked equation (SIMPLE) and its variants are popular in finite volume discretizations (Noll, 1993; Ferziger and Peric, 1996). For finite element discretizations multi-level methods with a multiplicative Vanka type smoother belong to the fastest methods. They are based on a block Gauss–Seidel type smoother and use a hierarchy of grid levels. An advanced multi-level multi-discretization solver which combines the advantage of high accuracy of higher-order finite element discretizations with the computational efficiency of multi-grid methods for low-order finite elements has been developed and numerically tested by (John *et al.*, 2002).

2.6 Implementation in FLUENT, MooNMD, COMSOL and ProMoT

Powerful tools of process simulation and analysis, modern mathematical methods and efficient algorithms are essential for the successful treatment of membrane reactors. In the present work, simulations based on CFD have been performed by FLUENT, MooNMD and COMSOL Multiphysics (COMSOL, 2006). The inhouse FEM software package MooNMD (John and Matthies, 2004) turned out to have advantages in case of special fluid–dynamic investigations due to the accessibility of its source code. For solving reduced 1-D + 1-D models toolbox software packages such as ProMoT/Diva and MATLAB have been applied. The use of FLUENT, MooNMD and ProMoT is discussed in the following sections.

2.6.1 Application of FLUENT

The solution of the partial differential equations in FLUENT is based on the control–volume technique. FLUENT 6.2 allows to choose either of two numerical solvers: segregated or coupled solver (Fluent, 2006). The two numerical methods employ a similar discretization process, but the approach used to linearize and solve the discretized equations is different (Fluent, 2006). For the simulations the segregated solver was used, by which the governing equations are solved sequentially. At first, fluid properties are updated, based on the current solution. Each of the momentum equations is solved using the current values for pressure and face mass fluxes, in order to update the velocity fields. If obtained velocities do not satisfy the continuity equation, the pressure correction equation is then solved to obtain the necessary corrections to the pressure and velocity fields and the face mass fluxes. In consequence, the balances for the scalars are solved with the updated values for velocities and pressure. These steps are continued until the convergence criteria are fulfilled (Fluent, 2006).

Referring to the simulation parameters, the first-order upwind discretization was chosen for the convective fluxes. For the approximation of the pressure gradients from the control volume center to the volume faces, the standard pressure interpolation scheme was used, taking into account the momentum equation coefficients. The pressure–velocity coupling was achieved by using the SIMPLE algorithm. The total and component mass balances and energy balance were controlled for the whole computational domain for each simulation and on the basis of the conservative character of the control–volume method always fulfilled. The source terms in the momentum and mass balances and the material properties of gases and porous media were defined for the different reactor zones in separate equations using user-defined functions (UDF).

2.6.2
Application of MooNMD

The C++ finite element package MooNMD (John and Matthies, 2004) was developed at the Institute of Analysis and Numerics at the University of Magdeburg as a result of *M*athematics and *o*bject-*o*riented *N*umerics in *M*ag*D*eburg. It is based on the discretization of partial differential equations by mapped finite elements. In contrast to most commercial software products the MooNMD code is open and fully extendable. This allows one to study the properties of advanced numerical schemes and to implement newly developed concepts for the robust and efficient solution of the nonlinear algebraic systems of equations. The package is in particular suited for the solution of incompressible flow problems in two- and three-dimensional domains decomposed into triangular/tetrahedral and quadrilateral/hexahedral meshes, respectively. Several benchmark computations have been performed using MooNMD which show its high accuracy and wide range of applicability (John, 2002; Lavrova *et al.*, 2002; Ganesan and Tobiska, 2008; Iliescu *et al.*, 2003). The most important features of MooNMD are:

- Higher-order finite elements of conforming and non-conforming type based on the concept of a family of reference mappings (Matthies and Tobiska, 2002, 2005)

- Isoparametric finite elements for better approximation of domains with curved boundaries
- Stabilization techniques like SUPG (Matthies and Tobiska, 2001; John *et al.*, 1998) or LPS (Matthies, Skrzypacz, and Tobiska, 2007, 2008) to handle instabilities caused by dominated convection and/or unstable finite element pairs
- Various shock capturing methods to suppress spurious oscillations (John and Knobloch, 2007, 2008)
- Postprocessing tools to enhance accuracy by superconvergence (Matthies, Skrzypacz, and Tobiska, 2005)
- State-of-the-art iterative solvers like flexible GMRES (Saad 2003), multiple-discretization multi-level approach with Vanka type smoothers for mixed problems (John *et al.*, 2002) and fast direct solver UMFPACK (Davis, 2004)
- Built-in output to graphical tools, for example, VTK (Schröder, Martin, and Lorensen, 2006)

The flexibility and adaptivity of MooNMD to user-defined problems allow full control over the design of decoupling strategies and the solution process for systems of nonlinear partial differential equations.

2.6.3
Application of ProMoT

Both FLUENT and MoonMD are simulation tools in the sense that they expect a fully developed model equation system from the user and provide an efficient numerical solution of this system. The approach of the process modeling tool ProMoT (Tränkle *et al.*, 2000) presented in this subsection is different: instead of solving a given model, ProMoT aims at supporting the model development process. The numerical solution of the models generated by ProMoT is done in other tools like DIVA (Köhler *et al.*, 2001), MATLAB, or DIANA (Krasnyk *et al.*, 2006). The need for computer-aided modeling of membrane reactors results from the complexity of the model approaches described above. A membrane reactor model has to meet demands on validity of the model on the one hand, but also on implementation issues on the other hand. Meeting the first demand, that is, developing a realistic membrane reactor model, is an iterative process: a model typically requires a lot of refinement until it proves to be adequate for the solution of a given problem. Once a model has been validated and found to be applicable to a design or a control problem, the right way to implement the model becomes important. Traditionally, the model of a chemical apparatus is implemented in a monolithic way without much internal structuring. This makes the understanding and debugging of the model difficult. Monolithically structured models are not very transparent and are hardly reusable for another modeler. Furthermore, the implementation of complicated differential equations in a flow-sheet simulator is tedious and error prone. Finally, in most simulation tools it is the responsibility of the modeler to formulate his models in a manner suitable for numerical treatment, for example, to avoid a higher differential index of a differential algebraic system.

From the requirements on model formulation and on model implementation, the main objectives of a computer-aided modeling tool follow immediately: a modeling tool should: (a) let a user concentrate on the physical modeling task and relieve him from mechanical coding work, (b) increase the reusability and transparency of existing models, (c) simplify the debugging process during model development and (d) provide libraries of predefined building blocks for standard modeling tasks such as reaction kinetics, physical properties, or transport phenomena.

The process modeling tool ProMoT is a software environment that possesses all the properties listed above. It is used as a framework for a model library of structured membrane reactor models. The development of such a library consists of two main steps: (a) the choice of a suitable structuring methodology for the models and (b) the implementation of the structured models in ProMoT. Both steps are discussed in the following.

The usefulness of a model library strongly depends on the structure of the implemented models. Typically, process models in flow-sheet simulators are formulated in terms of process unit models. However, process unit models are not suitable elementary units of a model library, because they are too complex and contain too much information on a specific process. In view of the many existing variants of chemical reactors, a library of reactor models will always be incomplete. It seems more promising to divide the process unit models further into smaller subunits and to store the subunits in the model library. The network theory of chemical processes (Gilles, 1998; Mangold, Motz, and Gilles, 2002) gives a guideline for the internal structuring of process unit models. The basic idea is to describe a process model by two types of elementary functional units: *components* and *coupling elements*. Components possess a hold-up for physical quantities like energy, mass, or momentum. They are described by a thermodynamic state or a state vector. The state of a component may be changed by fluxes or flux vectors. The task of the second class of functional units, the coupling elements, is to determine these flux vectors. In accordance with the principles of irreversible thermodynamics, it is assumed that the flux vector is an algebraic function of potential differences and potential gradients. As an example, consider the general balance (same as Equation 2.5):

$$\frac{\partial(\rho\varphi)}{\partial t} = -\nabla\cdot(\rho\varphi\vec{v}) - \nabla\vec{j}_{\varphi} + \dot{s}_{\varphi} \tag{2.66}$$

In the methodology of the network theory, the left-hand side of the above equation can be seen as a storage for field quantity $\rho\varphi$, and the three terms on the right-hand side can be represented by coupling elements for convective transport, non-convective transport and internal sinks and sources, respectively. Figure 2.4 shows a formal graphical representation of Equation 2.66, which uses the symbols introduced in (Gilles, 1998).

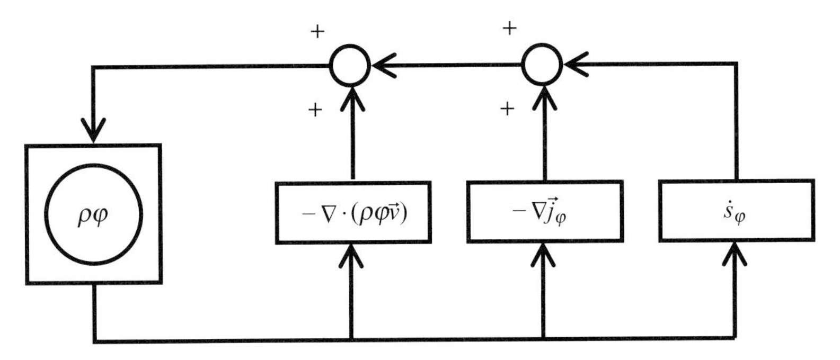

Figure 2.4 Graphical representation of the general balance by the symbols of the network theory (Gilles, 1998).

The storage element $\rho\varphi$ delivers information on its internal state to the three coupling elements, which in turn use this information to determine the fluxes that change the component's state. One should note that Figure 2.4 is simplified, because in general it is not only the field quantity φ but the complete thermodynamic state vectors that determine the fluxes.

The decomposition of a process model into components and coupling elements can be carried out on different hierarchical levels. A connected system of components and coupling units on one level can always be considered as an aggregated component on the next higher level. For example, models of process units can be considered as components of a plant model. This structures the plant on the level of process units. However, the model of a process unit can be decomposed into models of thermodynamic phases interacting via phase boundaries. Therefore, thermodynamic phases can be considered as components on a level of phases; the phase boundaries are the corresponding coupling elements. Finally, a thermodynamic phase is able to store macroscopic thermodynamic quantities like mass or internal energy, the components on the level of macroscopic thermodynamic storages. The coupling elements on that level are mass and energy transport phenomena inside a phase as well as chemical reactions. The decomposition of the general balance (2.66) shown in Figure 2.4 occurs on the level of storages.

By applying the described structuring concept, a model library of distributed dynamic packed-bed and membrane reactor models has been implemented in the process modeling tool ProMoT. The model library was described in detail in (Mangold, Ginkel, and Gilles, 2004). Only a brief overview is given here.

The library contains modules for the formulation of models for heterogeneous catalytic reactors with gas phase reactions. Dynamic homogeneous, heterogeneous and pseudo-homogeneous models with one space coordinate are considered. The model library in ProMoT differs from a library of process units in a traditional flow-sheet simulator mainly in two respects. The first difference is the deep structuring below the level of process units. A membrane reactor model is decomposed into models of several thermodynamic phases, for example, for the membrane, the sweep gas side and the tubular side. Each phase model is built up from modules for mass storage and energy storage, from modules for transport phenomena inside a thermodynamic phase like convection, diffusion, or heat conduction, as well as from modules describing sinks and sources in the phase due to chemical reaction. These modules on the level of storages are the elementary modeling entities in the model library. The advantage of this fine granulation of the model structure is that a comparatively small amount of elementary modeling units or building blocks suffices to describe a wide variety of membrane reactor models.

The second difference is that object oriented concepts like aggregation and inheritance can be used to formulate a model. Aggregation can be seen as the software representation of the decomposition levels described above. Inheritance stands for the ability of a model to adopt properties like equations or variable definitions from another model, the so-called super-class. In ProMoT it is possible to use multiple inheritance, that is, a model can inherit from several super-classes.

In the following, the simulation of a composite membrane as described by (Hussain, 2006) is discussed. The membrane consists of four layers of varying thickness and porosity. A model of the membrane in a testing environment as shown in Figure 2.5 is to be developed in ProMoT.

In a first step, a model of a single membrane layer is constructed from predefined building blocks in the model library. The isothermal dusty gas model is used to describe the membrane layer. A screenshot of the single layer model in ProMoT is shown in Figure 2.6.

It can be seen that the model contains two spatially distributed storage elements for total mass and component masses corresponding to total mass and component mass balances. As the membrane is assumed to be isothermal, a reservoir with constant temperature is added instead of an energy storage. A connecting element,

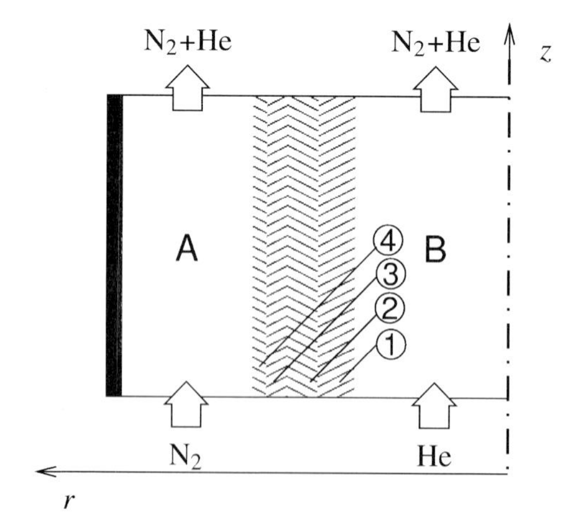

Figure 2.5 Scheme of a testing environment for a composite membrane with four layers denoted as (1)–(4).

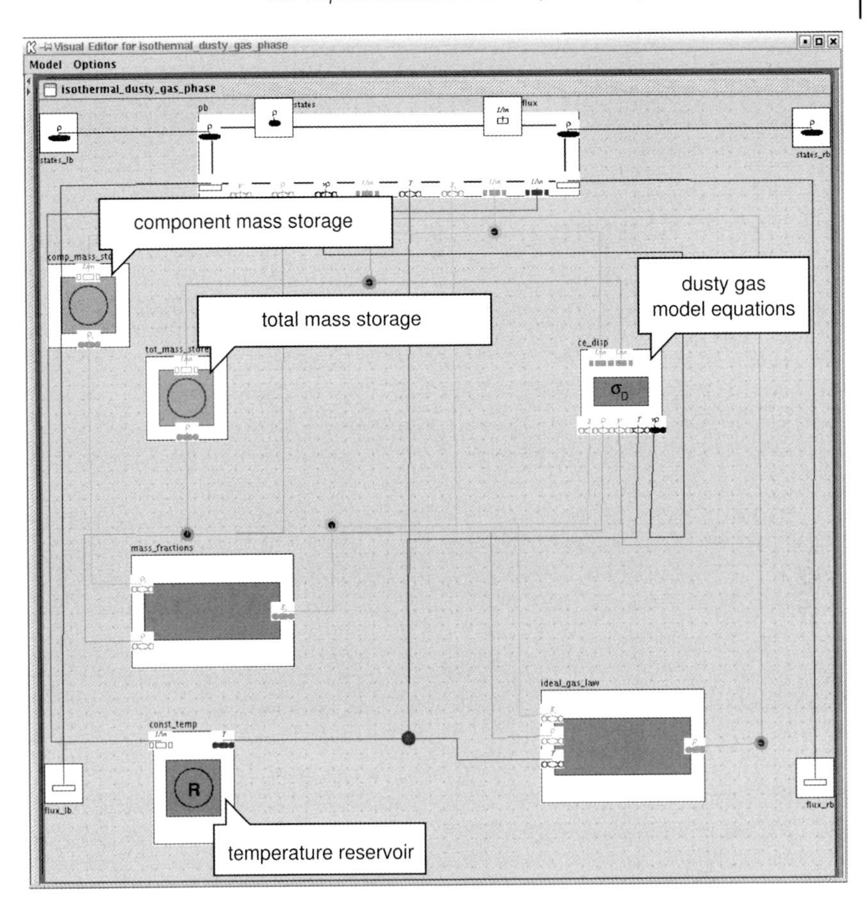

Figure 2.6 Structured ProMoT model of a single layer of the composite membrane.

which contains phenomenological relations of the mass fluxes according to the dusty gas model, describes the mass transport through the membrane. Two additional building blocks, one of them converting partial densities into mass fractions and the other computing the total pressure from the ideal gas law, complete the model. Extensions of the single layer model are straightforward, for example, the extension to a nonisothermal model by substituting the constant temperature reservoir with an energy storage, or the extension to a reactive membrane layer by adding a reactive connecting element.

Using aggregation, the assembly of connected modules in Figure 2.6 can be fused to a new modeling entity, which describes a single membrane layer. In order to model the composite membrane in Figure 2.5, four of those aggregated modules are necessary. Each of the four modules possesses identical model equations, but a different set of model parameter values to account for the different properties of the four membrane layers.

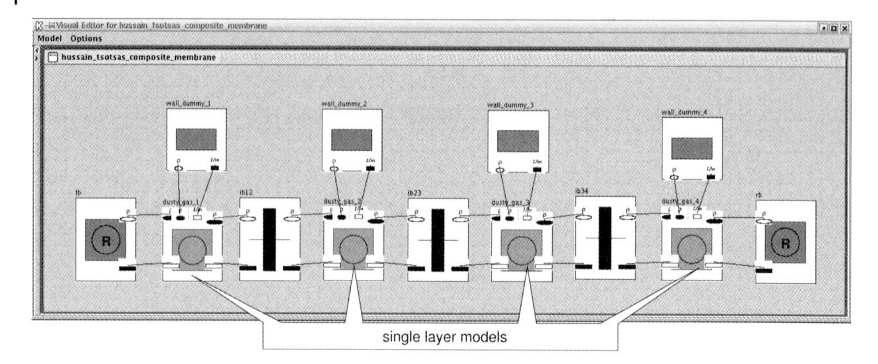

Figure 2.7 Structured ProMoT model of the membrane process shown in Figure 2.5. The single layer models can be decomposed into the structure shown in Figure 2.6.

The ProMoT model of the complete membrane is shown in Figure 2.7. In addition to the building blocks for the membrane layers it contains connecting elements for the internal boundary conditions between the layers and modules for the description of the gas bulks on the left-hand side and on the right-hand side of the membrane.

2.7 Conclusion

The diversity concerning configuration and operation of membrane reactors requires a general, common frame for developing and solving respective models. After presenting this frame, we can proceed in the following chapters with the identification of model parameters and the detailed treatment of specific membrane reactor configurations. Hereby, selected parts of the theoretical models introduced and the tools described are used.

Notation used in this Chapter

Latin Notation

A	m^2	area
B_0	m^2	permeability parameter
C	m^2/s	diffusion coefficient matrix in DGM
c_p	J/(kg K)	specific heat capacity
$\tilde{c}$	mol/m^3	molar concentration
D_{ij}	m^2/s	diffusion coefficient
d	m	diameter
e	J/kg	specific energy

F	Pa/m	driving force vector in DGM
F_0		diffusion parameter
$\vec{f}$	N/m^3	force source term
f_1	$Pa\,s/m^2$	viscous resistance factor
f_2	$Pa\,s^2/m^3$	inertial resistance factor
g	m/s^2	gravitational acceleration
h	J/kg	specific enthalpy
$\tilde{h}$	J/mol	molar enthalpy
$\tilde{h}^0$	J/mol	molar enthalpy of formation
$\tilde{h}_i^*$	J/mol	partial molar enthalpy
I		identity matrix
K_0	m	knudsen parameter
l	m	mean free path
L	m	length
M	kg	mass
$\tilde{M}$	kg/kmol	molar mass
$\vec{m}$	$kg/(m^2 s)$	mass flux
$\dot{n}$	md/s	permeate flux of the membrane
$\vec{n}$	$mol/(m^2 s)$	molar flux
p	Pa	pressure
$\dot{q}$	W/m^2	heat flux
$\tilde{R}$	J/(mol K)	universal gas constant
R_i	$kg/(m^3 s)$	reaction source term
r_j	mol/(kg s)	rate of reaction
r	m	radial coordinate
$\dot{s}$	$mol/(m^2 s)$, $kg/(m^3 s)$	source term
T	K, °C	temperature
t	s	time
$\vec{u}$	m/s	volume average velocity
$\vec{u}_i$	m/s	velocity of component i
V	m^3	volume
$\vec{v}$	m/s	mass average velocity, superficial velocity
v_f	m/s	interstitial velocity
$\tilde{v}$	m^3/mol	molar volume
$\tilde{v}_i^*$	m^3/mol	partial molar volume of component i
y		mass fraction
$\tilde{y}$		mole fraction
z	m	axial coordinate

Greek Notation

α		molar flux ratio parameter
Γ		matrix of non-ideal coefficients

ε		porosity
η	Pa s	dynamic viscosity
κ	Pa s	dilatational viscosity
λ	W/(m K)	thermal conductivity
μ	J/mol	chemical potential
ν_{ij}		stoichiometric coefficient
ρ	kg/m^3	density
ρ_i^*	kg/m^3	partial density of component i
$\bar{\rho}_f$	kg/m^3	average density
τ		tortuosity
$\bar{\bar{\tau}}$	Pa	stress tensor
φ		arbitrary field quantity

Super- and Subscripts

b	boundary
cat	catalyst
cor	corrected, correction
D	total diffusion
Diff	molecular diffusion
e, eff	effective
f	fluid
g	gas
i	inner (tube side)
i,j	component index, reaction index
K	Knudsen
M	Stefan–Maxwell
m	membrane
mol	molecule
n	number of components, inert species
nr	number of reactions
o	outer (shell side)
p	at constant pressure
p	particle
pb	pseudo-binary
pore	pore
r	radial
ref	reference
s	solid
SS	shell side
T	at constant temperature

TS	tube side
t	transposed
v	viscous flow
x	in x-direction
y	in y-direction
z	axial
φ	for arbitrary field quantity
φ	in φ-direction (circumferential)
0	initial value
∞	in the core of the bed
*	partial

References

Amann, H. (2001) On the strong solvability of the Navier-Stokes equations. *J. Math. Fluid Mech.*, **2**, 16–98.

Baranowski, B. (1975) Nichtgleichgewichts-Thermodynamik in der physika-lischen Chemie, Dt. Verl. für Grundstoffindustrie, Leipzig.

Bird, R.B., Stewart, W.E., and Lightfoot, E.N. (2002) *Transport Phenomena*, John Wiley & Sons, Inc., New York.

Bosanquet, C.H. (1944) British TA Rept., BR-507:770.

Brezzi, F., and Fortin, M. (1991) *Mixed and Hybrid Finite Element Methods*, Springer, Berlin.

Brooks, A.N., and Hughes, T.J.R. (1982) Streamline upwind/Petrov-Galerkin formulation for convection dominated flows with particular emphasis on the incompressible Navier-Stokes equations. *Comput. Methods Appl. Mech. Engrg.*, **32**, 199–259.

Chou, S.H., and Ye, X. (2007) Unified analysis of finite volume methods for second order elliptic problems. *SIAM J. Numer. Anal.*, **45** (4), 1639–1653.

COMSOL (2006) *Handbook, COMSOL Multiphysics 3.2*, 1997–2007 COMSOL, Inc., All Rights Reserved.

Cussler, L. (1997) *Diffusion, Mass Transfer in Fluid Systems*, Cambridge University Press.

Davis, T.A. (2004) Algorithm 832: UMFPACK V4.3–an unsymmetric-pattern multifrontal method. *ACM Trans. Math. Softw.*, **30** (2), 196–199.

de Groot, S.R., and Mazur, P. (1984) *Non-Equilibrium Thermodynamics*, Dover Publ., New York.

Deriagin, B.V., and Bakanov, S.P. (1957) Theory of gas flow in a porous body near the Knudsen region: pseudomolecular flow. *Sov Phys. Dokl.*, **2**, 326.

Dixon, A., and Nijemeisland, M. (2001) CFD as a design tool for fixed-bed reactors. *Ind. Eng. Chem. Res.*, **40**, 5246–5254.

Dixon, A., Nijemeisland, M., and Stitt, H. (2003) CFD simulation of reaction and heat transfer near the wall of a fixed bed. *Int. J. Chem. React. Eng.*, **1**, A22.

Eckhaus, W. (1973) *Matched Asymptotic Expansions and Singular Perturbations, Mathematic Studies* 6, North Holland/ American Elsevier, Amsterdam, New York.

Elnashaie, S.S.E.H., and Elshishini, S. (1993) *Modelling, Simulation and Optimization of Industrial Fixed Bed Catalytic Reactors*, Gordon and Breach, London.

Ergun, S. (1952) Fluid flow through packed columns. *Chem. Eng. Prog.*, **48**, 9–94.

Evans, R., Watson, G., and Maso, E. (1962) Gaseous diffusion in porous media at uniform pressure. *J. Chem. Phys.*, **35**, 2076–2083.

Ferziger, J.H., and Peric, M. (1996) *Computational Methods for Fluid Dynamics*, Springer, Berlin.

FLUENT (2006) FLUENT 6.2 Help.

Forchheimer, P. (1901) Wasserbewegung durch Goden. *Zeits. V. deutsch. Ing.*, **45**, 1782–1788.

Fuller, E., Schettler, P., and Giddings, J. (1966) New method for prediction of binary gas-phase diffusion coefficients. *Ind. Eng. Chem. Res.*, **58**, 18–27.

Galdi, G.P. (1994) *An Introduction to the Mathematical Theory of the Navier-Stokes Equations,Vol. I, Linearized Steady Problems, Vol. II, Nonlinear Steady Problems*, Springer, New York.

Ganesan, S., and Tobiska, L. (2008) An accurate finite element scheme with moving meshes for computing 3D-axisymmetric interface flows. *Int. J. Numer. Methods Fluids*, **57** (2), 119–138.

Ganesan, S., Matthies, G., and Tobiska, L. (2008) Local projection stabilization of equal order interpolation applied to the Stokes problem, Math. Comp., electr. published May 9, 2008.

Georgieva, K., Mednev, I., Handtke, D., and Schmidt, J. (2005) Inflence of the operating conditions on yield and selectivity for partial oxidation of ethane in a catalytic membrane reactor. *Catal. Today*, **101**, 168–176.

Gilles, E.D. (1998) Network theory for chemical processes. *Chem. Engng. Technol.*, **21**, 121–132.

Girault, V., and Raviart, P.-A. (1986) *Finite Element Methods for Navier-Stokes Equations*, Springer, Berlin.

Haase, R. (1963) *Thermodynamik Der Irreversiblen Prozesse*, Steinkopff, Darmstadt.

Hunt, M., and Tien, C. (1990) Non-Darcian flow, heat and mass transfer in catalytic packed-bed reactors. *Chem. Eng. Sci.*, **45** (1), 55–63.

Hussain, A. (2006) Heat and mass transfer in tubular inorganic membranes, PhD thesis, Otto-von-Guericke-Universität, Magdeburg.

Iliescu, T., John, V., Layton, W., Matthies, G., and Tobiska, L. (2003) A numerical study of a class of LES models. *Int. J. Comput. Fluid Dynamics*, **17** (1), 75–85.

Jackson, R. (1977) *Transport in Porous Catalysts*, Elsevier, Amsterdam.

John, V. (2002) Higher order finite element methods and multigrid solvers in a benchmark problem for the 3D Navier-Stokes equations. *Int. J. Numer. Methods Fluids*, **40** (6), 775–798.

John, V., and Knobloch, P. (2007) On spurious oscillations at layer diminishing (SOLD) methods for convection-diffusion equations, Part I – a review. *Comput. Methods Appl. Mech. Engrg*, **196**, 2197–2215.

John, V., and Knobloch, P. (2008) On spurious oscillations at layer diminishing (SOLD) methods for convection-diffusion equations, Part II – Analysis for P_1 and Q_1 finite elements. *Comput. Methods Appl. Mech. Engrg*, **197**, 1997–2014.

John, V., and Matthies, G. (2004) MooNMD – a program package based on mapped finite element methods. *Comput. Vis. Sci.*, **6** (2–3), 163–169.

John, V., Matthies, G., Schieweck, F., and Tobiska, L. (1998) A streamline-diffusion method for nonconforming finite element approximations applied to convection-diffusion problems. *Comput. Methods Appl. Mech. Engrg*, **166** (1–2), 85–97.

John, V., Knobloch, P., Matthies, G., and Tobiska, L. (2002) Non-nested multi-level solvers for finite element discretizations of mixed problems. *Computing*, **68**, 313–341.

Kerkhof, P., and Geboers, M. (2005) Towards a unified theory of isotropic molecular transport phenomena. *AIChE J.*, **51** (1), 79–121.

Knobloch, P., and Tobiska, L. (2008) *On the Stability of the Finite Element Discretization of Convection-Diffusion-Reaction Equations, Preprint 08-11*, Otto-von-Guericke-University Magdeburg.

Köhler, R., Mohl, K.D., Schramm, H., Zeitz, M., Kienle, A., Mangold, M., Stein, E., and Gilles, E.D. (2001) Method of lines within the simulation environment DIVA for chemical processes, in *Adaptive Method of Lines* (eds A. van de Wouver, P. Saucez, and W.E. Schiesser), Chapman & Hall, London, pp. 371–406.

Kracmar, S., and Nestupa, J. (2001) A weak solvability of a steady variational inequality of the Navier-Stokes type with mixed boundary conditions. *Nonlinear Anal.*, **47**, 4169–4180.

Krasnyk, M., Bondareva, K., Milokhov, O., Teplinskiy, K., Ginkel, M., and Kienle, A. (2006) The ProMoT/DIANA simulation environment, in *Proceedings of the 16th European Symposium on Computer Aided*

Process Engineering (eds W. Marquardt and C. Pantelides), Elsevier, London, pp. 445–450.

Krishna, R., and van Baten, J.M. (2009) An investigation of the characteristics of Maxwell-Stefan diffusivities of binary mixtures in silica nanopores. *Chem. Eng. Sci.*, **64**, 870–882.

Lavrova, O., Matthies, G., Mitkova, T., Polevikov, V., and Tobiska, L. (2002) Finite element methods for coupled problems in ferrohydrodynamics, in *Challanges in Scientific Computing–CISC* (ed. E. Bänsch), Springer, pp. 160–183, Lect. Notes Comput. Sci. Eng. 35, 2003.

Lions, P.-L. (1996) *Mathematical Topics in Fluid Mechanics, Vol. 1 Incompressible Models, Vol. 2 Compressible Models*, Oxford Science Publications.

Lucas, K., and Luckas, M. (2002) *Berechnungsmethoden Für Stoffeigenschaften, Section Da in: VDI-Wärmeatlas*, 9th edn, Springer, Berlin.

Mackey, M.C. (1971) Kinetic theory model for ion movement through biological membranes. *Biophys. J.*, **11**, 75–90.

Mangold, M., Motz, S., and Gilles, E.D. (2002) Network theory for the structured modelling of chemical processes. *Chem. Engng. Sci.*, **57**, 4099–4116.

Mangold, M., Ginkel, M., and Gilles, E.D. (2004) A model library for membrane reactors implemented in the process modeling tool ProMoT. *Comput. Chem. Eng.*, **28**, 319–332.

Manjhi, N., Verma, N., Salem, K., and Mewes, D. (2006) Lattice Boltzmann modelling of unsteady-state 2D concentration profiles in adsorption bed. *Chem. Eng., Sci.*, **61**, 2510–2521.

Masel, R. (2001) *Chemical Kinetics and Catalysis*, John Wiley & Sons, Inc., New York.

Mason, E.A., Malinauskas, A.P., and Evans, R.B. (1967) Flow and diffusion of gases in porous media. *J. Chem. Phys.*, **46**, 3199–3216.

Matthies, G., Skrzypacz, P., and Tobiska, L. (2005) Superconvergence of a 3D finite element method for stationary Stokes and Navier-Stokes problems. *Numer. Methods Part. Diff. Equat.*, **21**, 701–725.

Matthies, G., Skrzypacz, P., and Tobiska, L. (2007) A unified convergence analysis for local projection stabilization applied to the Oseen problem. *ESAIM: M2AN*, **41** (4), 713–742.

Matthies, G., Skrzypacz, P., and Tobiska, L. (2008) Stabilization of local projection type applied to convection-diffusion problems with mixed boundary conditions. *ETNA*, **32**, 90–105.

Matthies, G., and Tobiska, L. (2001) The streamline-diffusion method for conforming and nonconforming finite elements of lowest order applied to convection-diffusion problems. *Computing*, **66** (4), 343–364.

Matthies, G., and Tobiska, L. (2002) The inf-sup condition for the mapped Q_k-P_k-1^disc element in arbitrary space dimensions. *Computing*, **69**, 119–139.

Matthies, G., and Tobiska, L. (2005) Inf-sup stable nonconforming finite elements of arbitrary order on triangles. *Numer. Math.*, **102**, 293–309.

Matthies, G., and Tobiska, L. (2007) Mass conservation of finite element methods for coupled flow-transport problems. *Int. J. Comput. Sci. Math.*, **1**, 293–307.

Melin, T., and Rautenbach, R. (2007) *Membranverfahren: Grundlagen der Modul- und Anlagenauslegung*, Springler, Berlin.

Noll, B. (1993) *Numerische Strömungsmechanik*, Springer, Berlin.

Poling, B.E., Prausnitz, J.M., and O'Connell, J.P. (2001) *The Properties of Gases and Liquids*, McGraw-Hill, New York.

Reid, R.C., Prausnitz, J.M., and Sherwood, T.K. (1977) *The Properties of Gases and Liquids*, 3th edn, McGraw-Hill, New York.

Russo, R., and Simader, C. (2006) A note on the existence of solutions to the Oseen system in Lipschitz domains. *J. Math. Fluid Mech.*, **8** (1), 64–76.

Saad, Y. (2003) *Iterative Methods for Sparse Linear Systems*, SIAM, Philadelphia.

Sani, R.L., and Gresho, P.M. (1994) Resume and remarks on the open boundary condition minisymposium. *Int. J. Numer. Methods Fluids*, **18**, 983–1008.

Schröder, W., Martin, K., and Lorensen, B. (2006) *The Visualization Toolkit: An Object-Oriented Approach to 3D Graphics*, Kitware.

Tabata, M. (1977) A finite element approximation corresponding the upwind differencing. *Mem. Numer. Math.*, **1**, 47–63.

Taylor, R., and Krishna, R. (1994) *Multicomponent Mass Transfer*, John Wiley & Sons, Inc., New York.

Tota, A., Hlushkou, D., Tsotsas, E., and Seidel-Morgenstern, A. (2007) Packed bed membrane reactors, Chapter 5, in *Modeling of Process Intensification* (ed. F.J. Keil), Wiley-VCH Verlag GmbH, Weinheim, pp. 99–148.

Tränkle, F., Zeitz, M., Ginkel, M., and Gilles, E.D. (2000) ProMoT: a modeling tool for chemical processes. *Math. Comp. Model. Dyn. Syst.*, **6**, 283–307.

Tsotsas, E. (1991) *Über Die Wärme- Und Stoffübertragung in Durchströmten Festbetten: Experimente, Modelle, Theorien, VDI Fortschr*, Ber., Ser. 3, No. 223, VDI Verlag, Düsseldorf.

Varma, A., and Morbidelli, M. (1997) *Mathematical Methods in Chemical Engineering*, Oxford University Press.

Wesselingh, J.A., and Krishna, R. (2000) *Mass Transfer in Multi-Component Mixtures*, Delft University Press.

Young, J.B., and Todd, B. (2005) Modelling of multi-component gas flows in capillaries and porous solids. *Int. J. Heat Mass Transf.*, **48**, 5338–5353.

Zajaczkowski, W.M. (2007) Some global regular solutions to Navier-Stokes equations, *Math. Methods Appl. Sci.*, **30** (2), 123–151.

3
Catalysis and Reaction Kinetics of a Model Reaction

Frank Klose, Milind Joshi, Tanya Wolff, Henning Haida, Andreas Seidel-Morgenstern, Yuri Suchorski, and Helmut Weiß

3.1
Introduction

The development of new and the improvement of existing reactor concepts is difficult without understanding the reaction mechanism and – if a catalyst is used – its operation principles. In the ideal case an accurate kinetic model with detailed structure–activity relations would be available. However, in industrial practice in general only simplified strategies are applied in order to minimize experimental effort and to reduce the complexity of the corresponding mathematical expressions. Often simplified kinetic models of the power law type are used to describe reaction rates and catalytic performance for a limited parameter range. Nevertheless, for the understanding and the correct prediction of the performance of alternative reactor concepts, as the membrane reactors treated in this book, a more detailed understanding of the kinetic relations is essential. In membrane reactors the concentration profiles of the reactants and the contact time profiles in the catalyst bed are influenced simultaneously. A reliable kinetic model has to describe also the reaction behavior for conditions which are far from those present in conventional packed-bed reactors.

In order to deliver reliable data for the comparison between conventional and membrane reactors on the one hand, and between different membrane reactor concepts on the other, it is important to study the same reaction system and to apply the same catalyst. In this and the following chapters, the focus is set on the catalyzed oxidative dehydrogenation of ethane (ODHE). Although this reaction is currently far from a wide industrial application, it has several advantages for the purpose of this conceptual study:

- The expected reaction network is limited to a rather low number of main products and byproducts, which facilitates mechanistic as well as kinetic studies.
- There are no condensable products under reaction conditions, which is beneficial for monitoring the reactor performance with high accuracy.

Membrane Reactors: Distributing Reactants to Improve Selectivity and Yield.
Edited by Andreas Seidel-Morgenstern

ISBN: 978-3-527-32039-4

- The reactants and products can be efficiently and reliably studied by gas chromatography.
- The reaction temperature is high enough to evaluate possible difficulties in the practical applications of the reactors, but also low enough to realize and to operate them safely on a pilot plant scale.
- The performance of conventional packed-bed reactors is comparably poor, and positive effects of new reaction engineering strategies should be well detectable.
- It should be possible to transfer major aspects of the ODHE to the selective oxidation of other hydrocarbons, for example, the oxidative dehydrogenation of propane (ODHP, see Chapter 5) or the selective oxidation of butane to maleic anhydride (Chapter 7).

Next to the choice of the reaction itself, the selection of a suitable catalyst is of importance. In particular, it should allow investigating the interplay between reactants, intermediates, products and the catalyst surface itself. With respect to the oxidative dehydrogenation of ethane, a wide variety of catalytic systems has already been studied and a considerable number of reviews published (Banares, 1999; Cavani and Trifiro, 1999; Grasselli, 1999; Bhasin *et al.*, 2001; Dai and Au, 2002). Based on the reaction mechanism, the catalytically active components can be divided into four main classes, namely: (a) reducible oxides of non-noble transition metals, including perovskites and related catalysts, (b) non-reducible oxides of the elements from the Ia, IIa and IIIb groups including lanthanoids, (c) platinum-based catalysts, usually applied for deep oxidation and (d) all catalysts not included in the previous classes. Beside from the reaction mechanism, these classes differ also in their operation temperatures and the maximum accessible ethylene yields.

In a study of ODHE for different membrane reactors of the distributor type and a comparison with conventional reactors it is important to use the same active catalyst in all experiments. The goal of the work presented here was not to find an optimized catalyst, but to use one which: (i) can be prepared in an easy and reproducible way and in larger quantities, (ii) is rather cheap in preparation and (iii) has proven before to be catalytically active in ODHE. All these prerequisites are fulfilled for some of the class (a) catalysts, that is, reducible oxides of non-noble transition metals. In particular, it is possible to produce catalytically active layers of these materials on different supports like, for example, alumina, silica or titania. For this study, supported vanadia catalysts were chosen for reasons which are explained in more detail below.

This chapter is organized as follows. We start with a short description of the reaction network, which was deduced from results of experimental studies using vanadia catalysts. A preliminary power law analysis shows that a distributed-feed concept should have the potential to improve the selectivity towards the desired hydrocarbon, for example, ethylene. We then further describe the preparation, characterization and properties of the vanadia catalysts used. Finally, we provide results of a more detailed analysis and suggest a quantitative model for the reaction kinetics of the reaction network valid for the supported vanadium oxide catalyst.

3.2 The Reaction Network of the Oxidative Dehydrogenation of Ethane

Using the above mentioned vanadia catalysts supported on γ-alumina (VO_x/γ-Al_2O_3) which are described in detail below, the reaction network of the oxidative dehydrogenation of ethane has been investigated in depth performing experiments in conventional packed-bed reactors. In these experiments, the feed concentrations of both hydrocarbon and oxygen, the temperature, the gas hourly space velocity (GHSV) and the vanadium loading on the catalyst support have been varied in a systematic manner. Here only essential findings of these investigations are given. The results of a more detailed kinetic analysis are described in Section 3.4.

Summarizing the experimental findings, the following main reactions were identified (Klose *et al.*, 2004):

$$C_2H_6 + 0.5O_2 \rightarrow C_2H_4 + H_2O \tag{3.1}$$

$$C_2H_6 + 3.5O_2 \rightarrow 2CO_2 + 3H_2O \tag{3.2}$$

$$C_2H_4 + 2O_2 \rightarrow 2CO + 2H_2O \tag{3.3}$$

$$C_2H_4 + 3O_2 \rightarrow 2CO_2 + 2H_2O \tag{3.4}$$

$$CO + 0.5O_2 \rightarrow CO_2 \tag{3.5}$$

$$C_2H_6 + 2.5O_2 \rightarrow 2CO + 3H_2O \tag{3.6}$$

Reaction 3.3 was formulated due to the fact that no acetaldehyde could be detected in the measurements reported below. In this reaction are lumped the acetaldehyde formation from ethylene, its oxidation to CO and the direct ethylene oxidation to CO:

$$\begin{array}{c} C_2H_4 + 0.5O_2 \rightarrow CH_3CHO \\ \underline{CH_3CHO + 1.5O_2 \rightarrow 2CO + 2H_2O} \\ C_2H_4 + 2O_2 \rightarrow 2CO + 2H_2O \end{array} \tag{3.7}$$

Under oxygen-free conditions, a number of side reactions contributing to ethylene formation and to its consumption have also to be taken into account. Such side reactions are:

$$\begin{array}{l} C_2H_6 + CO_2 \rightarrow C_2H_4 + CO + H_2O \\ C_2H_4 + 4CO_2 \rightarrow 6CO + 2H_2O \\ C_2H_4 \rightarrow 2C + 2H_2 \\ 2CO \rightarrow C + CO_2 \end{array} \tag{3.8}$$

A major challenge is the accurate measurement of the extent of these side reactions. Especially, the reliable quantification of carbon deposits on the catalyst surface by ethylene pyrolysis and Boudouard reactions is very complicated.

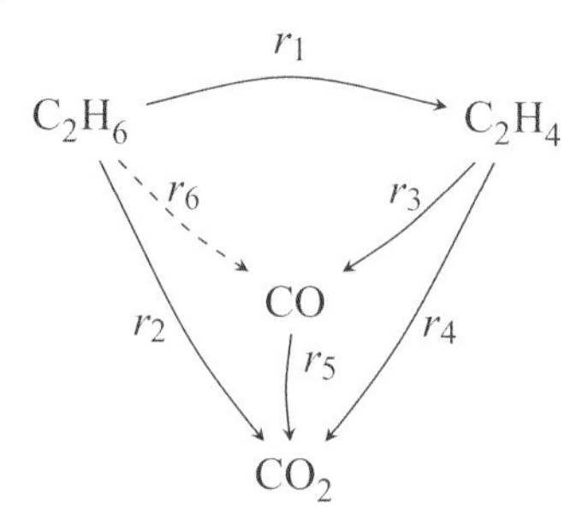

Figure 3.1 The proposed reaction network for the oxidative dehydrogenation of ethane (ODHE) over VO_x/Al_2O_3 catalysts. The formation of carbon monoxide from ethane (r_6) was neglected in the kinetic modeling.

Table 3.1 Reaction rate orders of Reactions 3.1 to 3.5 with respect to C_2H_6, C_2H_4 and CO (α_j) and oxygen (β_j) (Equation 3.9, Tóta *et al.*, 2004), VO_x-Cat.

	r_1	r_2	r_3	r_4	r_5
α_j	0.88	0.75	0.84	0.87	1.13
β_j	0.02	0.24	0.11	0.20	0.13

Already in the first experimental investigations using the laboratory reactor described later, it was found that Reaction 3.6 was not significant and could be neglected. Thus, the five reactions shown in Figure 3.1 represent the reaction network considered below.

The most important question regarding a successful membrane reactor concept is: can a distributed feed of one of the reactants improve the reactor performance? To answer this question simplified power law kinetics (Levenspiel, 1999) were assumed and the orders were estimated in a first analysis of the experimental data described later. Hereby, the rate laws assumed were:

$$r_j = k_j \tilde{c}_{HC/CO}^{\alpha_j} \tilde{c}_{O_2}^{\beta_j} \quad j = 1 \ldots 5 \tag{3.9}$$

The results obtained revealed that the order of the reaction of ethane to ethylene (3.1) with respect to oxygen, β_1, is approximately zero, while this order is higher for the other four steps (β_2 ... β_5; Table 3.1; Tóta *et al.*, 2004). Referring to Section 1.6, it can be stated that the prerequisite for a successful operation of a membrane reactor distributing oxygen is fulfilled.

3.3
Catalysts and Structure–Activity Relations

Due to the necessity to find a catalyst well suited for the study of the membrane reactor concept, several non-noble transition metal oxides have been tested.

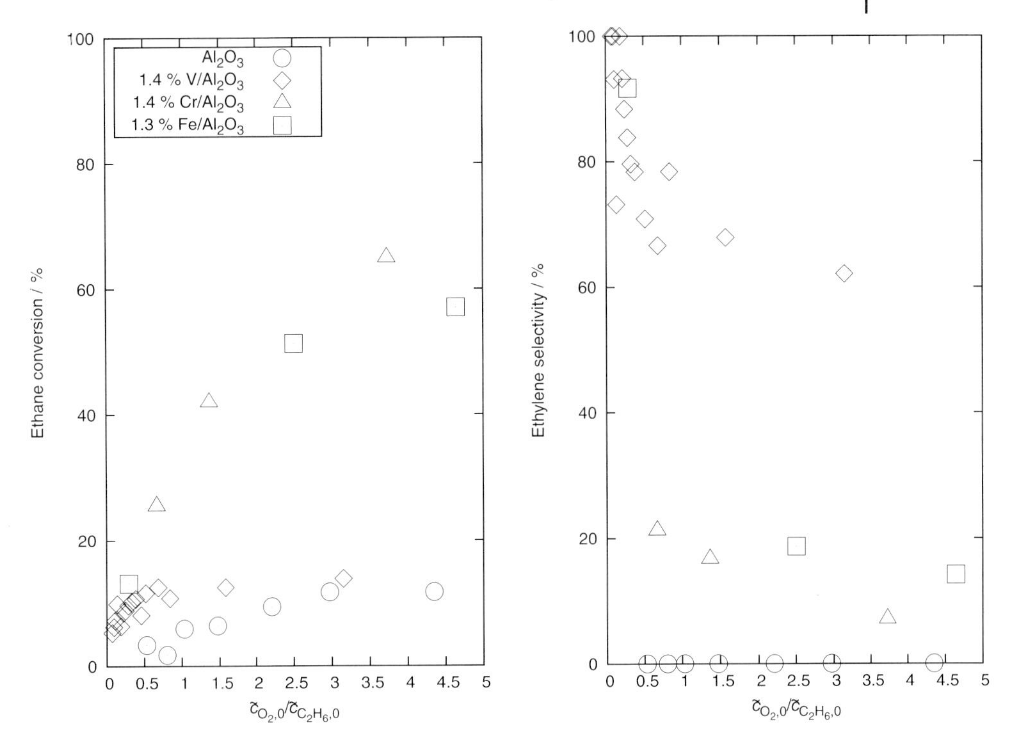

Figure 3.2 Ethane conversion (left) and selectivity towards ethylene (right) as a function of the oxygen to hydrocarbon ratio in the reactor feed for different catalytically active species. The most active species was chromium (Δ), but the best tradeoff between activity and selectivity was achieved with the vanadium doped catalyst (◇). The experimental conditions were $M_{Cat}/\dot{V} = 558\,kg\,s/m^3$, $T = 863\,K$, $M_{cat} = 3.1\,g$.

Figure 3.2 shows measured activities of vanadium, iron and chromium oxides on γ-Al_2O_3 as well as those of the bare γ-alumina support in the oxidation of ethane and the corresponding selectivity towards the desired product ethylene. As can be seen, CrO_x and FeO_x are both characterized by high conversion rates, but only low selectivities. In contrast, the conversion of ethane above VO_x spread over γ-Al_2O_3 is rather low, but has the highest selectivity towards ethylene. It should be noted that the pure support material has a comparable activity but a selectivity of 100% towards the thermodynamically stable final product CO_2. Evaluating these findings vanadia-based catalysts on a γ-Al_2O_3 support were used in all subsequent studies.

The following briefly summarizes the state of knowledge on vanadia catalysts in the oxidative dehydrogenation of ethane. ODHE over vanadia catalysts belongs to the class of redox catalysis reactions following the Mars–van Krevelen mechanism (Mars and van Krevelen, 1954), which is found also in the oxidation of higher hydrocarbons (Busca, 1996; Magagula and van Steen, 1999). In the first step of this mechanism, the hydrocarbon is oxidized by the catalyst, which itself is reduced and in a second step re-oxidized by an oxidant from the surrounding gas phase, for example, O_2 or CO_2. In general, a reaction network consisting of: (a) the selective oxidation yielding olefins or oxygenates, (b) the parallel "deep oxidation" of the primary hydrocarbon to CO_x and (c) the consecutive oxidation of the olefins or oxygenates to CO_x is to be expected. Thus, selective oxidation is a kinetically controlled reaction. It is evident that the formation of carbon oxides is not desired as it minimizes selectivity and yields of the oxidation products of interest, in this case ethylene.

The pure vanadium oxide (V_2O_5) is known to be rather inactive (Oyama and Somorjai, 1990; Le Bars *et al.*, 1992) and has also unfavorable mechanical properties when stable pellets or monoliths are needed (Chary *et al.*, 2003; Reddy and Varma, 2004). For these reasons, vanadia is typically deposited onto high-surface-area supports like γ-alumina, silica, zeolites, titania or zirconia. Usually, the support is loaded with a suitable vanadium compound (precursor), for example, by impregnation, grafting or chemical vapor deposition (CVD) techniques and afterwards treated at elevated temperatures under an oxidizing environment. During this calcination step, the vanadium precursor converts to the oxide, which undergoes a chemical interaction with the support yielding the formation of M-O–V bonds (M = metal cation of the support material). At low and moderate loadings, this results in the formation of a two-dimensional vanadate "monolayer" with a high dispersion of V species. Only if the surface is overloaded the excess vanadia agglomerates and forms three-dimensional V_2O_5 crystallites (Blasco and López Nieto, 1997; Mamedow and Cortés Corberán, 1995; Banares, 1999). The monolayer capacity for vanadia and the strength of the support–vanadia interaction depend on the type of the support material.

The nature of the vanadia species formed on different support materials is already the subject of several reviews (e.g., Weckhuysen and Keller, 2003; Haber, Witko, and Tokarz, 1999; Deo, Wachs, and Haber, 1994; Grzybowska-Swierkosz, 1999). The most prominent concept postulates the existence of monovanadates and divanadates, polyvanadates and bulk-like V_2O_5 structures. Monovanadates (VO_4^{3-}) as the most dispersed vanadia species should appear predominantly at low V loadings, where the tetrahedral VO_4 base unit is assumed to consist of one V=O double bond and three M-O–V bonds to the support. With moderately increased V loadings the formation of divanadate and two-dimensional tetrahedral polymeric vanadate species is postulated. Beside the V=O double bond and the M-O–V bonds, these species should contain additional V–O–V bonds. Furthermore, the existence of a three-dimensional dispersed polyvanadate phase with up to five overlayers is postulated (García-Bordejé *et al.*, 2004), in which the three-dimensional $(V_2O_5)_n$ crystallites have an octahedral structure with the base unit VO_6. The latter are reported to be less active and less selective towards ODHE than the dispersed vanadate species (Pieck *et al.*, 2004; Liu *et al.*, 2004).

It should be noted that within this prevailing concept all vanadia species are assumed to be in the oxidation state +5. However, there are also a number of studies, in which significant amounts of V(IV) species are found in the catalysts even after oxidizing treatment (Matralis *et al.*, 1995; Enache *et al.*, 2004; Pieck, Banares, and Fierro, 2004; Reddy and Varma, 2004; Reddy *et al.*, 2004). This is important when "active sites" are discussed, since it is directly related to the reaction kinetics in a Mars–van Krevelen mechanism.

In the following, we present results regarding the catalyst characterization and structure–activity relations.

3.3.1 Catalyst Preparation and Characterization

Structure–activity relations in heterogeneous catalysis reflect relations between catalytic activity and structural properties of the catalysts such as doping degree, porosity, dispersion of the active species, reducibility and acid–base properties. In order to study these for ODHE over supported vanadia catalysts a large number of samples were prepared by impregnation of γ-alumina, silica and titania supports either with $VO(acac)_2$ as V(IV) or with NH_4VO_3 as V(V) precursors (Klose, 2008). The V loading was varied between 1 wt% and 16 wt%.

All samples were characterized by numerous techniques like atomic absorption spectroscopy (AAS), surface area determination (BET), N_2 porosimetry, differential thermal analysis and thermal gravimetry (DTA/TG), X-ray diffraction (XRD), diffuse reflection infrared Fourier transform spectroscopy (DRIFTS), nuclear magnetic resonance (NMR) spectroscopy of ^{51}V, X-ray photoelectron spectroscopy (XPS), temperature-programmed reduction (TPR), temperature-programmed oxidation (TPO) and temperature-programmed desorption of pyridine (TPD). Catalytic tests were carried out under oxygen-lean and oxygen-excess conditions (0.35% and 21% O_2, respectively) in the oxidation of ethane (0.7%), ethylene and CO (0.7%, 1% and 21% O_2) as the expected intermediates in the reaction network. The complete experimental information is described in detail by (Klose, 2008) and provides the base for the upcoming discussion.

Consistent with (Weckhuysen and Keller, 2003) it was found by XRD, DRIFTS and ^{51}V-NMR spectroscopy that on alumina and titania vanadate monolayers are formed up to V densities of 6–7 atoms/nm^2, but only 1–2 atoms/nm^2 on silica. The appearance of crystalline vanadia goes parallel with a dropdown of the specific surface area by up to one order of magnitude. In the case of titania, crystalline vanadia catalyzes the conversion of anatase to rutile and leads to a collapse of the porous framework of this support material (Djerad *et al.*, 2004; Reddy *et al.*, 2004).

In tests of the catalytic activity in ODHE under lean oxygen conditions, it was found that alumina-supported vanadia catalysts provided the highest ethylene yields among the studied supports. Ethylene yields of up to more than 20% were obtained, in the same range as reported in the literature (Le Bars *et al.*, 1996; Argyle *et al.*, 2002). In these experiments, at ~35% ethane conversion, complete oxygen consumption was reached. The silica-supported catalysts were less active, with ethylene yields up to 10%. The titania-supported catalysts were found to be rather

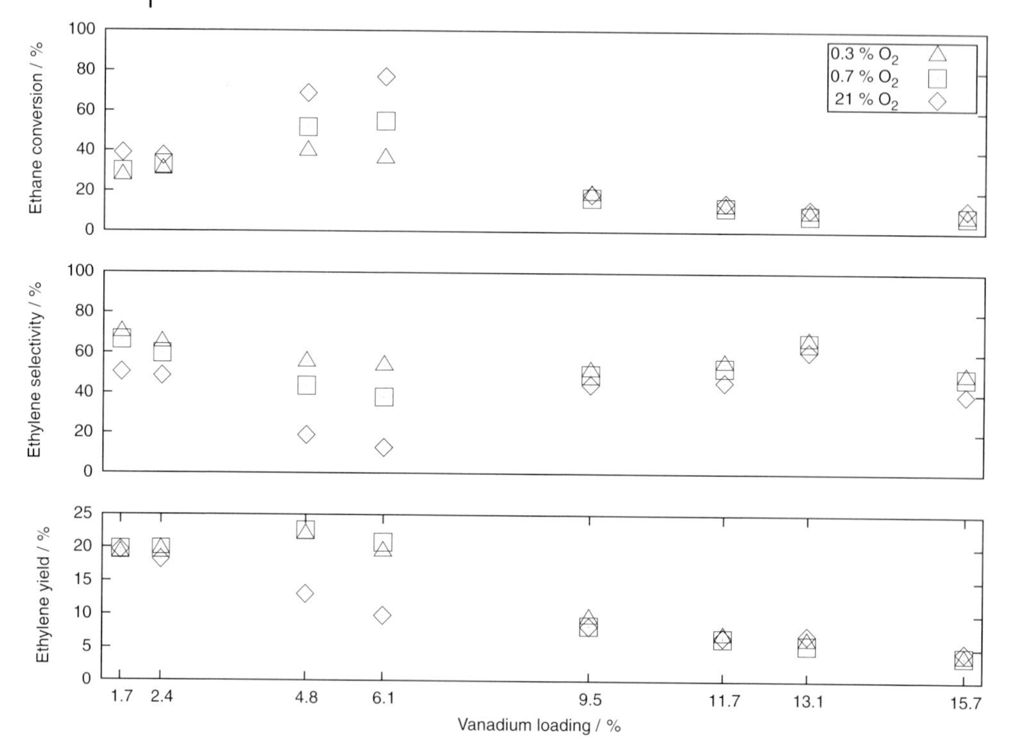

Figure 3.3 Ethane conversion (upper), ethylene selectivity (center) and ethylene yield (lower) against the vanadium loading on the γ-Al_2O_3 for various oxygen mole fractions in the feed (0.3% Δ, 0.7% □, 21% ◇, respectively, with nitrogen as balance). The experimental conditions were $M_{Cat}/\dot{V} = 75\ kg\,s/m^3$, $T = 873\ K$, $M_{Cat} = 0.2507\,g$.

active but poorly selective, so that the ethylene yields were found at the same low level. In all cases the vanadia precursor material had only minor impact on the catalytic performance; the samples prepared from NH_4VO_3 gave comparable results to those prepared from $VO(acac)_2$.

Due to the higher ethylene yields and the selectivity the following studies were focused on vanadia on the γ-Al_2O_3 support material. Figure 3.3 shows the correlation of ethane conversion and ethylene selectivity to the oxygen supply and visualizes the importance of a low oxygen concentration in the reaction: While the total conversion of ethane increases with the percentage of oxygen in the feed gas up to a vanadium loading of about 6 wt%, the selectivity towards ethylene decreases significantly in this range. The decrease in ethane conversion above a vanadium loading of 6 wt% is due to the formation of three-dimensional vanadia crystallites, which is connected to clogging of the pores and a significant reduction of the available surface area (from more than 160 m^2/g to less than about 13 m^2/g).

These results demonstrate clearly:

- A local low oxygen concentration and a distributed oxidant dosing should help in ODHE.
- High vanadium loadings of the support material lead to the formation of crystalline vanadia and to a decrease in ethane conversion and ethylene production.
- Highest selectivity towards ethylene is found for lowest vanadium loadings.

For these reasons, all following kinetic experiments were performed at low vanadia coverages.

3.3.2 Mechanistic Aspects: Correlation Between Structure and Activity

In the following, some chemical and mechanistic aspects of the catalysts in the oxidative dehydrogenation of ethane are discussed in order to answer the major question: What is the active species, and what is the reaction path? Here we focus on main aspects only. More details can be found elsewhere (Klose *et al.*, 2007).

Regarding ethylene formation from ethane over supported VO_x catalysts, a Mars–van Krevelen redox mechanism is commonly accepted to describe the reaction behavior. Assuming an original vanadium oxidation state of +5 in these catalysts, the prevailing models would suggest that all vanadia species in the vanadate monolayer should contribute to the catalytic performance, resulting in comparable turnover frequencies (TOF) when normalized to the surface area of the catalyst and to the amount of vanadium deposited on the support. However, normalized TOF rate constants calculated from power law expressions of the first order with respect to C_2H_6, C_2H_4 or CO and orders of 0.5 and zero with respect to oxygen for oxygen-lean and oxygen-excess conditions, respectively, show surprisingly that the TOF for agglomerated vanadia species (i.e., above about 6 wt% V) seems to be comparable or even higher than those of dispersed ones, apparently in contradiction to literature where the opposite order with respect to the performance of vanadates and V_2O_5 phases has been reported (for details see Klose *et al.*, 2007). In addition, at vanadate coverages up to ~6 wt% V only under oxygen-excess conditions is a constant TOF level observed, while under oxygen-lean conditions the TOF decreases with increasing V loading. These trends were discovered not only for ethylene formation but also for the deep oxidation reactions.

An explanation is found when the initial vanadium oxidation states of the individual catalysts are taken into account. Based on extensive TPR and XPS experiments with fresh as well as used catalysts it was found that the dispersed vanadate monolayer phases contain only ~30% V(V) species, the residual fraction being V(IV). This V(V) percentage cannot be enlarged even after exposure of the catalysts to air at temperatures up to 650 °C. In contrast, the catalysts with agglomerated vanadia species have an oxidation state ratio of V(V) : V(IV) = 80 : 20. It should be noted that, under time on stream in ODHE, a considerable part of the less reducible agglomerated vanadia phases is converted into a kind of multilayer polyvanadate phase with reducibility comparable to the dispersed monolayer vanadates (Suchorski *et al.*, 2005; Klose *et al.*, 2007).

It should be mentioned that in a study of the reoxidation dynamics of highly dispersed VO_x species supported on γ-Al_2O_3 Frank *et al.* found a maximum vanadium oxidation state +5 at temperatures around 700 K (Frank *et al.*, 2009). Starting from V(III), these authors observed maximum average oxidation states of only 4.3–4.5 for their catalysts at temperatures up to ~550 K. In our studies, and for the catalysts used in the different membrane reactor experiments, reoxidation of the highly dispersed VO_x to average vanadium oxidation states of more than 4.3–4.4 was not observed up to temperatures of above 900 K.

For the identification of the active V species the turnover frequencies were normalized to either the V(IV) or the V(V) fractions, respectively. A meaningful correlation is only obtained for those related exclusively to V(V): here the ethylene formation follows the expected trend, with a constant TOF level for monolayer catalysts and a decrease with the appearance of three-dimensional crystallites for both oxygen-lean and oxygen-excess conditions, respectively. This strongly suggests that the outermost V(V) species are able to catalyze oxidative dehydrogenation, while V(IV) is not active for this reaction [but is reduced by H_2 to V(III) in TPR experiments].

In the attempt to find a structural model for the catalysts, it has to be kept in mind that an average oxidation state of about 4.3, corresponding to a V(IV) : V(V) ratio of 2 : 1, could not be exceeded in the mono- and polyvanadates used herein. In consequence, two different types of V cations with different redox behavior should exist in the monolayer vanadates of our catalysts, only one of which is active. Both the existence of two types of V cations within the vanadate structure and the maximum oxidation state of 4.3 were found to be in significant contradiction to present concepts on the nature of supported vanadate phases. Thus, a new structural model was required, which had to be consistent with previous experimental findings in literature and with all the results reported herein, as well. In this chapter we can only describe the essential idea. Details of the model and additional supporting experiments are described by (Klose *et al.*, 2007).

In this model, the first building block is the V=O unit, the presence of which is indicated by the corresponding absorption bands in DRIFT spectra of the catalysts. Furthermore, M-O–V units should be included as the crucial sites determining performance, reducibility and acid–base properties of the catalysts since these functionalities are strongly affected by the used support. Because of the significant support effect on ODHE activity and reducibility and the strong correlation between both, it is consequent to assign M-O–V units to the active V(V) species.

As another prerequisite, the presence of 2/3 of V(IV) and 1/3 of V(V) cations has to be reflected in the vanadate structures. There is no indication for three-dimensional growth in this coverage region, thus the possibility of core-shell structures of extremely small clusters of oxidation state +5 on the surface and +4 inside can be excluded. It has to be concluded that the V(IV) and V(V) cations are uniformly distributed within the overall vanadate phases. This is also supported by TEM pictures of the polyvanadate phase, which show well ordered regions with chain-like structures for the coverage regime attributed to polyvanadates and no indication for three-dimensional growth (Klose *et al.*, 2007).

The smallest unit which fulfills all the requirements described above is $[V_3O_8]^{3-}$, bound by three oxygens to the support, with one central V(V) and two adjacent V(IV) cations connected by oxide anion bridges and an additional V=O double bond on each vanadium. It is assumed that this structure represents the amorphous "isolated vanadate" species, which is predominantly formed at low V densities (<2 V atoms/nm^2). However, the existence of real "monovanadates" at ultra-low V loadings cannot completely be ruled out.

For the polyvanadate phase, which is predominantly formed on γ-alumina at V densities higher than 2 cations/nm^2 but below the appearance of crystalline vanadia species and which is characterized by a well ordered structure in the range of tens of nanometers, chains in which a V(V) cation is followed by two V(IV) cations are proposed (Klose *et al.*, 2007). Again, the V cations are bridged by oxygen anions. Herein, oxygen bonds to the support are assigned to the V(V) cations only, and the M-O–V(V) unit is suggested to be the active site. In this model, the ratio of 2 : 1 for V(IV):V(V) is given, too. This polyvanadate phase exhibits reducibility similar to that of the "isolated vanadates" regarding reduction temperature and oxidation states, but differs from the latter by a lower strength of Broenstedt acidity, as was found in the experiments.

These proposed structures on dispersed vanadate phases do not contradict the spectroscopic findings previously reported in literature: V=O, M–O–V and V–O–V bonds causing, for example, the various known IR and RAMAN absorptions are included. It should be remarked that the existence of V(IV)–V(V) mixed oxides like V_6O_{13} is also known for bulk species (Weckhuysen and Keller, 2003). Although at the present state a number of questions still remain open and further research is required to confirm these structural proposals, they are able to explain and understand the obtained experimental results, in contrast to other models.

3.4 Derivation of a Kinetic Model

Reaction rates are usually described by mechanistic models as a function of the concentrations of the components present in the system and adjustable free parameters as, for example, reaction rate constants, activation energies and adsorption equilibrium constants. Despite intensive research these rate laws can hardly be predicted theoretically. Therefore, a detailed kinetic modeling requires parallel experimental investigations, that is, a correlation of experimental results with proposed kinetic models followed by model discrimination (Froment, 1975; Levenspiel, 1999).

3.4.1 Experimental

3.4.1.1 Catalyst

Detailed investigations on the reaction kinetics of the ODHE were carried out with a 1.4% VO_x/γ-Al_2O_3 catalyst. This catalyst was used later in various studies of membrane reactor concepts at laboratory scale and pilot scale. It was selected because of a sufficient activity, a rather high initial selectivity to ethylene and a comparable simple and well reproducible preparation. This reference catalyst was prepared by wet impregnation of a γ-Al_2O_3 spherical support (Condea Chemie, Germany) with a solution of vanadyl acetylacetonate in acetone, followed by calcination at 700 °C for 4 h. The vanadium content was measured by means of AAS analysis after a microwave extraction in HNO_3. The properties of the reference catalyst were:

- Surface area of 184.5 m^2/g (by BET analysis),
- Average pore diameter of 10.9 nm,
- Particle sizes of 0.4 mm, 1.0 mm and 1.8 mm.

This catalyst was used to derive a detailed kinetic model for the reaction network shown in Figure 3.1 based on experimental data collected in a set-up described below.

3.4.1.2 Set-Up

The laboratory plant consisted of a packed-bed reactor, a catalytic afterburner and analytic equipment. Electronic mass flow controllers were used to mix different feed compositions from the pure gases ethane, ethylene, CO, CO_2, air and nitrogen. By means of a heated eight-port multiposition valve both the feed and the product stream were analyzed by gas chromatography with a mass selective detection and a thermal conductivity detection (GC-MSD/TCD).

The laboratory-scale packed-bed reactor (inner diameter 12 mm, length 19 mm) was a quartz glass tube filled with 0.25–0.50 g catalyst. The reactants were preheated inside the reactor in an inert entrance zone. The catalyst-filled reaction zone was equipped with a thermocouple to measure the reaction temperature. In addition, two sample capillaries were placed at the inlet and the outlet to measure the concentrations of the reactants and the products, respectively. The quantitative analysis was performed with a GC-MSD/TCD, as mentioned above. First, water was removed from the samples in a *Haysep* column (Chrompack) and then a *HP PLOT/Q* column separated hydrocarbons and CO_2. In a 5 Å Molsieve column permanent gases and CO were separated. Finally, oxygenates were separated in a *DB-FFAP* column. Furthermore, the permanent gases, hydrocarbons and carbon oxides were quantified by means of a TCD. The trace components like oxygenates were detected by the MSD. The relative errors in the concentrations were estimated to be <3%.

Table 3.2 Summary of experiments performed in a laboratory packed-bed reactor for kinetic modeling and parameter estimation (Joshi, 2007).

Measurements	Total flow rate $\dot{V}$(L·h^{-1})	Catalyst mass M_{Cat} (g)	Temperature T (°C)	Overall inlet concentrations c (vol%)			
				C_2H_x	O_2	CO	CO_2
C_2H_6 feed							
A1	12.05–24.12	0.2507–0.5052	480–570	0.6–1.2	6.0	–	–
A2	12.05–24.12	0.2507–0.5052	480–570	1.2	0.2–3.0	–	–
A3	15.05–24.12	0.2507–0.5052	480–570	0.7	0.4–1.0	–	–
A4	19.18	0.2013	530–620	0.3–0.8	0.67	0.28–1.22	–
A5	19.18	0.2013	530–620	0.3–0.8	0.67	–	0.28–0.9
C_2H_4 feed							
B1	12.05	0.2507	480–570	0.6–1.2	6.0	–	–
B2	12.05	0.2507	480–570	1.26	0.4–4.0	–	–
B3	19.18	0.2013	480–620	0.15–1.0	1.2–6.0	0.5–3.2	–
B4	19.18	0.2013	480–620	0.15–1.0	1.2–6.0	–	0.5–3.2
CO feed							
C1	12.05	0.2507	480–570	–	6.0	0.6–1.2	–
C2	12.05	0.2507	480–570	–	0.5–2.6	1.2	–

3.4.1.3 **Procedures**

Preliminary experiments were devoted to evaluate possible mass transfer restrictions. For this purpose experiments with three particle sizes (0.4 mm, 1.0 mm, 1.8 mm) were conducted. To analyze the network three types of experiments using the smallest particles (0.4 mm) were performed: (group A) the oxidation of ethane, (group B) the oxidation of ethylene and (group C) the oxidation of CO (see, respectively, A1–A5, B1–B4 and C1–C2 in Table 3.2). To enable the comparison of these experiments the ratio of the catalyst mass to the total flow rate was kept constant. The temperature was varied between 480 °C and 620 °C. In all experiments the feed contained oxygen from 0.2 vol% up to 6.0 vol% diluted in nitrogen. The hydrocarbons were varied between 0.15 vol% to 1.26 vol% C_2H_4 and C_2H_6 (experiments of group A and group B, respectively). Furthermore, some experiments were carried out with products (CO, CO_2) mixed into the inlet stream. Moreover, the experiments of group C contained only oxygen, CO and nitrogen in the feed.

Within these parameter ranges a so-called cube design (Box, Hunter, and Hunter, 1978) was applied where the effects of the following variables were studied: partial pressures of the main reactants (C_2H_6, C_2H_4, CO, respectively), the partial pressure of oxygen and the temperatures (Joshi, 2007). The experimental database consisted of about 500 observations obtained in the tubular laboratory reactor for various feed concentrations. A summary of the conditions is given in Table 3.2.

3.4.2 Qualitative Trends

Before performing a quantitative kinetic analysis the overall performance of the catalyst and the influence of mass transfer resistances were evaluated qualitatively.

3.4.2.1 Overall Catalyst Performance

Here, selected trends regarding the conversion and selectivity as a function of temperature are shown. The results presented were obtained for a flow rate of 12.05 L/h ($M_{Cat}/\dot{V}$=75 kg s/m^3). Figure 3.4 illustrates an increase of ethane conversion and CO selectivity with increasing temperature. In contrast, ethylene selectiv-

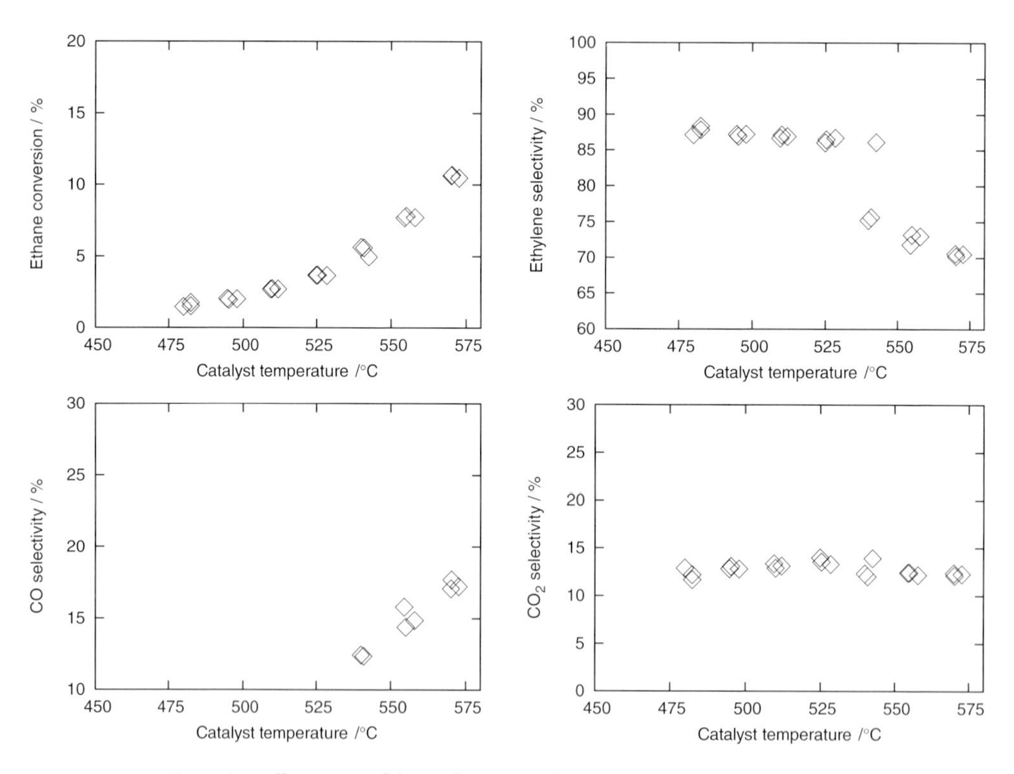

Figure 3.4 Illustration of the performance of the catalyst characterization: ethane conversion (upper left), ethylene selectivity (upper right), CO selectivity (lower left) and CO_2 selectivity (lower right) as a function of temperature. Experimental conditions: $M_{Cat}/\dot{V}$= 75 kg s/m^3, 1.22 vol% ethane and 1.67 vol% oxygen (balance nitrogen), M_{Cat} = 0.2507 g.

ity decreases with temperature. The formation of CO_2 remains constant in the observed temperature region. Selectivities of ethylene, CO and CO_2 were found at 540 °C to be 80.5%, 8.4% and 11.1%, respectively. In general, the catalyst was very selective towards ethylene for lower temperatures. These findings are in a good agreement with publications from other authors (e.g., Argyle *et al.*, 2002; Grabowski, 2006; Waku *et al.*, 2003).

3.4.2.2 Evaluation of Intraparticle Mass Transfer Limitations

For investigating reaction kinetics the experimental data should not be influenced by mass transfer limitations to allow transferring results to other reactor concepts. To evaluate the influence of possible mass transfer limitations additional experiments were performed with two different catalyst particle sizes (1.0 mm and 1.8 mm) in the range of 580 °C to 670 °C. A mixture of 2.16 vol% ethane and 2.10 vol% oxygen (balance nitrogen) was fed to the reactor. Measured overall reaction rates of ethane and of ethylene with respect to the temperature are shown in Figure 3.5. The overall consumption of ethane is for the conditions mentioned not influenced by mass transfer limitations. Regarding ethylene

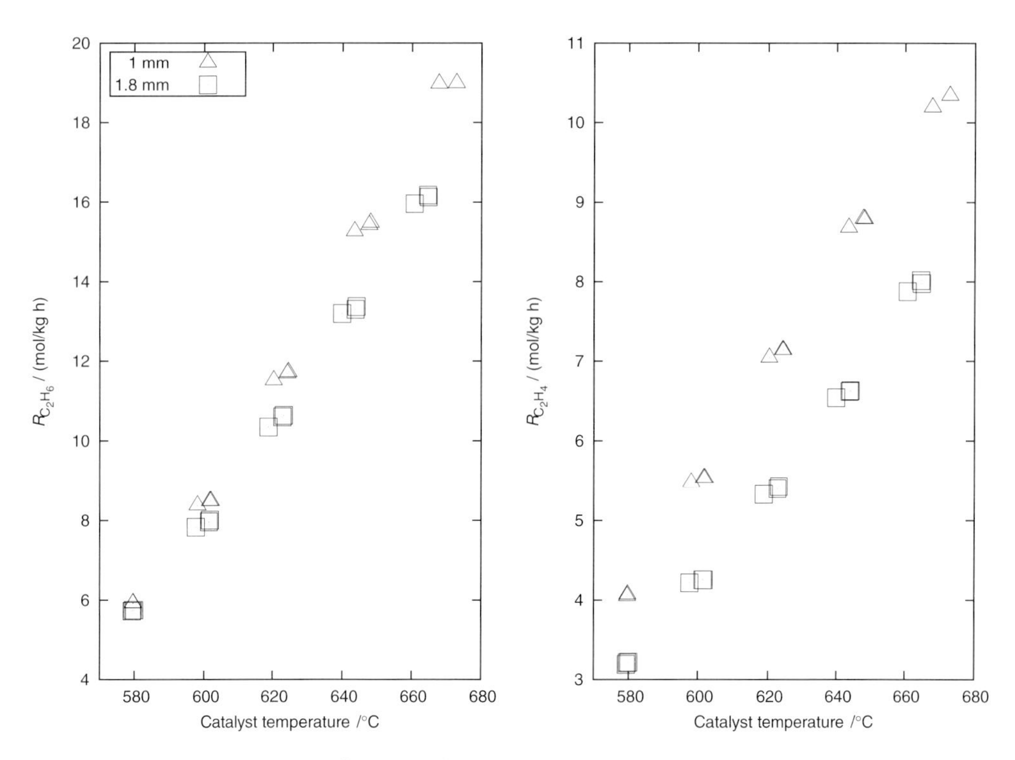

Figure 3.5 Overall reaction rates of ethane (left) and ethylene (right) against catalyst temperature for catalyst particles of 1 mm (Δ) and 1.8 mm (□) size, respectively. The feed consisted of 2.16 vol% ethane and 2.1 vol% oxygen (balance nitrogen).

formation, lower rates were observed with the 1.8 mm particles than with the 1 mm particles. This result led to the decision to perform the experiments used in the kinetic analysis exclusively with the smallest catalyst particle size (0.4 mm).

3.4.3 Quantitative Evaluation

3.4.3.1 Simplified Reactor Model and Data Analysis

To extract kinetic data from measurements performed in tubular reactors, in principle an integration of the mass and energy balances is required. This allows matching theoretical and experimental outlet concentrations. If, however, conversion and deviations from isothermal conditions are low, a much simpler analysis can be applied. In this case averaged reactor concentrations can be determined for all components from the arithmetic means of the inlet and outlet concentrations. Further, component specific overall reactions rates R_i can be estimated using the following simplified steady-state mass balance:

$$R_i = \frac{(\tilde{c}_{i,\mathrm{in}} - \tilde{c}_{i,\mathrm{out}})\dot{V}}{M_{\mathrm{Cat}}}, \tag{3.10}$$

where $\tilde{c}_{i,\mathrm{in}}$ and $\tilde{c}_{i,\mathrm{out}}$ are inlet and outlet concentrations, respectively, $\dot{V}$ is the volumetric flow rate, M_{Cat} is the mass of catalyst. The overall reaction rate R_i as used here has the dimension: $\mathrm{mol/(kg_{Cat}\,h)}$. In accordance with the reaction network under investigation the overall rates R_i are linked to the individual reaction rates r_j using the corresponding stoichiometric coefficients ν_{ij},

$$R_i = \sum_j^M \nu_{ij} r_j \tag{3.11}$$

The approach described was used to analyze the subset of the available experimental data characterized by ethane conversions below 20% providing flat temperature profiles and almost isothermal conditions in the reactor. The remaining experiments yielding larger conversions were used for model validation, as described in Section 3.4.4.

3.4.3.2 Kinetic Models

The structure of suitable rate equations – the models of $r_j(T, \tilde{c}_1, \ldots, \tilde{c}_N)$ – is not known *a priori*, although physico-chemical insight and preliminary knowledge about the mechanism of the involved reaction steps limit the spectrum of possible models (Froment, 1975). Beside other formulations, the Langmuir–Hinshelwood and Hougen–Watson (LHHW) models (Hougen and Watson, 1943) and the Mars–van Krevelen (Mars and van Krevelen, 1954) model are widely known and used in heterogeneous catalytic processes. The latter is especially suited for catalytic oxidations, such as the oxidative dehydrogenation of ethane (Reaction 3.1). As mentioned above, it assumes that oxygen is provided by the lattice of the solid catalyst and the reduced catalyst is then re-oxidized by gas phase oxygen. The model reads:

$$r_1 = \frac{k_{\text{red}}\tilde{c}_{\text{C}_2\text{H}_{n+2}} k_{\text{ox}}\tilde{c}_{\text{O}_2}^{\beta}}{k_{\text{red}}\tilde{c}_{\text{C}_2\text{H}_{n+2}} + k_{\text{ox}}\tilde{c}_{\text{O}_2}^{\beta}}, \tag{3.12}$$

where β is the order with respect to oxygen. It is set to $\beta = 0.5$, regarding the assumption of a dissociative oxygen adsorption. The well known LHHW models assumes reactants and products can be adsorbed on the catalyst surface and that all the elementary steps of an overall reaction are close to the equilibrium except the rate-determining step (e.g., the surface reaction). Depending on the decisions of the rate-determining step different formulations of LHHW models are possible. Details regarding the model formulation can be found in the literature, for example, (Hougen and Watson, 1947; Froment and Bischoff, 1979).

Usually various possible model formulations are tested in a regression analysis. Afterwards model discrimination is applied. Here, for Reactions 3.2 to 3.5 various LHHW models were derived based on different assumptions on the mechanism and the rate-determining step of the elementary reactions (Yang and Hougen, 1950; Doraiswamy and Sharma, 1984; Joshi, 2007).

3.4.3.3 Parameter Estimation

The free parameters of the various kinetic models evaluated were the reaction rate constants, the activation energies and the adsorption equilibrium constants. Initial estimates of the adsorption equilibrium constants were extracted from the literature (Argyle *et al.*, 2002). All initial estimates were refined in a regression analysis using a Levenberg–Marquardt algorithm (Moré, 1978) minimizing the following objective function:

$$\text{OF} = \sum_{i}^{N_{\text{components}}} \sum_{k}^{N_{\text{exp}}} \left(\frac{R_{ik}^{\text{exp}} - R_{ik}^{\text{mod}}}{R_{ik}^{\text{exp}}} \right)^2 \tag{3.13}$$

Based on the magnitude of the OF values obtained model discrimination was performed and a model was identified, which is described in the next section.

3.4.4 Suggested Simplified Model

The oxidative dehydrogenation of ethane to ethylene was quantified using a reaction network, which was derived from experiments on oxidation of ethane and ethylene and CO as intermediates. The reaction network consists of five partial reactions: (r_1) ethane to ethylene and water, (r_2) ethane to carbon dioxide and water, (r_3) ethylene to carbon monoxide and water, (r_4) ethylene to carbon dioxide and water and (r_5) carbon monoxide to carbon dioxide (Figure 3.1, Equations 3.1–3.5). To quantify the rates of the Reactions 3.1 to 3.5 for a $VO_x/\gamma\text{-}Al_2O_3$ catalyst with 1.4% V the expressions summarized in Table 3.3 are suggested. Reaction 3.1 is described by a Mars–van Krevelen mechanism. Reactions 3.2

Table 3.3 Suggested kinetic model of the oxidative dehydrogenation of ethane to ethylene on a $VO_x/\gamma\text{-}Al_2O_3$ catalyst with 1.4% V. The corresponding parameters are listed in Table 3.4.

$$r_1 = k_1(T)\frac{k_{red}\tilde{c}_{C_2H_6}k_{ox}\tilde{c}_{O_2}^{0.5}}{k_{red}\tilde{c}_{C_2H_6}+k_{ox}\tilde{c}_{O_2}^{0.5}}$$

$$r_2 = \frac{k_2(T)K_{C_2H_6}\tilde{c}_{C_2H_6}K_{O_2}^{0.5}\tilde{c}_{O_2}^{0.5}}{(1+K_{C_2H_6}\tilde{c}_{C_2H_6}+K_{CO_2}\tilde{c}_{CO_2})(1+K_{O_2}^{0.5}\tilde{c}_{O_2}^{0.5})}$$

$$r_3 = \frac{k_3(T)K_{C_2H_4}\tilde{c}_{C_2H_4}K_{O_2}^{0.5}\tilde{c}_{O_2}^{0.5}}{(1+K_{C_2H_4}\tilde{c}_{C_2H_4}+K_{CO}\tilde{c}_{CO})(1+K_{O_2}^{0.5}\tilde{c}_{O_2}^{0.5})}$$

$$r_4 = \frac{k_4(T)K_{C_2H_4}\tilde{c}_{C_2H_4}K_{O_2}^{0.5}\tilde{c}_{O_2}^{0.5}}{(1+K_{C_2H_4}\tilde{c}_{C_2H_4}+K_{CO_2}\tilde{c}_{CO_2})(1+K_{O_2}^{0.5}\tilde{c}_{O_2}^{0.5})}$$

$$r_5 = \frac{k_5(T)K_{CO}\tilde{c}_{CO}K_{O_2}^{0.5}\tilde{c}_{O_2}^{0.5}}{(1+K_{CO}\tilde{c}_{CO}+K_{O_2}^{0.5}\tilde{c}_{O_2}^{0.5}+K_{CO_2}\tilde{c}_{CO_2})^2}$$

Table 3.4 Optimized parameter values for the favorite kinetic model as presented in Table 3.3.

k value [mol (kg h)$^{-1}$]	*E* value (kJ mol^{-1})	*K* value (L mol^{-1})
$k_{1,0} = 1.0$	$E_{A,1} = 94$	$K_{C_2H_6} = 4770$
$k_{2,0} = 1.6 \times 10^7$	$E_{A,2} = 114$	$K_{C_2H_4} = 3026$
$k_{3,0} = 2.0 \times 10^4$	$E_{A,3} = 51$	$K_{CO} = 3456$
$k_{4,0} = 1.0 \times 10^3$	$E_{A,4} = 51$	$K_{CO_2} = 3234$
$k_{5,0} = 1.1 \times 10^7$	$E_{A,5} = 118$	$K_{O_2} = 1003\,L^{0.5}\,(mol^{0.5})^{-1}$
$k_{red} = 4.3 \times 10^9\,L\,(kg\,h)^{-1}$	$k_{ox} = 1.1 \times 10^8\,mol^{0.5}L^{0.5}\,(kg\,h)^{-1}$	

to 3.4 are described by two-site Langmuir–Hinshelwood equations. In contrast Reaction 3.5 is quantified by a Langmuir–Hinshelwood mechanism based on single-site competitive adsorption. In each kinetic equation the oxygen reaction order is 0.5, which assumes that oxygen participates in all reactions in dissociated form. The temperature dependencies of the five pre-factors k_j are described by Arrhenius' law:

$$k_j = k_{j,0}\exp[-E_{A,j}/(\tilde{R}T)] \quad j = 1\ldots5 \tag{3.14}$$

To reduce the number of free model parameters and to allow for convergence of the optimization procedure the adsorption equilibrium constants K_i were assumed not to depend on temperature. All parameters estimated are summarized in Table 3.4.

Finally, in order to validate the derived kinetic model with the available experiments leading to conversions above 20%, numerical simulations were performed using a non-isothermal pseudo-homogeneous one-dimensional plug-flow model (Froment and Bischoff, 1979; Elnashaie and Elshishini, 1994) implemented in

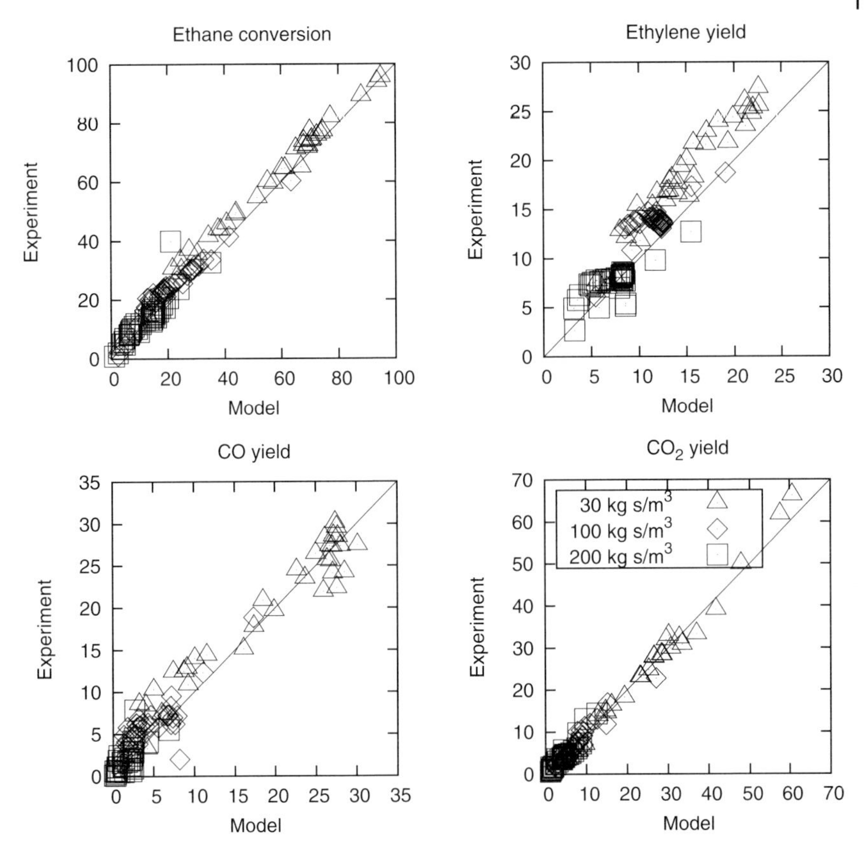

Figure 3.6 Comparison of experimentally determined and modeled ethane conversion (upper left), ethylene yield (upper right), CO yield (lower left) and CO_2 yield (lower right) over $VO_x/\gamma\text{-}Al_2O_3$ catalyst for three experimental conditions: $M_{Cat}/\dot{V} = 30\,kg\,s/m^3$ (Δ), $100\,kg\,s/m^3$ ($\diamond$) and $200\,kg\,s/m^3$ ($\square$).

COMSOL Multiphysics and PROMOT/DIVA (see also Chapter 2). Theoretical reactor outlet concentrations were generated and compared with the corresponding data. Figure 3.6 compares experimental and theoretical conversion, yield and selectivity data. The first subset of theoretical data (conversion <20%) is based on the application of the averaged reactor concentrations in combination with the determined rate laws. The second subset (>20%) is based on the numerical simulations of the more detailed reactor model. A relatively good agreement for the whole parameter range studied can be observed. More detailed results of the mentioned study are not given here and the interested reader is referred to the literature (Joshi, 2007).

Despite the relative large data basis acquired and the systematic model generation and discrimination approach followed, the kinetic model suggested does not represent a description of the real reaction mechanism. It rather can be considered as a formal description of the reaction network valid in the temperature and concentration space covered. This kinetic model is valid for hydrocarbon concentrations between 0.15% and 1.0%, for oxygen concentrations between 0.15% and 20.5% and for a temperature range from 450 °C to 650 °C. It should be of value for the analysis and design of conventional and new reactor concepts as, for example, the membrane reactors of the distributor type considered in this book (Chapters 4–8).

Special Notation not Mentioned in Chapter 2

Latin Notation

$E_{A,j}$	J/mol	activation energy of reaction j
k_j	mol/(kg s)	rate constant of reaction j
$k_{j,0}$	mol/(kg s)	pre-factor of reaction j
k_{ox}	$mol^{0.5} L^{0.5}/(kg\,s)$	rate constant of oxidation step in Mars–van Krevelen mechanism
k_{red}	L/(kg h)	rate constant of reduction step in Mars–van Krevelen mechanism
K_i	L/mol	adsorption equilibrium constant of component i
M_{Cat}	kg	mass of catalyst
OF		objective function
R_i	mol/kg s	overall reaction rate of component i
$\dot{V}$	m^3/s	volumetric flow

Greek Notation

α	reaction rate order with respect to hydrocarbon
β	reaction rate order with respect to oxygen

References

Argyle, M.D., Chen, K., Bell, A.T., and Iglesia, E. (2002) Effect of catalyst structure on oxidative dehydrogenation of ethane and propane on alumina-supported vanadia. *J. Catal.*, **208**, 139–149.

Banares, M.A. (1999) Supported metal oxide and other catalysts for ethane conversion: a review. *Catal. Today*, **51**, 319–348.

Bhasin, M.M., McCain, J.H., Vora, B.V., Imai, T., and Pujadó, P.R. (2001) Dehydrogenation and oxydehydrogenation of paraffins to olefins. *Appl. Catal. A Gen.*, **221**, 397–419.

Blasco, T., and López Nieto, J.M. (1997) Oxidative dehydrogenation of short chain alkanes on supported vanadium oxide catalysts. *Appl. Catal. A Gen.*, **157**, 117–142.

Box, G.E.P., Hunter, W.G., and Hunter, J.S. (1978) *Statistics for Experimenters: An Introduction to Design, Data Analysis, and Model Building*, John Wiley & Sons, Inc.

Busca, G. (1996) Infrared studies of the reactive adsorption of organic molecules over metal oxides and of the mechanisms of their heterogeneously-catalyzed oxidation. *Catal. Today*, **27**, 457–496.

Cavani, F., and Trifiro, F. (1999) Selective oxidation of light alkanes: interaction between the catalyst and the gas phase on different classes of catalytic materials. *Catal. Today*, **51**, 561–580.

Chary, K.V.R., Kishan, G., Kumar, C.P., and Sagar, G.V. (2003) Structure and catalytic properties of vanadium oxide supported on alumina. *Appl. Catal. A Gen.*, **246**, 335–350.

Dai, H.X., and Au, C.T. (2002) The oxidative dehydrogenation of ethane to ethene. *Curr. Topics Catal.*, **3**, 33–80.

Deo, G., Wachs, I.E., and Haber, J. (1994) Supported vanadium oxide catalysts: molecular structural characterization and reactivity properties. *Crit. Rev. Surf. Chem.*, **4**, 141–187.

Djerad, S., Tifouti, L., Crocoll, M., and Weisweiler, W. (2004) Effect of vanadia and tungsten loadings on the physical and chemical characteristics of V_2O_5-WO_3/TiO_2 catalysts. *J. Mol. Catal. A Chem.*, **208**, 257–264.

Doraiswamy, L.K., and Sharma, M.M. (1984) *Heterogeneous Reactions Analysis, Examples and Reactor Design: Gas-Solid and Solid-Solid Reactions*, John Wiley & Sons, Inc.

Elnashaie, S.S.E.H., and Elshishini, S.S. (1994) *Modelling, Simulation, and Optimization of Industrial Fixed Bed Catalytic Reactors*, CRC Press.

Enache, D.I., Bordes-Richard, E., Ensuque, A., and Bozon-Verduraz, F. (2004) Vanadium oxide catalysts supported on zirconia and titania: I. Preparation and characterization. *Appl. Catal. A Gen.*, **278**, 93–102.

Frank, B., Fortrie, R., Hess, Ch., Schlögl, R., and Schomäcker, R. (2009) Reoxidation dynamics of highly dispersed VO_x species supported on γ-alumina. *Appl. Catal. A Gen.*, **353**, 288–295.

Froment, G.F. (1975) Model discrimination and parameter estimation in heterogeneous catalysis. *AIChE J.*, **21**, 1041–1057.

Froment, G.F., and Bischoff, K.B. (1979) *Chemical Reactor Analysis and Design*, John Wiley & Sons, Inc.

García-Bordejé, E., Lázaro, J., Moliner, R., Galindo, J.F., Sotres, J., and Baro, A.M. (2004) Morphological characterization of vanadium oxide supported on carbon-coated monoliths using AFM. *Appl. Surf. Sci.*, **228**, 135–142.

Grabowski, R. (2006) Kinetics of oxidative dehydrogenation of C2–C3 alkanes on oxide catalysts. *Catal. Rev.*, **48**, 199–268.

Grasselli, R.K. (1999) Advances and future trends in selective oxidation and ammoxidation catalysis. *Catal. Today*, **49**, 141–153.

Grzybowska-Swierkosz, B. (1999) Active centres on vanadia-based catalysts for selective oxidation of hydrocarbons. *Appl. Catal. A Gen.*, **157**, 409–420.

Haber, J., Witko, M., and Tokarz, R. (1999) Vanadium pentoxide I. Structures and properties. *Appl. Catal. A Gen.*, **157**, 3–22.

Hougen, O.A., and Watson, K.M. (1943) Solid catalysts and reaction rates. *Ind. Eng Chem.*, **35** (5), 529–541.

Hougen, O.A., and Watson, K.M. (1947) *Chemical Process Principles: Part 3. Kinetics and Catalysis*, John Wiley and Sons, Inc.

Joshi, M. (2007) Statistical analysis of models and parameters in chemical and biochemical reaction networks, Dissertation, Otto von Guericke University.

Klose, F. (2008) Structure-Activity Relations of Supported Vanadia Catalysts and the Potential of Membrane Reactors for the Oxidative Dehydrogenation of Ethane, Habilitation thesis, Otto von Guericke University Magdeburg.

Klose, F., Joshi, M., Hamel, C., and Seidel-Morgenstern, A. (2004) Selective oxidation of ethane over a VO_x/γ-Al_2O_3 catalyst – investigation of the reaction network. *Appl. Catal. A Gen.*, **260**, 101–110.

Klose, F., Wolff, T., Lorenz, H., Seidel-Morgenstern, A., Suchorski, Y., Piórkowska, M., and Weiss, H. (2007)

Active species on γ-alumina-supported vanadia catalysts: nature and reducibility. *J. Catal.*, **247**, 176–193.

Le Bars, J., Védrine, J.C., Auroux, A., Pommier, B., and Pajonk, G.M. (1992) Calorimetric study of vanadium pentoxide catalysts used in the reaction of ethane oxidative dehydrogenation. *J. Phys. Chem.*, **96**, 2217–2221.

Le Bars, J., Auroux, A., Forissier, M., and Vedrine, J.C. (1996) Active Sites of V2O5/γ-Al2O3Catalysts in the Oxidative Dehydrogenation of Ethane. *J. Catal.*, **162**, 250–259.

Levenspiel, O. (1999) *Chemical Reaction Engineering*, 3rd edn, John Wiley & Sons Inc.

Liu, Y.-M., Cao, Y., Yi, N., Feng, W.-L., Dai, W.-L., Yan, S.-R., He, H.-Y., and Fan, K.-N. (2004) Vanadium oxide supported on mesoporous SBA-15 as highly selective catalysts in the oxidative dehydrogenation of propane. *J. Catal.*, **224**, 417–428.

Magagula, Z., and van Steen, E. (1999) Time on stream behaviour in the (amm) oxidation of propene/propane over iron antimony oxide: cyclic operation. *Catal. Today*, **49**, 155–160.

Mamedow, E.A., and Cortés Corberán, V. (1995) Oxidative dehydrogenation of lower alkanes on vanadium oxide-based catalysts. The present state of the art and outlooks. *Appl. Catal. A Gen.*, **127**, 1–40.

Mars, P., and van Krevelen, D.W. (1954) Oxidations carried out by means of vanadium oxide catalysts. *Chem. Eng. Sci. (Spec. Suppl.)*, **3**, 41–59.

Matralis, H.K., Ciardelli, M., Ruwet, M., and Grange, P. (1995) Vanadia catalysts supported on mixed TiO2-Al2O3 supports. Effect of composition on the structure and acidity. *J. Catal.*, **157**, 368–379.

Moré, J.J. (1978) The Levenberg-Marquardt algorithm: Implementation and theory, Numerical Analysis, 630/1978, pp. 105–116, in Lecture Notes in Mathematics, Springer.

Oyama, S.T., and Somorjai, G.A. (1990) Effect of structure in selective oxide catalysis: oxidation reactions of ethanol and ethane on vanadium oxide. *J. Phys. Chem.*, **94**, 5022–5028.

Pieck, C.L., Banares, M.A., and Fierro, J.L.G. (2004) Propane oxidative dehydrogenation on VOx/ZrO2 catalysts. *J. Catal.*, **224**, 1–7.

Reddy, B.M., Ganesh, I., and Khan, A. (2004) Stabilization of nanosized titania-anatase for high temperature catalytic applications. *J. Mol. Catal. A Chem.*, **223**, 295–304.

Reddy, E.P., and Varma, R.S. (2004) Oxidative dehydrogenation of short chain alkanes on supported vanadium oxide catalysts. *J. Catal.*, **221**, 93–104.

Suchorski, Y., Rihko-Struckmann, L., Klose, F., Ye, Y., Alandjiyska, M., Sundmacher, K., and Weiss, H. (2005) Evolution of oxidation states in vanadium-based catalysts under conventional XPS conditions. *Appl. Surf. Sci.*, **249**, 231–237.

Tóta, A., Hamel, C., Thomas, S., Joshi, M., Klose, F., and Seidel-Morgenstern, A. (2004) Theoretical and experimental investigation of concentration and contact time effects in membrane reactors. *Chem. Eng. Res. Des.*, ISMR3-CCRE18, **82**, 236–244.

Waku, T., Argyle, M., Bell, A., and Iglesia, E. (2003) Effects of O_2 concentration on the rate and selectivity in oxidative dehydrogenation of ethane catalyzed by vanadium oxide: implications for O_2 staging and membrane reactors. *Ind. Eng. Chem. Res.*, **42**, 5462–5466.

Weckhuysen, B.M., and Keller, D.E. (2003) Chemistry, spectroscopy and the role of supported vanadium oxides in heterogeneous catalysis. *Catal. Today*, **78**, 25–46.

Yang, K.H., and Hougen, O.A. (1950) Determination of mechanism of catalyzed gaseous reactions. *Chem. Eng. Prog.*, **46**, 146–157.

4
Transport Phenomena in Porous Membranes and Membrane Reactors

Katya Georgieva-Angelova, Velislava Edreva, Arshad Hussain, Piotr Skrzypacz, Lutz Tobiska, Andreas Seidel-Morgenstern, Evangelos Tsotsas, and Jürgen Schmidt

4.1
Introduction

Membrane reactors offer the possibility to integrate dosing, separation and reaction processes in a single apparatus. Their efficient operation depends on the effectiveness of the catalyst and the corresponding reaction kinetics as well as on the transport processes in the reactor. The diffusive and convective mass and energy transfer can be systematically influenced by the reactor geometry, the membrane morphology and the operation conditions. Hence, there is a high optimization potential for these reactors. Due to the large number of parameters, the numerical simulation is increasingly used for reactor sizing and design. The applied models and tools for this purpose are presented in Chapter 2, where different modeling depths are taken into account. Chapter 3 provides details about the kinetics of an example reaction (oxidative dehydrogenation of ethane), which are necessary for the study of the catalysts, membranes and processes in the framework of this project. The aim of the current chapter is the analysis of the transport phenomena, in particular the superposition of convection and diffusion processes. It focuses on fixed-bed or packed-bed reactors (PBR), packed-bed membrane reactors (PBMR) and catalytic membrane reactors (CMR), which are investigated numerically by a pseudo-homogeneous model approach. Concerning the superposition of convective and diffusive processes, some aspects of the numerical solution of the corresponding differential equations are discussed from a mathematical point of view in Section 4.2.

The complexity of the simulation model and the numerical effort depend on the model dimension and especially on the manner of description of the velocity profiles. Through meaningful simplifications in the modeling, based on a balance between expenses and benefits, computing effort can often be significantly reduced. Plug flow conditions or developed velocity profiles are frequently used as model assumptions. For example in PBRs, developed velocity profile can be reliably used for the momentum analysis without great deviation to the full-order model, (Hein, 1999), due to the setting of this profile after a very short entrance length. But, no developed profile can be obtained in the membrane reactor due to

Membrane Reactors: Distributing Reactants to Improve Selectivity and Yield.
Edited by Andreas Seidel-Morgenstern

ISBN: 978-3-527-32039-4

the axial reactant dosing. Only a few publications consider this phenomenon in detail. Therefore, Section 4.3 deals with the particularities of the velocity fields in membrane reactors and focuses on:

- The need to consider compressibility in the equation of motion
- The developing flow behavior of the fluid and the possibilities of formulating a scaled velocity profile, independent from the reactor length
- The effect of the pressure drop in the reactor tube on the axial dosing (Hamel, 2008).

Another studied point is the analysis of the transport processes in the membranes. In this work, two types of membrane (Al_2O_3 membranes, sintered metals) are considered experimentally and theoretically. There are different transport mechanism in these two types of membrane and an experimental identification of the effective morphological and transport coefficients is essential for a correct model description. The experimental set-up and the analysis of the results based on the dusty gas model (DGM) are briefly described and discussed in Section 4.4. Furthermore the summarized results for the different membrane layers are used for a modeling study in this chapter as well as in Chapters 5 and 8. The application of the derived DGM model parameters in CFD simulations requires a link between the different approaches and subsequently the model validation. The implementation procedure is discussed regarding a 2-D model.

The combination of convective and diffusive transport in the different reactor zones and in conjunction with the reaction kinetics is the subject of Section 4.5. Ethane and products can diffuse through the membrane into the shell side, causing an undesired mass loss. To avoid or limit this process, both operating conditions like mass flux through the membrane and membrane structure parameters like pore diameter, thickness and ε/τ ratio can be used. Due to the smaller mass flux through the membrane in a CMR compared to a PBMR, the risk of product loss is much greater in a CMR. The increase in the mass flux of a mixture of inert gas/oxygen through the membrane can be kept only in narrow limits because ethane is transported into the catalyst layer only by diffusion and a reaction inhibition can arise. The developing concentration profiles in the reactor tube and the oxygen dosage determine the activity of the catalyst layer. In a PBMR, oxygen reaches the catalyst particles at the reactor axis only after a given length, so that near the inlet an inactive region is formed. A sensible matching of the reaction and transport processes is required and an optimal operation of the membrane reactor can be achieved only by using appropriate parameters.

Furthermore, an analysis of the influence of the membrane geometry and structural parameters as well as of operating conditions is performed for the CMR in Section 4.6. Yield and selectivity of the desired product are sensitively affected in this reactor by the transport processes. A few results from an experimental investigation were achieved for the CMR, revealing that the preparation of the catalytic active layer did not provide a sufficient thickness. Therefore, numerical simulations were carried out to evaluate the reactor performance. The results obtained contribute to a better understanding of the coupled processes occurring in the reactor.

4.2 Aspects of Discretizing Convection-Diffusion Equations

The balance equations for the energy and species derived in Chapter 2 are of the convection–diffusion type in which convection is often dominant due to the smallness of thermal conductivity and diffusion coefficients, respectively. This causes several difficulties when solving these equations numerically; for an overview see (Roos, Stynes, and Tobiska, 2008). The problems and basic ideas to handle them can be illustrated for the simple convection–diffusion equation:

$$-D\Delta y + \vec{v}\cdot\nabla y = f \quad \text{in } \Omega, \qquad y = y_D \quad \text{on } \Gamma_D, \quad D\frac{\partial y}{\partial n} = 0 \quad \text{on } \Gamma_N, \tag{4.1}$$

where y denotes the concentration of a chemical species, D is the diffusion coefficient, $\vec{v}$ is the velocity field, f is a source term, Ω is a two- or three-dimensional domain and y_D is the concentration given at the Dirichlet part Γ_D of the boundary, and Γ_N is the Neumann outflow part of the boundary. The numerical results of a standard Galerkin and an upwind approach for the 2-D test case on a 33 × 33 cartesian triangular grid corresponding to 1089 degrees of freedom are presented for the data $D = 10^{-7}$, $f = 0$, $\vec{v} = [8xy(1-x), -4(2x-1)(1-y^2)]$, $\Gamma_N = \{(x, y) \in \partial\Omega : 1/2 < x < 1, y = 0\}$, $\Gamma_D = \partial\Omega \backslash \Gamma_N$, $y_D = 1$ for $1/4 \le x \le 1/2$, $y = 0$ and $0 \le y \le 1$, $x = 1$, $y_D = 0$ otherwise on Γ_D. The in- and out-flow parts can be seen in vector plot of the velocity, see Figure 4.1. As one can see in Figure 4.2, the solution obtained by

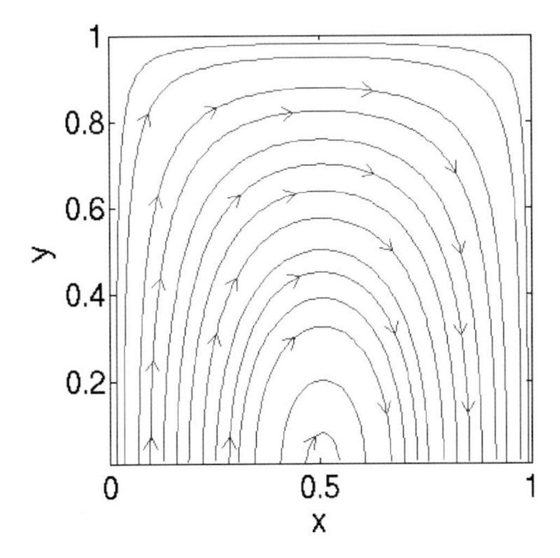

Figure 4.1 Velocity field.

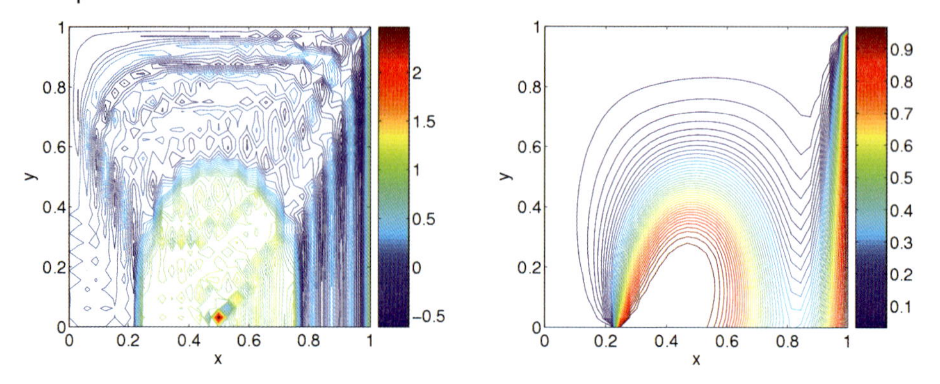

Figure 4.2 Galerkin solution (left), upwind solution (right).

the standard Galerkin finite element method exhibits spurious nonphysical oscillations and is therefore completely useless. Applying a first-order upwind finite element method improves the solution quality considerably but the sharp inflow front at $x = 1/4$, $y = 0$ is smeared out.

The consistent streamline–diffusion method has been developed for convection–diffusion reaction problems in order to combine stability with high accuracy, see for example, (Roos, Stynes, and Tobiska, 2008). This type of stabilization is based on adding weighted residuals to:

$$S_{\mathrm{SD}}(y,\psi)=\sum_K \tau_K\left(-D\Delta y+\vec{v}\cdot\nabla y-f,\vec{v}\cdot\nabla\psi\right)_K \tag{4.2}$$

the standard Galerkin formulation where the summation goes over all finite element cells K and τ_K denotes a user chosen positive stabilization parameter. However, applied to systems of convection–diffusion reaction equations additional nonphysical couplings are introduced by the consistency requirement. Therefore, a new stabilization method based on local projections has been developed, mathematically investigated and successfully implemented (Matthies, Skrzypacz, and Tobiska, 2007; Matthies, Skrzypacz, and Tobiska, 2008). Instead of weighted residuals only weighted fluctuations $\kappa_h(\nabla y)$ are added to the Galerkin formulation, resulting in a stabilizing term of the form:

$$S_{\mathrm{LP}}(y,\psi)=\sum_K \tau_K\left(\kappa_h(\nabla y),\kappa_h(\nabla\psi)\right)_K. \tag{4.3}$$

Applied to systems, such symmetric stabilization terms avoid nonphysical additional couplings between the balance equations and lead to block diagonal matrices in the algebraic system. The stabilizing effect of the local projection stabilization (LPS) can clearly be seen in Figure 4.3 (left). Compared to the Galerkin method the oscillations are damped and localized in the direct neighborhood of sharp fronts. This stability behavior similar to the streamline–diffusion method has been mathematically established by (Knobloch and Tobiska, 2008), however the LPS does not introduces artificial couplings when applied to systems of equations.

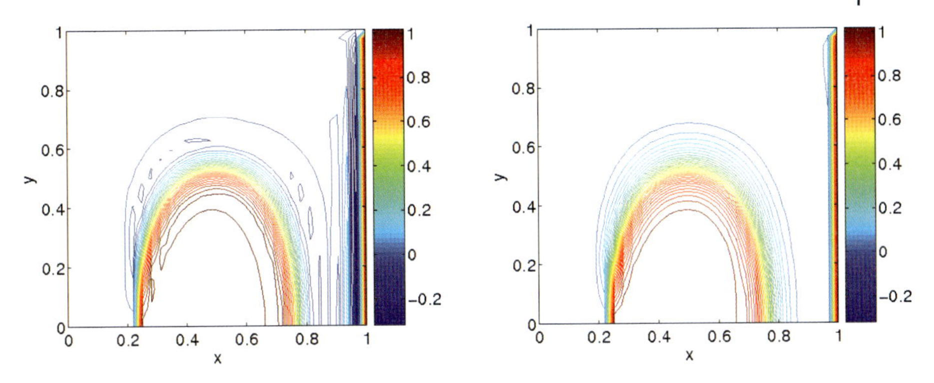

Figure 4.3 LPS solution (left), LPS solution with additional shock-capturing (right).

Although the new developed LPS combines good stability properties with high accuracy, it does not satisfy – just like the streamline–diffusion stabilization – the discrete maximum principle. Thus, there is no guarantee that physical properties like the positivity and boundedness of the solution of the balance equations are preserved in the discrete case. To improve the quality of the discrete solution the stabilized schemes can be modified by adding a nonlinear shock-capturing term. In Figure 4.3 (right) this improvement can be clearly seen for an edge-oriented shock-capturing term. The numerical results for the LPS and LPS with shock capturing have been computed for an enriched piecewise linear finite element method on a triangular mesh and a piecewise constant projection space, cf. (Matthies, Skrzypacz, and Tobiska, 2008). The new approach has been successfully applied to the various coupled transport flow problems.

4.3 Velocity Fields in Membrane Reactors

The numerical effort of the simulation mainly depends on the used model for the velocity field. Often, the velocity and pressure fields are governed by the Navier–Stokes equations which can be formulated for a compressible or incompressible fluid. In order to estimate the need for a compressible formulation, the influence of the gas compressibility on the yield and transmembrane pressure difference was investigated in preliminary studies. For compressible flows, the gas density is a function of the pressure changes in the reactor; otherwise the density depends only on the predefined operating pressure. For membranes with a thin catalyst layer (3 μm thickness), minor variations in the ethylene yield and transmembrane pressure drop are observed varying the SS/TS = $\dot{V}_{SS}/\dot{V}_{TS}$ ratio. In these cases, an incompressible formulation (ρ, η = const) of the momentum balances equation is sufficient to study the transport phenomena.

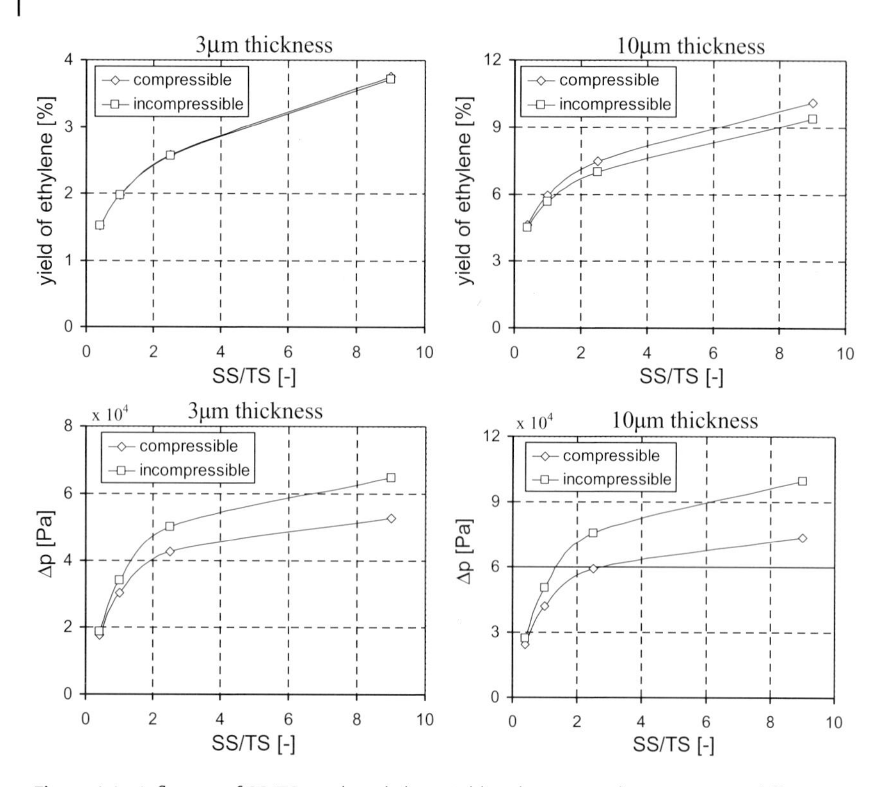

Figure 4.4 Influence of SS/TS on the ethylene yield and transmembrane pressure difference for 3 and 10 μm thicknesses of the catalyst layer for GHSV = 27 736 L/h, T = 620 °C.

However, increasing the membrane layer thickness as well as reducing the pore diameter and porosity to tortuosity ratio, it is expected that the gas compressibility strongly influences the reactor performance. As shown in Figure 4.4, where simulations are carried out also with a 10 μm catalyst layer, obvious differences appear in the yield values, especially for high-volume flow rates at the shell inlet. The deviations of the transmembrane pressure drop are significantly larger and reach 26%. Therefore, obtaining results for the transmembrane pressure drop under the given conditions requires simulations based on compressible fluid behavior.

The transport processes in the reactor are influenced by the different types of velocity profiles (plug-flow, parabolic profile, etc.) used and the development of the flow behavior. In particular, for large entrance zones a detailed computation of the velocity fields is needed and a fully developed velocity profile cannot be assumed. This is also the case in membrane reactors of the distributor type where a developed velocity profile cannot be expected due to the axial reactant dosing.

Based on the combination of reactor and membrane functions, there are differences in the formation of velocity, concentration and temperature profiles compared to conventional reactors. This is demonstrated in the following example.

The axial and radial velocity profiles in two types of CMR, calculated with the 2-D model under isothermal conditions, are compared and presented in Figure 4.5 in dimensionless form. The inlet velocity at the tube side and the reactor diameter are used as a scale basis. The velocity fields are calculated on the basis of macroscopic average superficial values. In the first type reactor, the tube is empty and, in the second one, the tube is filled with inert particles. By these simulations, identical operating conditions are defined for both reactors.

In the case of CMR without packed bed, parabolic profiles of axial velocity are characteristic for the laminar flow in the channels. If a packed bed is posed in the reactor tube, the velocity profiles are formed by a sink in the momentum equations. The pressure loss is described by usage of the Ergun equation, Equation 2.21, and as a result a flat profile of the axial velocity is observed. Furthermore, a radial dependence of bed porosity is considered and calculated according to Equation 2.22. At a low tube : particle diameter ratio, the consideration of the porosity profile in the reactor has a strong influence not only on the flow maldistribution, but also on the concentration and temperature profiles (Keil, 2007). The porosity rises rapidly close to the membrane wall and the axial velocity increases significantly causing undesirable flow channeling, (Schlünder and Tsotsas, 1988).

In the membrane, the values of the axial velocity decrease significantly as a result of the low permeability coefficients of the porous media and they approach zero for both reactors. Moreover, a convective radial flow through the membrane is characteristic for the membrane reactors of the distributor type. The development of the radial velocity profile depends on the intensity and direction of the pressure gradients. The axial pressure gradients in the annulus, the packed bed or the tube are small compared to the pressure drop between the tube and shell side in the considered cases. Consequently, qualitative identical radial velocity profiles with the reactor length are obtained in both CMRs. In case of a significant pressure loss caused by the packing, an influence of the velocity field on the axial distribution of the dosing is given (Hamel, 2008). The radial velocity profiles go in both reactors through a minimum. This occurs in the CMR with a packed bed in the vicinity of the wall. After that the radial velocity increases linearly towards the core. In the CMR, the minimum lies near the core, caused by the well known laminar profile.

If the dimensionless axial velocities in both reactors are scaled to the local average cross-sectional velocity in reactor tube, an uniform self-similar distribution can be predicted along the reactor, see Figure 4.5e,f. In the PBMR, the similar profiles are observed after a very short distance, whereas the flow in the CMR needs to develop along about 10% of the reactor length. Figure 4.6 shows the development of the scaled maximal velocity as a function of the axial coordinate of the CMR.

The different velocity distributions in the case of a CMR with and without inert particles influence significantly the reactor performance for fixed operating conditions (Georgieva-Angelova, 2008).

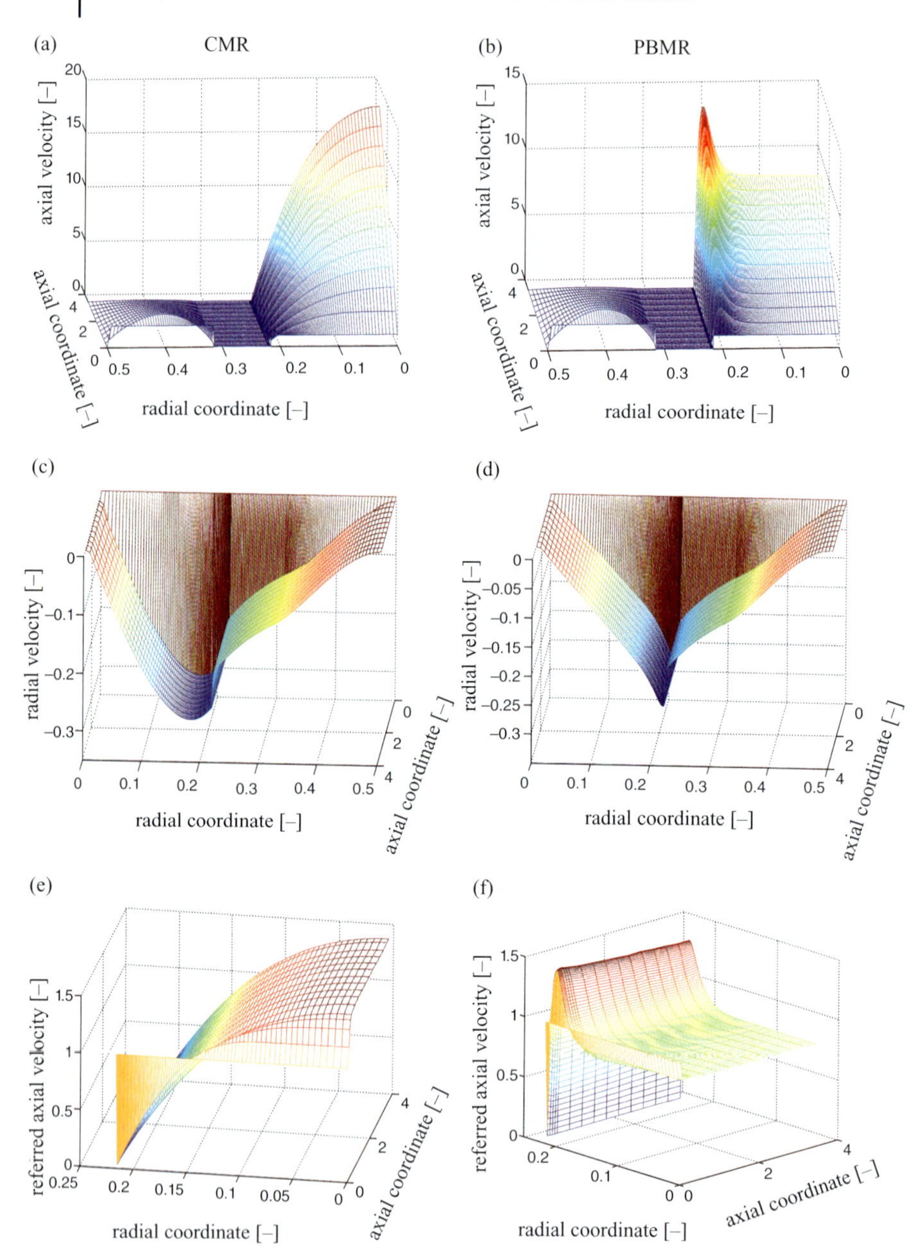

Figure 4.5 Velocity profiles in CMR with and without packed bed for GHSV = 27 736 L/h, SS/TS = 9, T = 600 °C.

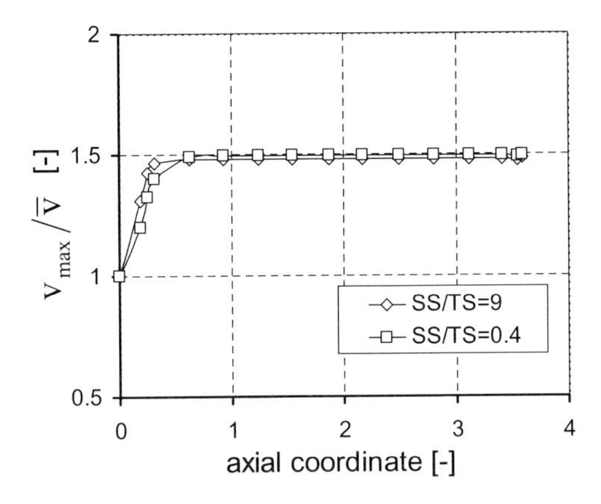

Figure 4.6 Development of the scaled maximal velocity.

For PBRs the velocity fields have been computed using the Brinkman–Forchheimer equation:

$$-\nabla\cdot\left(\frac{\varepsilon}{\mathrm{Re}}\nabla\vec{v}-\varepsilon\vec{v}\otimes\vec{v}\right)+\varepsilon\nabla p+\sigma(\vec{v})=0,\qquad \nabla\cdot(\varepsilon\vec{v})=0, \tag{4.4}$$

where $\varepsilon = \varepsilon(r)$ denotes the radial symmetric porosity, Re is the Reynolds number and σ is a friction force. The existence and uniqueness of a weak solution for a linear friction term $\sigma(\vec{v})=\sigma\,\vec{v}$ under appropriate boundary conditions has been shown by (Skrzypacz and Tobiska, 2005). In the simplest case of a plane channel of gap L with a constant porosity $\varepsilon(r) = \varepsilon =$ const, the analytical solution:

$$\vec{v}=\left(\frac{p_1}{\sigma}\left(1-\frac{\cosh\lambda(y-L/2)}{\cosh\lambda L/2}\right),0\right),\quad p=p_0-p_1x,\quad \lambda^2=\sigma\mathrm{Re}/\varepsilon \tag{4.5}$$

for the fully developed flow can be used to evaluate the accuracy of different numerical solutions. We found that the Q_2/P_1^{disc} solution computed by the in-house code MooNMD approximates the analytical solution already on a coarser mesh with the same accuracy as the P_2/P_1 solution computed by the COMSOL package (Chapter 2). In case that the porosity is not constant there is no longer an analytical expression for the fully developed velocity profile available.

Therefore, the profiles have been computed by MooNMD with the accurate Q_2/P_1^{disc} finite element for different Reynolds numbers. Figure 4.7 shows clearly the tunneling effect due to decreasing porosity in the neighborhood of the reactor wall. Although there is no exact representation of the developed profile, the method of matched asymptotic expansions can be used to construct an analytical approximation in the case of a high Reynolds number. For this let the friction term be given by $\sigma(\vec{v})=\frac{\alpha}{\mathrm{Re}}\vec{v}+\beta\vec{v}|\vec{v}|$. Assuming a linear pressure drop $p = p_0 - p_1 z$ we get for the developed velocity profile $\Phi = \Phi(r)$ in a tube of radius R the two-point boundary value problem:

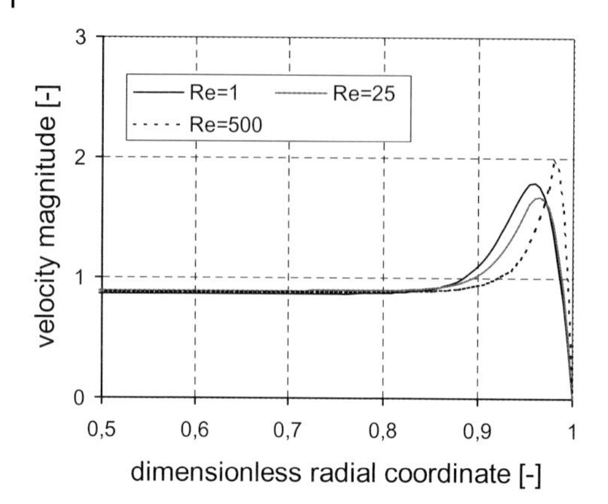

Figure 4.7 Fully developed velocity profiles in PBR for different Reynold numbers.

$$\frac{1}{\mathrm{Re}}\left[-\frac{1}{r}\frac{\partial}{\partial r}\left(r\varepsilon(r)\frac{\partial \Phi}{\partial r}\right)+\alpha(r)\Phi\right]+\beta(r)\Phi|\Phi|=\varepsilon(r)p_1,\quad \Phi'(0)=\Phi(R)=0, \tag{4.6}$$

which can be solved by a finite element method on layer-adapted grids (Roos, Stynes, and Tobiska, 2008) or for high Reynolds numbers by the matched asymptotic expansion method (Goering, 1977). The asymptotic expansion up to first-order terms is:

$$\Phi_{as}(r)=\sqrt{\frac{p_1\varepsilon(r)}{\beta(r)}}-\sqrt{\frac{p_1\varepsilon_R}{\beta_R}}\frac{3}{\cosh^2\left(\frac{(R-r)\sqrt{\mathrm{Re}}}{\sqrt{2}}\sqrt[4]{\frac{p_1\beta_R}{\varepsilon_R}}+\gamma\right)},\quad 3\cosh^2(\gamma)=1,\quad \gamma>0. \tag{4.7}$$

In Figure 4.8 we compare the first order asymptotic approximation for $\mathrm{Re} = 1000$ with the 1-D finite element solution which are in good agreement. Note that the asymptotic approximation is valid for the limit case $\mathrm{Re} \to \infty$ and therefore valid only for high Reynolds numbers. The 1-D finite element approximation can be used for the full range of Reynolds numbers as a method to avoid 2-D and 3-D computations whenever only developed profiles are of interest. Figure 4.9 presents the velocity profiles for a PBMR with a uniform dosing profile over the reactor length. We see that, with an appropriate scaling, one uses the concept of self-similar solutions also in this case.

Concerning an accurate numerical computation of velocity fields, particularly for PBMR with large velocity gradient near the wall, we performed several tests which show that the quality of the finite element solution of the incompressible Navier–Stokes equation strongly depends on the pair of finite elements used to discretize velocity and pressure. In particular, the local mass conservation – important when coupled fluid-transport phenomena are computed – is much better satisfied for discontinuous pressure approximations than for continous ones (Matthies

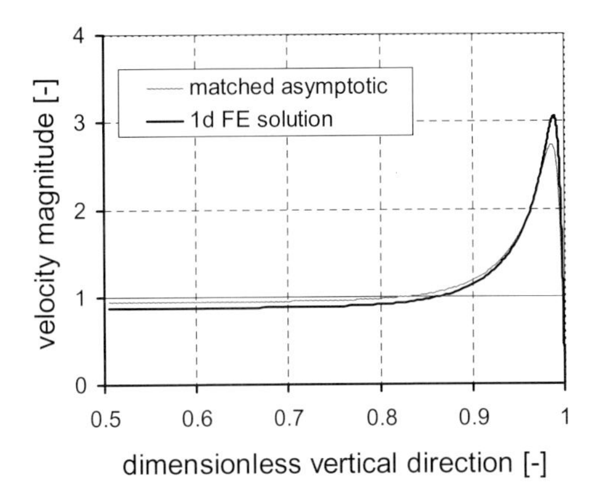

Figure 4.8 Developed velocity profiles for the PBR for Re = 1000.

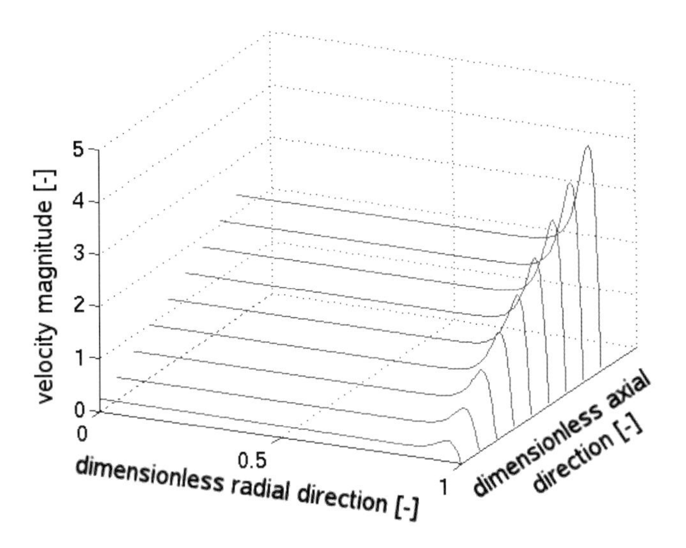

Figure 4.9 Self-similar velocity profiles in PBMR for Re = 50.

and Tobiska, 2007). Therefore, instead of using continuous, piecewise polynomials of degree $k \geq 2$ to approximate the velocity on triangular or tetrahedral meshes and continuous, piecewise polynomials of degree $k - 1$ for the pressure (Taylor–Hood element), we propose to approximate the pressure by piecewise, discontinuous polynomials of degree $k - 1$. However, combined with a continuous, piecewise polynomial approximation of degree k on triangular and tetrahedral meshes this would lead to an unstable finite element pair which does not satisfy the inf-sup stability condition already mentioned in Chapter 2. In order to fulfill the stability condition the velocity space can be enriched by additional bubble functions or the discontinuous pressure space can be used with a velocity space of continuous,

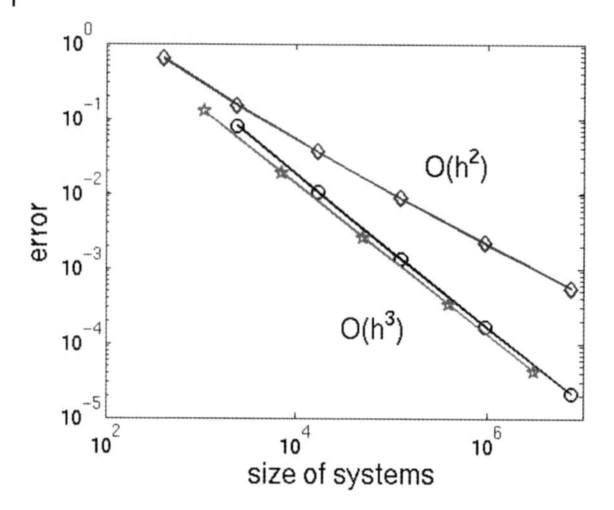

Figure 4.10 Errors of the (Q_2, P_1^{disc}) solution (diamonds), the (Q_3, P_2^{disc}) solution (stars) and the post-processed solution (circles).

piecewise polynomials of degree k in each variable on quadrilateral and hexahedral meshes, respectively (Girault and Raviart, 1986). We denote this inf-sup stable element pair briefly by $(Q_k, P_{k-1}^{\text{disc}})$.

Numerical experiments verify the theoretically predicted convergence rates of order k with respect to the (H^1,L^2)–norm. Moreover, (Matthies, Skrzypacz, and Tobiska, 2005) proved that, for $k = 2$, a special interpolant is superclose; more precisely the error of the computed discrete solution to the interpolated analytical solution is one order better than it is to the analytical solution. Based on this theoretical result a local postprocessing applied to the computed discrete solution leads to a discrete solution with enhanced accuracy. Figure 4.10 shows the convergence rates of the errors of the computed discrete solutions for the 3-D test example on the unit cube $\Omega = (0,1)^3$with the analytical solution:

$$\vec{u} = \begin{Bmatrix} \sin(\pi x)\sin(\pi y)\sin(\pi z) + x^4\cos(\pi y) \\ \cos(\pi x)\cos(\pi y)\cos(\pi z) - 3y^3 z \\ \cos(\pi x)\sin(\pi y)\cos(\pi z) + \cos(\pi x)\sin(\pi y)\sin(\pi z) - 4x^3 z\cos(\pi y) + \frac{9}{2}y^2 z^2 \end{Bmatrix}$$
$$p = 3x - \sin(y + 4z) - p_0 \tag{4.8}$$

and the Reynold number Re = 10. For a given mesh the next finer mesh is constructed by dividing each cube into eight subcubes such that the finest mesh corresponds to 6 440 067 velocity and 1 048 576 pressure degrees of freedom for the (Q_2, P_1^{disc}) discretization (Figure 4.10, diamonds). Note that the higher order element (Q_3, P_2^{disc}) (Figure 4.10, stars) is more accurate already on a coarser mesh level with

352 947 velocity and 40 960 pressure degrees of freedom. We see that the accuracy of the (Q_2, P_1^{disc}) solution on the finest mesh is almost achieved by the post-processed (Q_2, P_1^{disc}) solution on a mesh with only 107 811 velocity and 16 384 pressure degrees of freedom. The gain in accuracy by post-processing increases due to the different convergence rates $O(h^2)$ and $O(h^3)$, respectively. Thus, the computational effort for a certain accuracy can be drastically reduced, in particular if local post-processing is used only in regions of interest. For more details we refer to (Matthies, Skrzypacz, and Tobiska, 2007). It is important to note that the mixed finite element discretizations of the Stokes, Navier–Stokes and the extended Brinkmann equation lead to a nonlinear algebraic system of a saddle point type for which sophisticated algorithms have been developed (Turek, 1999; John *et al.*, 2002).

4.4
Determination of Transport Coefficients and Validation of Models

In the following we discuss mass transport in ceramic membranes based on results by (Hussain, Seidel-Morgenstern, and Tsotsas, 2006) and (Thomas, 2003) and mass transport in sinked metall membranes based on results by (Edreva *et al.*, 2009). Both types of membranes are porous and asymmetric. They differ not only in material and structure, but also in the fact that all precursors were available for the ceramic membranes, whereas just the final composite was available in the case of metallic membranes. This leads to different methodologies of parameter identification and validation in Sections 4.4.1 and 4.4.2. Section 4.4.3 shows that the quality of prediction of some mass transfer experiments can be further increased by the use of 2-D CFD models. This section is based on results summarized by (Georgieva-Angelova, 2008). The identification of thermal parameters of the membranes and the prediction of the results of experiments with combined mass and heat transfer are not discussed here with reference to work by (Hussain, Seidel-Morgenstern, and Tsotsas, 2006).

4.4.1
Mass Transport Parameters of Multilayer Ceramic Membranes – Precursors Available

4.4.1.1 Task and Tools

The dusty gas model (DGM) presented in Chapter 2 can be used (despite the already-mentioned criticism) to describe mass transport in porous membranes. As already discussed (see Equations 2.49 to 2.54), the model has three parameters (B_0, K_0, F_0) that quantify viscous flow, Knudsen diffusion and molecular diffusion, respectively, in a porous membrane. To translate this formal quantification into morphological features of the membrane, additional assumptions are necessary. Here we assume tortuous, monodispersed capillaries, which are neither interconnected, nor change their cross-sectional area with their length. Then, the three parameters of the DGM can be expressed as:

$$B_0 = F_0 \frac{d_{\text{pore}}^2}{32} \tag{4.9}$$

$$K_0 = F_0 \frac{d_{\text{pore}}}{4} \tag{4.10}$$

$$F_0 = \frac{\varepsilon}{\tau} \tag{4.11}$$

As Equations 4.9 to 4.11 reveal, in the frame of the selected simple model for membrane structure only two of the three parameters are independent from each other. K_0 and B_0 are selected as those two parameters. They correlate with the diameter of the assumed capillaries:

$$d_{\text{pore}} = \frac{8B_0}{K_0} \tag{4.12}$$

and with the ratio ε/τ:

$$\frac{\varepsilon}{\tau} = \frac{(K_0)^2}{2B_0} \tag{4.13}$$

between the porosity ε and tortuosity τ of the membrane.

Consequently, in order to describe mass transport with the DGM, the parameters K_0 and B_0 (or d_{pore} and ε/τ) must be identified with the help of experimental results. For a multilayer membrane, this identification should be done for every single layer. In the present section we refer to tubular multilayer ceramic membranes with available precursors. Such membranes consist of one support layer (α-Al_2O_3), two or three intermediate layers (also α-Al_2O_3) and one separation layer (γ-Al_2O_3). Availability of precursors means that–in the case of two intermediate layers–we have the support membrane, the membrane consisting of the support and the first intermediate layer, the membrane consisting of support and two intermediates and, finally, the complete multilayer membrane. This was possible by close cooperation with the producer of the ceramic membranes (Inocermic GmbH, Hermsdorf, Germany).

Identification of the mass transport parameters is carried out by means of single gas permeation experiments. The validity of parameters gained in this way is checked by additional experiments of isobaric diffusion and transient diffusion. The principles of all three mentioned isothermal experiments are summarized in Figure 4.11 and Table 4.1, where the index "*o*" is used for the shell-side (annulus) and the index "*i*" for the tube-side of the reactor. More explanations are given in subsequent sections.

All measurements can be conducted in the experimental set-up illustrated in Figure 4.12, which is an extended Wicke–Kallenbach cell for tubular specimens. Various valves and mass flow controllers (MFC) enable accurate dosing of gases at the tube and/or shell side of the membrane in the measuring cell. At both outlets the gas composition and gas flow rates can be measured by a thermal conductivity GC sensor and various instruments (depending on the absolute value

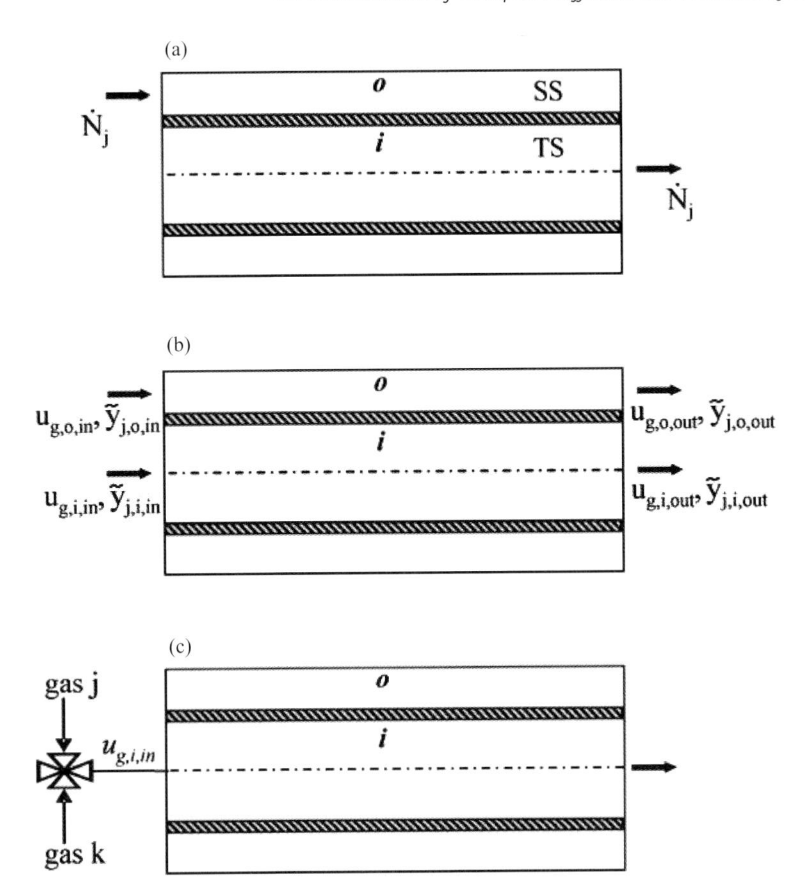

Figure 4.11 Principles of isothermal experiments used in order to identify and validate the mass transport parameters of porous membranes.

Table 4.1 Main features of the experiments from Figure 4.1.

Experiment	Conditions	Components	Principal measured quantity	Purpose
Single gas permeation	Steady state	Pure gas	$p_o - p_i$	Identification of K_0, B_0
Isobaric diffusion	Steady state	Two components	$y_{j,o,out}$, $u_{g,o,out1}$ $y_{j,i,out}$, $u_{g,i,out}$	Validation
Transient diffusion	Transient	Two components	$p_o(t) - p_i$	Validation

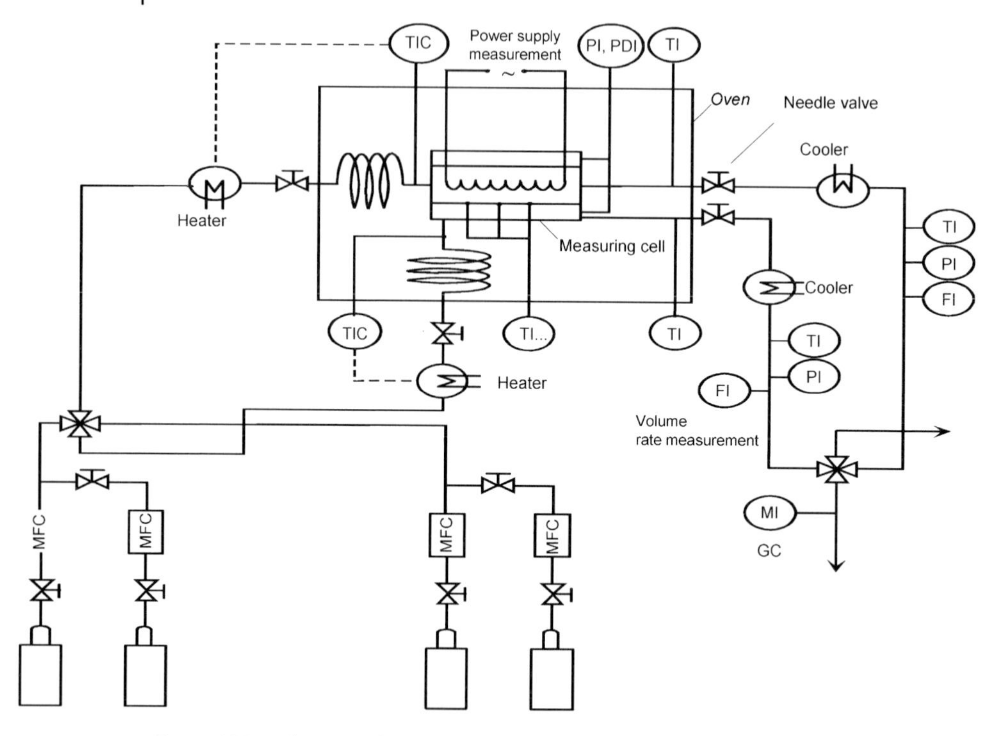

Figure 4.12 Schematic of experimental set-up.

of the gas flow rate), respectively. Additionally the pressure difference between tube and annulus and several absolute pressures are determined. If necessary, isobaric conditions can be attained by a fine adjustment of the needle valves at both gas outlets. Several methods of sealing have been tried out, including conventional O-rings and slightly conical graphite rings. The tightness between the tube- and shell-side of the membrane as well as tightness to the environment have been checked by a helium leakage detector and by pressure measurements. The reactor is placed in a controllable oven, so that different temperature levels can be set for isothermal experiments. An electrical heater that can be placed in the tube, gas pre-heaters, post-coolers, insulations and various measurements of temperature enable additional, non-isothermal modes of operation.

4.4.1.2 **Identification by Single Gas Permeation**

In the single gas permeation experiment according to Figure 4.11a a pure gas is introduced at the shell-side, is pressed through the membrane and leaves the cell at the end of the tube. Under such conditions, the general DGM equation for a homogeneous membrane (2.60) reduces to:

$$\dot{n}_j = -\frac{1}{\tilde{R}T}\left(\frac{4}{3}K_0\sqrt{\frac{8\tilde{R}T}{\pi\tilde{M}_j}} + \frac{B_0}{\eta_j}\bar{p}\right)\nabla p \tag{4.14}$$

For cylindrical coordinates and a relatively moderate membrane thickness, integration of Equation 4.14 leads to the expression:

$$\frac{\dot{N}_j}{\Delta p} = -\frac{2\pi L}{\tilde{R}T\ln\left(\frac{r_{m,o}}{r_{m,i}}\right)}\left(\frac{4}{3}K_0\sqrt{\frac{8\tilde{R}T}{\pi\tilde{M}_j}} + \frac{B_0}{\eta_j}\bar{p}\right) \tag{4.15}$$

Equation 4.15 means that the permeate flux of the membrane (the ratio $\dot{N}_j/\Delta p$) should be a linear function of the average pressure in the membrane ($\bar{p} = (p_o + p_i)/2$). Consequently, linear regression of respective experimental data gives the Knudsen coefficient K_0 from the intercept and the permeability constant B_0 from the slope of the resulting straight line. The measurements necessary in every single permeation experiment are those of pressure drop, $\Delta p = p_o - p_i$, one absolute pressure (p_o or p_i) and gas molar flow rate, $\dot{N}_j$.

After K_0 and B_0 of the support layer of a composite membrane have been determined in this way, single gas permeation experiments are conducted with the composite precursor consisting of the support and the first intermediate layer. Application of Equation 4.15 to the support with the previously determined values of K_0 and B_0 delivers the pressure at the interface between support and intermediate. In this manner, pressures and fluxes are known for the intermediate layer, so that K_0 and B_0 can be derived also for this layer. Recursively, the mass transport parameters of every layer of any composite membrane are obtained, provided that all precursors of the composite are available. Results of the application of this scheme to three ceramic membranes, one with a large diameter (membrane 1) and two with a small diameter (membranes 2a and 2b) are presented in Table 4.2. Herein, layer thicknesses have been taken over from producer information. Comparison between membranes 2a and 2b shows that the structural properties of membranes, which are nominally the same but belong to different production charges, may differ to a certain extent. Defects or fissures have an even stronger influence on membrane properties.

It is obvious from the foregoing discussion, that a certain variation of the pressure level is necessary in order to obtain mass transport parameters from single gas permeation experiments. This variation is in the range of 1–3 bar (1 bar = 100 kPa) in the present work. Variation of temperature or of the used gas is strictly not necessary. However, it can serve as a check for the consistency of the derived values of K_0 and B_0–which should not depend systematically on the temperature level or on the kind of the gas. For this reason, because temperature is an important operating parameter of membrane reactors, and because multicomponent gas mixtures appear in practice, an ample variation of temperature (20–500 °C) has been conducted for three different gases (air, N_2, He). Representative experimental results for the large membrane (membrane 1, see Table 4.2) are shown and compared with calculations in Figures 4.13–4.16.

First, single gas (N_2) permeation data gained with the (homogeneous) support of this membrane at different temperatures are depicted in Figure 4.13, illustrating that the linearity foreseen by Equation 4.15 is fulfilled. According to the same

Table 4.2 Mass transport parameters of three ceramic membranes.

Layer	Material	Thickness [μm]	K_0 [m]	B_0 [m²]	d_{pore} [nm]	ε/τ
Membrane 1, $L = 250\,mm$, $r_{m,i} = 10.5\,mm$, $r_{m,o} = 16\,mm$ (approx.) according to Hussain, 2006						
Support	α-Al_2O_3	5500	8.16×10^{-8}	2.96×10^{-14}	2900	0.112
First intermediate	α-Al_2O_3	25	7.99×10^{-8}	2.73×10^{-14}	2730	0.124
Second intermediate	α-Al_2O_3	25	2.98×10^{-9}	2.85×10^{-17}	76.5	0.156
Separation	γ-Al_2O_3	2	2.03×10^{-9}	7.47×10^{-18}	29.4	0.277
Membrane 2a, $L = 200\,mm$, $r_{m,i} = 3.5\,mm$, $r_{m,o} = 5\,mm$ (approx.) according to Hussain, 2006						
Support	α-Al_2O_3	1500	8.80×10^{-8}	3.32×10^{-14}	3010	0.117
First intermediate	α-Al_2O_3	25	7.95×10^{-8}	2.56×10^{-14}	2570	0.124
Second intermediate	α-Al_2O_3	25	7.71×10^{-9}	2.74×10^{-16}	284	0.109
Third intermediate	α-Al_2O_3	25	5.72×10^{-9}	7.88×10^{-17}	110	0.208
Separation	γ-Al_2O_3	2	8.83×10^{-11}	5.45×10^{-20}	4.94	0.072
Membrane 2b, $L = 300\,mm$, $r_{m,i} = 3.5\,mm$, $r_{m,o} = 5\,mm$ (approx.) according to Thomas, 2003						
Support	α-Al_2O_3	1500	9.34×10^{-8}	3.58×10^{-14}	3070	0.122
First intermediate	α-Al_2O_3	25	4.11×10^{-8}	9.47×10^{-15}	1840	0.089
Second intermediate	α-Al_2O_3	25	9.40×10^{-9}	2.24×10^{-16}	191	0.197
Third intermediate	α-Al_2O_3	25	5.97×10^{-9}	5.69×10^{-17}	76	0.313
Separation	γ-Al_2O_3	2	1.11×10^{-9}	2.18×10^{-18}	16	0.283

equation, the influence of temperature should be proportional to $T^{-1/2}$ in the Knudsen regime and very strong–proportional to approximately $T^{-1.75}$–in the viscous regime. The latter is a combination of the explicit, inverse proportionality on temperature of Equation 4.15 with the temperature dependence of viscosity ($\eta_j \sim T^{0.75}$ according to [Schlünder and Tsotsas, 1988]). As we see in Figure 4.13b,c, the intercepts and slopes of the straight lines of Figure 4.3a agree well with these theoretical expectations. However, the intercept (Knudsen diffusion) represents single gas permeation only at the limit of $\bar{p} \to 0$, and the slope (viscous flow) only at the limit of $\bar{p} \to \infty$. In reality, we have a certain finite value of average pressure and, thus, a mix of Knudsen diffusion and viscous flow that depends on the specific value of $\bar{p}$ and on the structure of the membrane. Consequently, the temperature dependence is characterized by an exponent somewhere between −0.5 and −1.75. This is illustrated in Figure 4.14. For moderate pressures in the rather permeable support the flow rate is approximately proportional to T^{-1} (Figure 4.14a). The flow rate through the composite membrane is always significantly lower than the flow rate through the support, due to the mass transport resistance of the additional layers. Since these layers have smaller pores, the influence of Knudsen diffusion increases, which is reflected in a weaker dependence of the flow rate upon temperature in the composite membrane. Results of $\dot{N}_j/\Delta P$ over pressure for every individual layer of membrane 1 are shown in Figure 4.15. We see small intercepts and large slopes for the support and the first intermediate

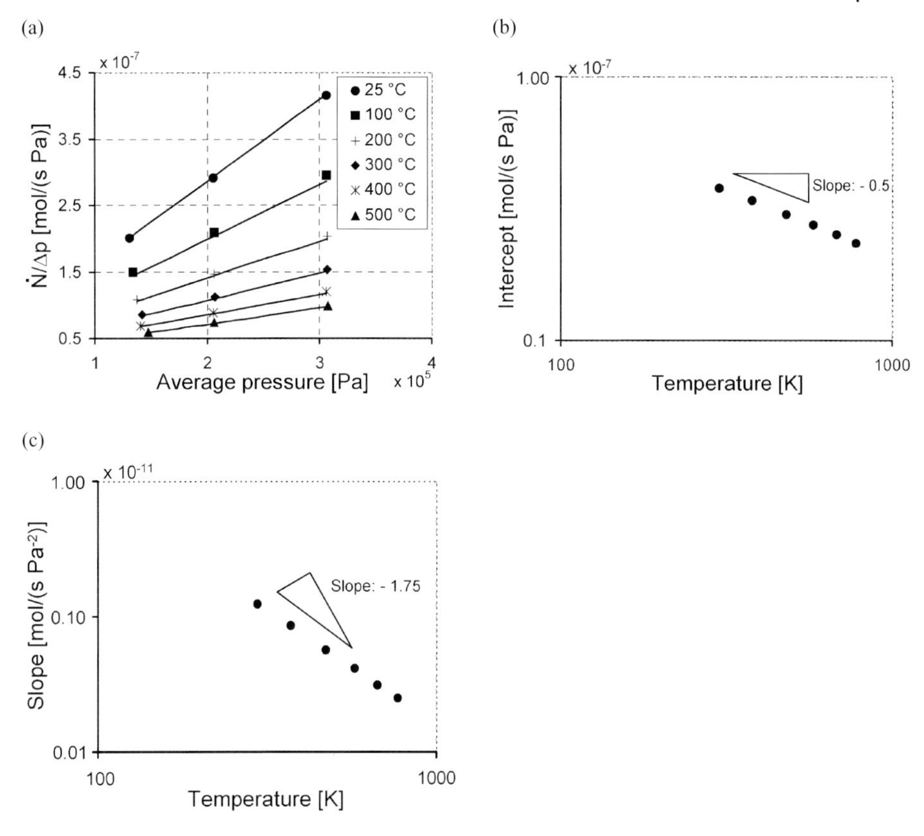

Figure 4.13 a) Results of single gas (N_2) permeation experiments for the support of membrane 1 at different temperatures. b) and c) Temperature dependence of the intercepts and slopes of the respective straight lines.

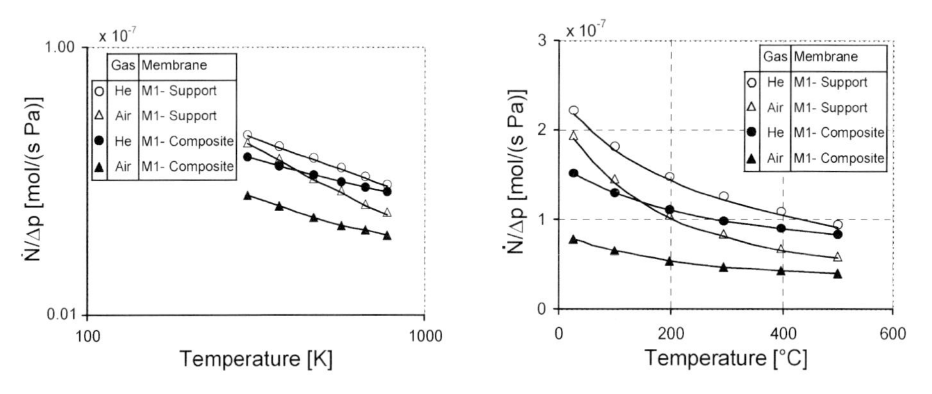

Figure 4.14 a) Logarithmic and b) linear plots for the influence of temperature on single gas permeation through the support and the composite membrane 1 at the lowest realized pressure level of approximately 1 bar.

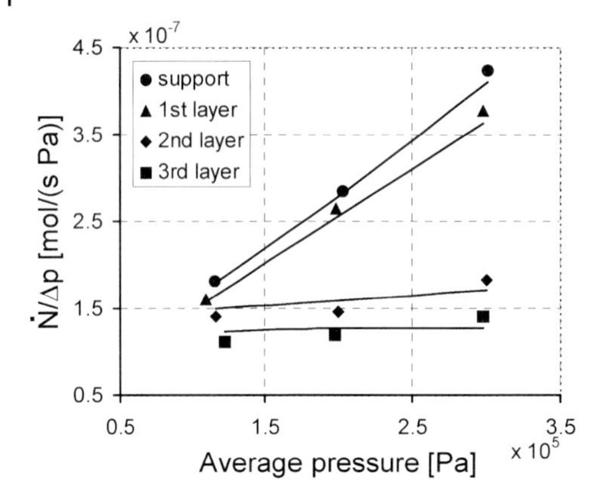

Figure 4.15 Single gas permeation (N_2, 25 °C) for every individual layer of membrane 1.

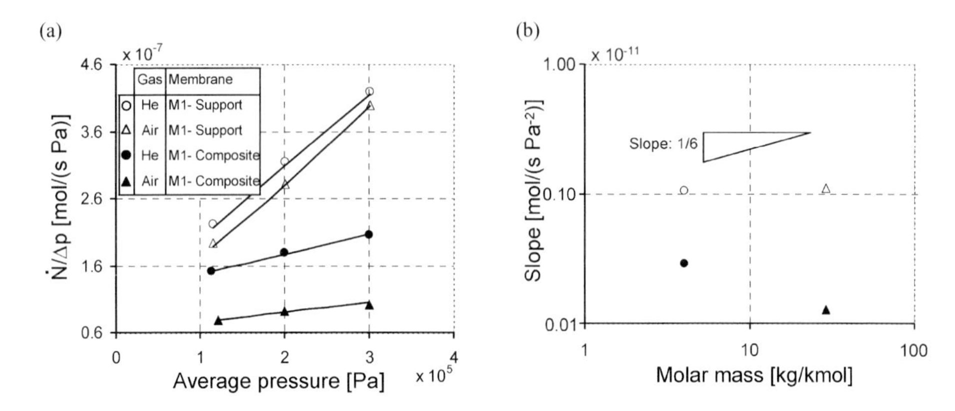

Figure 4.16 a) Results of permeation experiments with two different gases for the support of membrane 1 and for the composite membrane 1 at 25 °C. b) Dependence of the slopes of respective lines on gas molar mass.

layer, which means that viscous flow dominates. In contrast, Knudsen diffusion dominates for the second intermediate and the separation layer, so that large intercepts and very small slopes are obtained. Evaluated pore diameters reflect this behavior (Table 4.2).

As to the influence of the gas, it is clear that He permeates easier than air or N_2 (Figure 4.14). This is verified by Figure 4.16, which also reveals some additional interesting features. Equation 4.15 suggests a dependence on $\tilde{M}_j^{-1/2}$ in the intercept (Graham's law). Dynamic gas viscosity is proportional to approximately $\tilde{M}_j^{-1/6}$ (see, e.g., Schlünder and Tsotsas, 1988, p. 76), so that the slope should depend on $\tilde{M}_j^{1/6}$. Figure 4.16b shows explicitly that the expectation concerning the influence of molar mass on the slope is fulfilled by data for the support. Graham's law is also fulfilled by this data (open symbols), though not explicitly illustrated. The same is not true for data gained with the composite membrane 1 (full symbols). Specifically, the slope of single gas permeation data decreases with increasing molar mass for the composite, while it increases for the support (Figure 4.16b). This inversion in respect to the behavior of a homogeneous porous body is a result of complicated combination of several Knudsen and viscous contributions in the different layers. It can be used for diagnosis of composite membrane structure. Such diagnosis cannot be done on the basis of linearity in permeation plots, because plots of this kind (e.g., Figure 4.16a) are nearly linear for both homogeneous membranes and composites. The issue of diagnosis is important in connection with membranes of completely unknown structure and, thus, is further discussed in Section 4.4.2.1.

4.4.1.3 **Validation by Isobaric Diffusion and by Transient Diffusion**

Though the single gas permeation experiment discussed in the previous section primarily serves the purpose of identifying the transport parameters of the membrane, it also has validation features given by variations in the temperature and the kind of gas used. Additional validation is possible by experiments using isobaric diffusion and transient diffusion.

In isobaric diffusion according to the principle of Figure 4.11b, helium can be sent through the tube-side and nitrogen through the shell-side of the membrane. Helium is transported from the tube to the annulus by Knudsen and molecular diffusion, and viscous flow does not take place because of equal pressure in the two compartments. Consequently, at the outlet we have some helium in the annulus ($\tilde{y}_{He,o,out} > 0$) and some nitrogen in the tube ($\tilde{y}_{N_2,i,out} < 1$). The values of $\tilde{y}_{He,o,out}$ and $\tilde{y}_{He,i,out}$ can be measured and also calculated quite accurately, as Figure 4.17a shows. The calculation is conducted by reducing the dusty gas model (2.60) for isobaric conditions to the relationship:

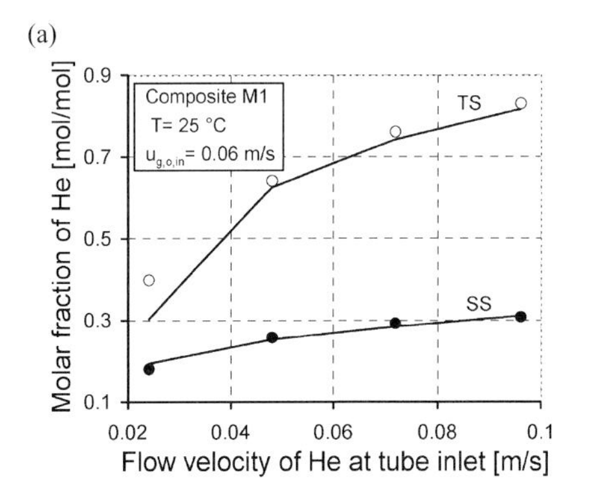

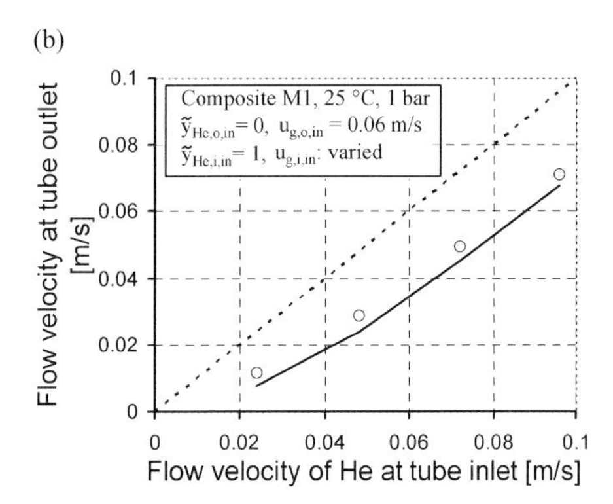

Figure 4.17 Measurement and prediction for isobaric diffusion through the composite membrane M1 (tube inlet: pure helium; annulus inlet: pure nitrogen): a) outlet molar fractions of helium, b) velocity at the outlet of the tube.

$$\sum_{k=1, k\neq j}^{N} \frac{\tilde{y}_k \dot{n}_j - \tilde{y}_j \dot{n}_k}{\varepsilon/\tau} \frac{D_{K,j}}{D_{jk}} + \dot{n}_j = -\frac{p}{\tilde{R}T} D_{K,j} \frac{\Delta \tilde{y}_j}{\bar{r} \ln(r_{m,o}/r_{m,i})} \tag{4.16}$$

with $\bar{r} = (r_{m,o} - r_{m,i})/2$, and applying this equation to every layer of a composite membrane. Equation 4.16 is coupled to the mass balances for the two compartments of the cell (SS, TS), assuming that axial dispersion can be neglected and using appropriate mass transfer coefficients (Hussain, Seidel-Morgenstern, and Tsotsas, 2006).

Figure 4.17a shows that the difference between the outlet molar fractions of helium decreases with decreasing velocity of incoming helium–that means with increasing residence time of helium in the tube. Because of Knudsen diffusion in the membrane, mass transfer between the compartments is not equimolar, so that the outlet and inlet velocities are different from each other. Specifically, the tube-side outlet velocity, $u_{g,i,out}$, will be smaller than the shell-side outlet velocity, $u_{g,o,out}$, because Knudsen diffusion favors the transfer of helium (the smaller molecule) from the tube to the annulus. The respective measured results can also be predicted well, see Figure 4.17b.

In the case of transient diffusion, the shell-side is kept closed and the cell is initially filled with nitrogen (gas "k", see Figure 4.11c). Then (at $t = 0$), the tube flow is switched to helium (gas "j"). Due to preferential diffusion of helium through the membrane the pressure increases in the closed shell-side (Figure 4.18). This trend is opposed by viscous flow, so that an equilibration of pressure and an atmosphere of pure helium are obtained throughout the entire cell after some time.

For calculation, the complete DGM must be used, because Knudsen diffusion, molecular diffusion and viscous flow take place simultaneously. The tube-side mass balance is treated in the same way as for isobaric diffusion. Spatially concentrated conditions can be assumed in the shell-side, reducing the respective mass balance–after application of the ideal gas law–to:

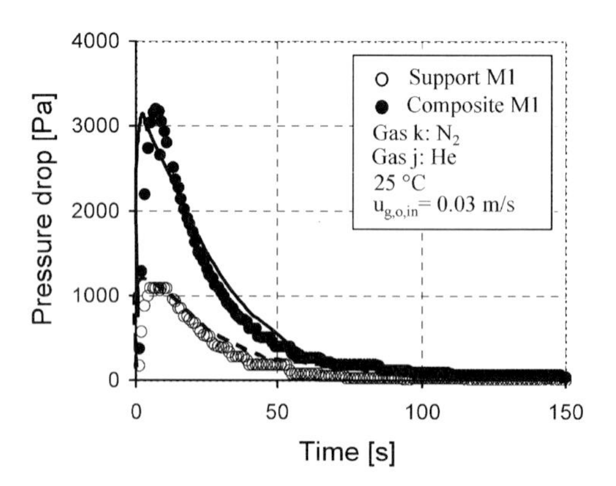

Figure 4.18 Measurement and prediction for transient diffusion (see Figure 4.1c) through the composite membrane M1 and its support.

$$\frac{dp_o}{dt} = \frac{\tilde{R}TA_{m,o}}{V_o}(\dot{n}_{j,m,o} + \dot{n}_{k,m,o}) \tag{4.17}$$

As Figures 4.15 and 4.16 show, the results of validation experiments can be satisfactorily predicted by the DGM using the previously identified parameters K_0 and B_0. The structural parameters d_{pore} and ε/τ, which can be derived from K_0 and B_0, have values within reasonable ranges (Table 4.2) and show reasonable trends when going from the support to the selective layer of the composite membrane. However, it would be a mistake to consider d_{pore}, ε and τ as the real values of average pore diameter, porosity and tortuosity in the porous media. The reason for this are the radical assumptions made in in order to connect K_0 and B_0 with d_{pore} and ε/τ by Equations 4.12 and 4.13. Additionally, defects and inhomogeneities are present in real membranes, transitions from one layer to the other are not sharp and inaccuracies occur in measurement, identification and determination of layer thicknesses.

4.4.2
Mass Transport Parameters of Metallic Membranes – Precursors not Available

Following the same procedure as in Section 4.4.1, our intention is to identify the DGM parameters of metallic membranes with the help of single gas permeation measurements and validate them by means of isobaric and transient diffusion experiments. The investigated tubular metallic membrane was produced by Fa. GKN Sinter Metal Filters GmbH. It has the same effective length as the ceramic membrane $L = 120\,\text{mm}$ and, with $r_{m,i} = 10.55\,\text{mm}$, it has approximately the same inner diameter; but it is thinner, with an outer radius of $r_{m,o} = 12.8\,\text{mm}$. No more information is available about this membrane.

4.4.2.1 Diagnosis

Single-gas permeation experiments have been conducted with air, argon and helium for different temperatures (26–500 °C) and pressures (1.5–3.0 bar). Without any further information about the membrane structure, these experiments have been evaluated by assuming the membrane to be homogeneous. In this way, the Knudsen diffusion coefficient, bulk flow coefficient, d_{pore} and ε/τ given in the upper part of Table 4.3 have been calculated.

Figure 4.19 shows permeate fluxes of He at different temperatures (full symbols) plotted against the mean pressure. It is evident that the single-layer model can describe the results of single-gas permeation experiments with good accuracy.

For validation, isobaric diffusion experiments have been carried out at a constant flow velocity ($u_{g,o,in} = 0.043\,\text{m/s}$) of pure nitrogen at the inlet of the annulus and various flow velocities of pure helium at the inlet of the tube. The respective experimental results are plotted in Figure 4.20.

Again, we expect the difference between the outlet molar fractions of helium ($\tilde{y}_{He,i,out}$ and $\tilde{y}_{He,o,out}$) to increase with $u_{g,i,in}$. However, the calculation conducted with

Table 4.3 Mass transport parameters of metallic membrane with different methods of identification.

Layer	Material	Thickness [μm]	K_0 [m]	B_0 [m^2]	d_{pore} [nm]	ε/τ
Membrane, $L = 250\,mm$, $r_{m,i} = 10.55\,mm$, $r_{m,o} = 12.8\,mm$ (approx.)						
One layer	Inconel 600	2250	1.48×10^{-7}	3.10×10^{-14}	1677	0.353
Two layers						
First layer	Inconel 600	1125	1.21×10^{-7}	1.69×10^{-14}	1119	0.432
Second layer	Inconel 600	1125	1.06×10^{-7}	1.39×10^{-13}	10490	0.040

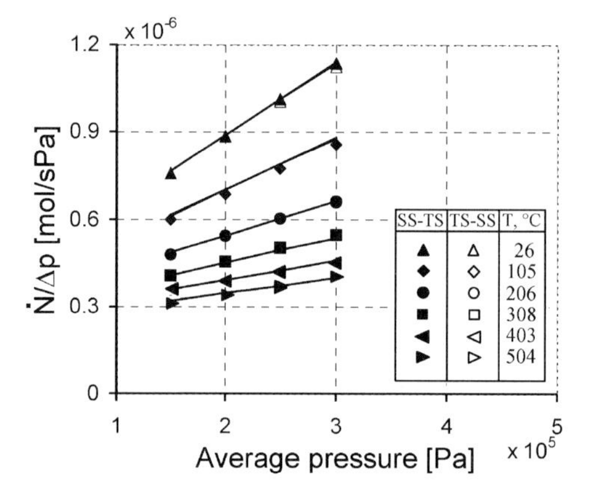

Figure 4.19 Results of single gas (He) permeation experiments for the metallic membrane at different temperatures and calculations based on the single-layer model. Full symbols: flow from SS to TS; empty symbols: inversed flow (Edreva *et al.*, 2009).

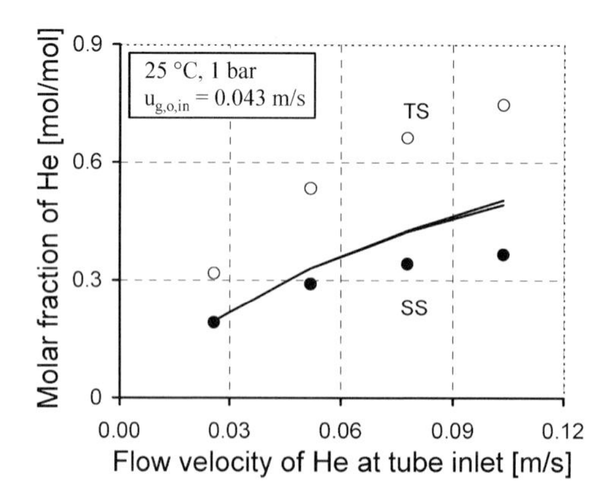

Figure 4.20 Measurement and calculation for isobaric diffusion (tube inlet: pure helium; annulus inlet: pure nitrogen) based on the assumption of a spatially homogeneous membrane with parameters gained from single gas permeation.

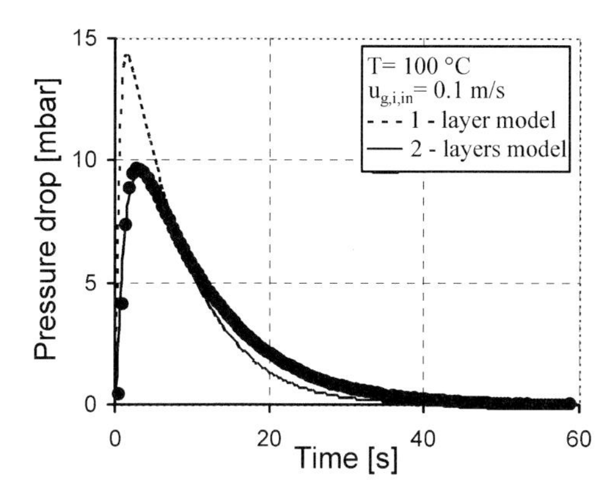

Figure 4.21 Measurement and calculations for transient diffusion.

the B_0 and K_0 from the upper part of Table 4.3 predicts conditions close to equilibrium at the outlet of the membrane, in contrast to the measurements. This means that Knudsen diffusion and molecular diffusion are significantly overestimated in the calculation.

Transient diffusion experiments have been conducted by using two gases, N_2 and He, with a significant difference in their molar masses. The comparison between experiment and simulations with the single-layer model in Figure 4.21 also shows an overestimation of the diffusion of helium from the tube to the closed annulus compartment, which leads to an overestimation of the maximum of shell-side pressure by the model (dashed curve).

The foregoing analysis shows that different modes of mass transport cannot be properly described when assuming a homogeneous membrane. This is a strong indication of a multilayer membrane structure, posing the question about additional methods for the non-destructive diagnosis of such structures. It is known from the literature that the permeate flux of asymmetric membranes may depend on the direction of flow (Uchytil, Schramm, and Seidel-Morgenstern, 2000). For this reason, the direction of flow, which had originally been from shell-side to tube-side, was inverted.

However, the respective results, which are represented by empty symbols in Figure 4.19, do not differ significantly from the previous results (full symbols). Therefore, flow inversion is not an adequate method for the diagnosis of the multilayer structure of the present membrane.

Another effect of structural inhomogeneity discussed in Section 4.4.1.2 refers to the influence of gas molar mass on the slope of permeate flux lines. This effect does occur with our present metallic membrane. As Figure 4.22 shows, the slope of the permeance line decreases by transition from helium to argon, instead of increasing.

The diagnosis of multilayer structure based on the failure of one-layer DGM to predict different mass transport experiments and the inversion of the molar mass effect was verified by breaking the metallic membrane. This revealed two layers

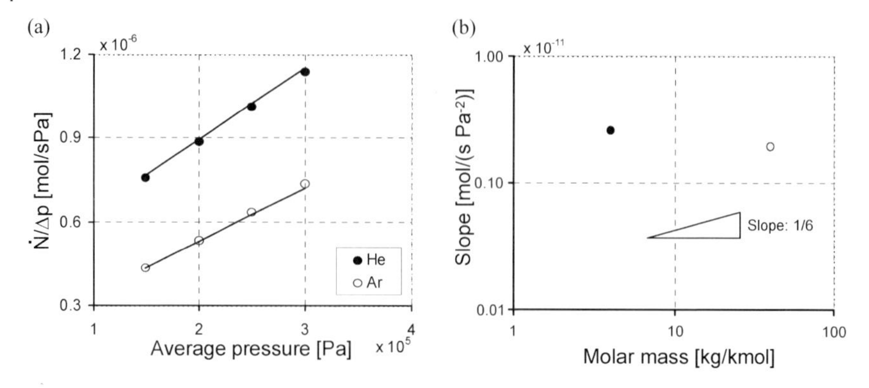

Figure 4.22 a) Results of permeation experiments with two different gases for the metallic membrane at 26 °C. b) Dependence of the slopes of respective lines on gas molar mass in double-logarithmic representation.

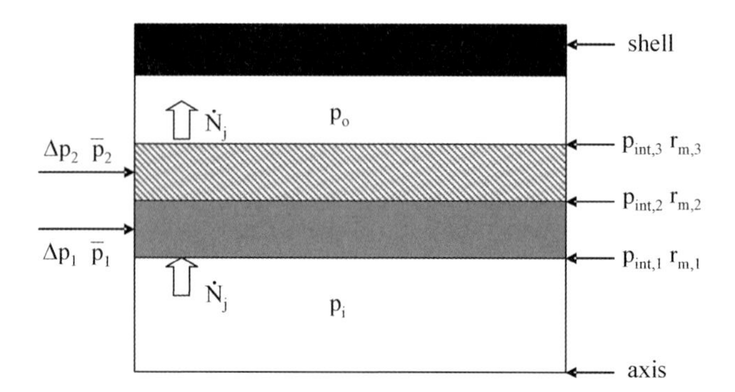

Figure 4.23 Sectional view for model of double layer membrane.

with approximately equal thickness. At the same time, it posed the problem of identifying B_0 and K_0 individually for each of the two layers of the membrane, when the precursors are not available. A solution to this problem is discussed in the following section.

4.4.2.2 **Identification**

In the next step, the same set of permeation experimental data with the metallic membrane is used to identify the parameters of a new two-layer model. These two layers (layer 1 and layer 2, with the same thickness) are shown schematically in Figure 4.23.

For modeling single-gas permeation through a multilayer membrane, Equation 4.15 can be written in the form:

$$\frac{\dot{N}_j}{\Delta p_i} = -\frac{2\pi L}{\tilde{R}T\ln\left(\frac{r_{m,i+1}}{r_{m,i}}\right)}\left(\frac{4}{3}K_{0,i}\sqrt{\frac{8\tilde{R}T}{\pi\tilde{M}_j}} + \frac{B_{0,i}}{\eta_j}\bar{p}_i\right) \tag{4.18}$$

where:

$$\sum_{i=1}^{n} \Delta p_i = p_i - p_o \tag{4.19}$$

$$\bar{p}_i = \frac{p_{\text{int},i} + p_{\text{int},i+1}}{2} \tag{4.20}$$

for $i = 1, 2, ..., n$
and:

$$p_{\text{int},1} = p_i \tag{4.21}$$

$$p_{\text{int},n+1} = p_o \tag{4.22}$$

Here n is the number of layers, and $p_{\text{int},i}$ is the pressure between two membrane layers. For each set of p_i and p_o a molar flow rate $\dot{N}_{\text{calc},j}$ can be calculated by solving the group of equations and compared with the measured value $\dot{N}_j$. The parameter identification problem is to find the best set of parameters of $B_{0,i}$ and $K_{0,i}$ which minimize the differences of model prediction and real observation by minimizing the objective function:

$$J = \sum_{j=1}^{m}\sum_{k=1}^{nj} w_{jk}\left(\dot{N}_{j,k} - \dot{N}_{\text{calc},j,k}\right)^2 \tag{4.23}$$

Here, w_{ij} is a weighting factor, set to unity in the present work.

To solve this optimization problem (see [Zhang, 2008] for more details) a genetic algorithm (GA) detecting the global optimum was used. After that, a local deterministic optimizer (LD) was applied to increase the accuracy of optimization. The combination of GA optimizer (Krasnyk *et al.*, 2006) and LD optimizer (Rodriguez-Fernandez, Mendes, and Banga, 2006) is called a hybrid optimizer (Byrd *et al.*, 1995). Both optimizers were incorporated in the DIANA simulation system (Teplinskiy, Trubarov, and Svjatnyj, 2005).

The results of this derivation for the investigated membrane are summarized in the lower part of Table 4.3. The identified transport parameters imply one loose layer with relatively small pores and a second, highly compacted layer with a few large pores. As Figures 4.24 and 4.25 show, the agreement between calculations with the two-layer model and measurements is, again, very good.

4.4.2.3 **Validation**

In Section 4.4.2.1 we discussed the single-layer model. The identified transport parameters of this model are capable of describing the results of single-gas permeation measurements even for a membrane that consists of two layers with significantly different structures. The deficiencies of the single-layer model are not visible without a broad variation of operating parameters, and even with such a

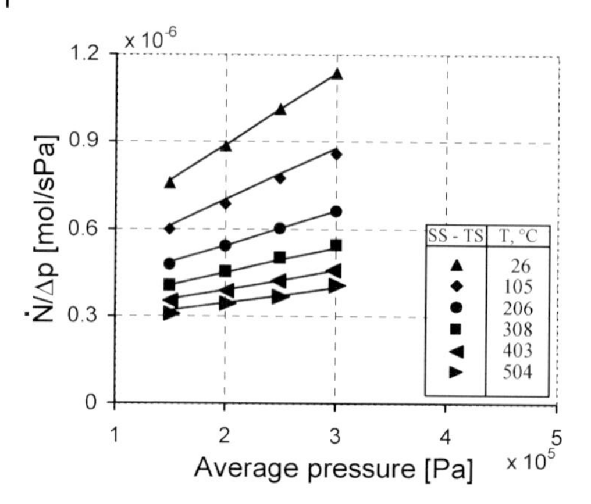

Figure 4.24 Results of single gas (He) permeation experiments (flow from SS to TS) for the metallic membrane at different temperatures and calculations based on the two-layer model.

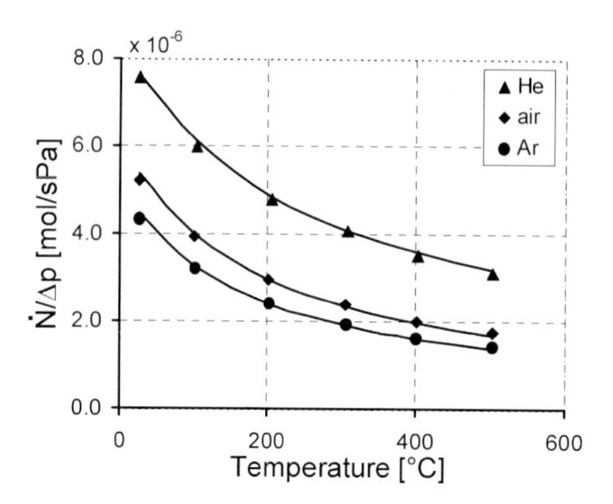

Figure 4.25 Results of single gas permeation experiments for three different gases at 1.5 bar (flow from SS to TS) for the metallic membrane and calculations based on the two-layer model.

broad variation they do not jeopardize the good overall agreement between simulation and permeation data. Therefore, the supremacy of the two-layer model from Section 4.4.2.2 can hardly be demonstrated statistically on measured permeation data.

Hence, the superiority of a model that realistically reflects the composite nature of the membrane (in our case of the two-layer model) becomes striking only when trying to validate the identified transport parameters with additional mass transfer experiments of another measuring principle, such as isobaric diffusion or transient diffusion.

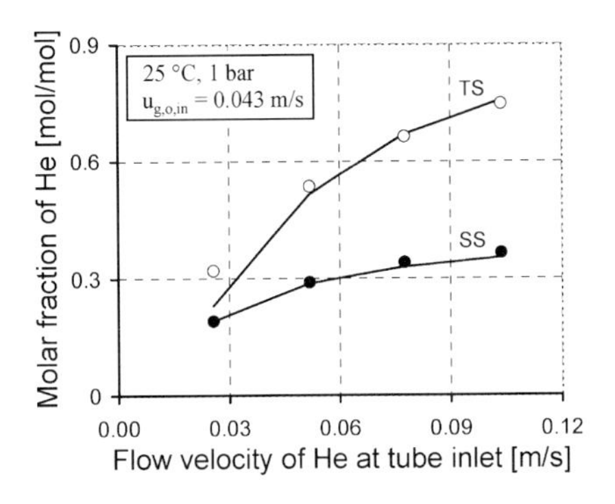

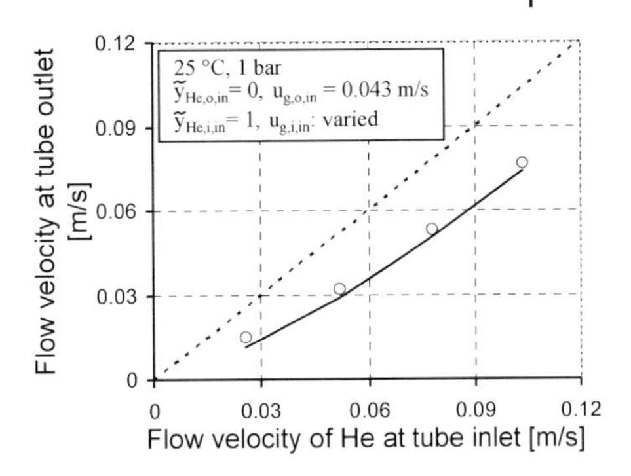

Figure 4.26 Measurement for isobaric diffusion and prediction by means of a two-layer model.

Results for isobaric diffusion are presented in Figure 4.26. We can see that the two-layer model enables a satisfactory prediction of helium molar fractions measured in the shell-side and in the tube-side at the outlet of the cell. The simulation is carried without any change of the previously identified mass transport parameters (lower part of Table 4.3) and contrasts radically the spectacular collapse of the one-layer model when compared with the same data (Figure 4.20). Figure 4.26b shows that the gas flow rate decreases in the tube along the membrane and gas velocities can also be predicted with satisfactory accuracy by means of the two-layer model. This results from preferential Knudsen diffusion of the smaller molecule (helium) from TS to SS and would not happen in the case of ordinary binary diffusion alone. The same is true for the transient diffusion experiment. The simulation results for the two-layer model are plotted by the solid line in Figure 4.21 and agree very well with the measured data, whereas calculation by the one-layer model (dashed line) does not describe the data well.

Since the one-layer model does not properly describe different types of mass transfer measurements, it is necessary to use multilayer models for multilayer membranes in membrane reactor simulations, based on careful diagnosis, parameter identification and validation.

Finally, it is interesting to compare the ceramic and metallic porous membranes investigated in the present work. The comparison of single-gas permeation data reveals a helium permeation of about 0.9×10^{-6} mol/(s Pa) at 2 bar and ambient temperature for the metallic membrane (Figure 4.24) in contrast to about 0.18×10^{-6} mol/(s Pa) at similar conditions for the ceramic membrane. A comparison of the isobaric diffusion (Figures 4.17 and 4.26) and transient diffusion (Figures 4.18 and 4.21) shows that equilibration is faster with the metallic than with the ceramic membrane.

Consequently, the thinner metallic membrane is characterized by higher permeability at comparable or even better selectivity. The metallic membranes are mechanically and thermally more stable. They would not easily get damaged by the relatively fast temperature changes occurring during start-up and shut-down of membrane reactors. Additionally, they can be more easily mounted and sealed, which facilitates membrane reactor design. In general, metallic membranes appear to be an attractive alternative to ceramics for membrane reactors.

4.4.3
Mass Transport in 2-D Models

Model descriptions with different complexity have been already presented in Chapter 2. The morphological parameters determined experimentally with the aid of the DGM are used for the 2-D simulations with FLUENT, after appropriate transformation. Validation can be conducted with the previously discussed steady-state experiments of single-gas permeation and isobaric diffusion. The consideration of transport phenomena such as diffusion and viscous slip in the membrane is particularly important in the 2-D CFD model. In case of a single-component flow, the non-slip boundary condition on the walls can be used reliably. In binary or multi-component mixtures, especially in capillary flows, this assumption is incorrect and may generate serious errors if the convective contributions are small and there is a large disparity in the molar masses of gas components (Young and Todd, 2005). It was derived by Kramers and Kistemaker (1943) that, in the presence of a concentration gradient parallel to the wall, the mixture would flow along the wall with non-zero mass average velocity. They called this phenomenon diffusion slip. Viscous slip occurs in tube flows when the molecular mean free path is a significant fraction of the tube radius (Young and Todd, 2005). In multi-component tube flow diffusion slip is important at all Knudsen numbers and exists even when the streamwise pressure gradient is zero (Young and Todd, 2005). In contrast, viscous slip is important in the Knudsen and transition regions.

By means of algebraic transformations of the analytical solution for single-gas permeation, the parameters of the DGM can be converted to a form appropriate for usage in CFD simulations, whereas the viscous slip phenomenon is taken into account on the basis of integral quantities. The definition of the viscous resistance factor f_1 has already been given in Equation 2.62. A comparison between experimental, analytical and numerical results is presented in Figure 4.27a for permeation of oxygen through the ceramic membrane 2a (see Table 4.2). A very good agreement for the permeation flux $\dot{N}/\Delta p$ can be observed between the analytical and numerical results with a maximal relative error of 0.08%. With increasing temperature and pressure the deviation between experimental and calculated values rises maximally to 10%.

Furthermore, the case of isobaric counter-diffusion of He versus N_2 is used with parameters of membrane 1 (Table 4.2) for the comparison of results. Because both components have very different molar masses, the diffusion slip cannot be omitted. Considering the velocity resulting from the diffusion process via a source term in the momentum transport equations, excellent agreement with the experimental data can be achieved, Figure 4.27b. The use of such a source term leads to a pressure drop through the membrane of just a few pascals. Despite its small value,

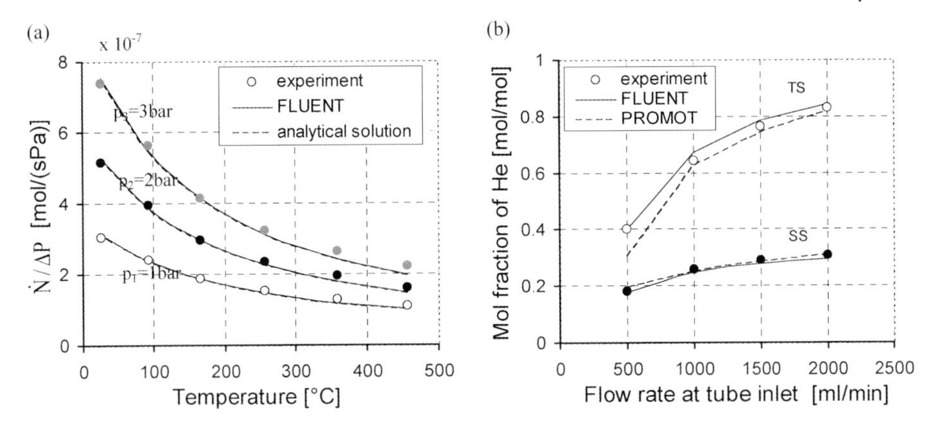

Figure 4.27 Validation of the 2-D model by transmembrane mass transport data.

this pressure difference is a key factor for obtaining close agreement between simulations and experiments. In the case of a dead-end reactor configuration, where the mass flow rate dosed at the shell inlet passes through the membrane, this pressure difference is negligible in comparison with the pressure drop related to convective flow. Therefore, only the viscous slip on the pore walls is taken into account in this kind of simulation.

4.5 Analysis of Convective and Diffusive Transport Phenomena in a CMR

This study aims at providing a better understanding of the subprocesses, especially the transport phenomena in the membrane zones. In the catalytic membrane reactor with dosing strategy, the diffusion process of C_2H_6, fed at the tube inlet, to the shell side of the membrane has an important influence on the reactor performance. The loss of ethane or products at the interface between membrane support and the shell side is known from literature as a back-diffusion phenomenon. In the PBMR, the back-diffusion of reactants has an undesirable effect to bypass the catalytic bed and leads to a reduction in the conversion of reactants and yield of the product. In the CMR, the diffusion problem of reactants and products is more complicated. Two contradictory requirements can be formulated: (a) enough diffusion rates of ethane in radial direction as well as in the membrane catalytic layer are needed and (b) further diffusion to the annulus is undesirable. Therefore, an optimization-based design for the CMR is required to obtain the best reaction conditions.

At first, the concentration profiles for reactants and products are presented in Figure 4.28 for both CMRs with and without a packed bed.

These results are related to the velocity fields already presented in the previous section.

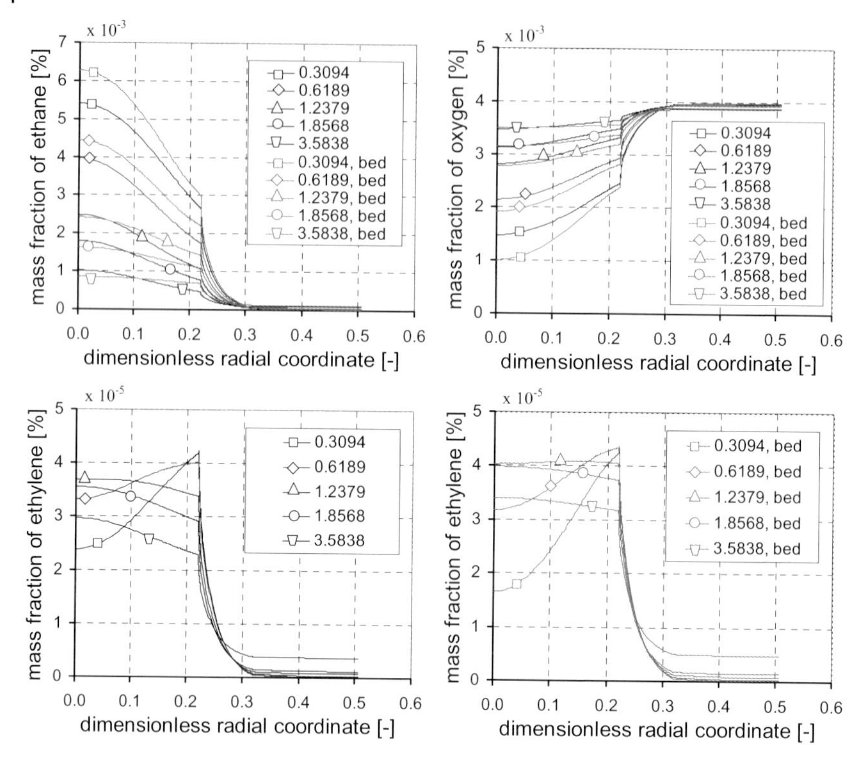

Figure 4.28 Mass fraction profiles for reactor (CMR) with and without packed bed at different axial positions z/D for GHSV = 27 736 L/h, SS/TS = 9, $T = 600\,°\text{C}$.

The mass fraction of ethane decreases in the tube side due to the reactions and system dilution caused by the dosed mass flow rate. It increases continuously in the shell side, depending on the mass flow rate at the tube inlet, membrane thickness and parameters as well as the total flow rate in the reactor. In contrast, the mass fraction of oxygen rises permanently in the tube and decreases in the shell. Because the reactions take place in the membrane catalyst layer and ethane is transported only by diffusion, pronounced profiles are formed especially in the first part of the reactor with and without a bed. For the CMR with an inert bed, the dispersion coefficients are calculated as a function of the axial velocity and porosity profiles. Therefore, almost flat profiles are developed near the reactor end with increasing the axial velocity in the tube. Ethylene is intensively generated near the inlet of both reactors and its mass fraction decreases with the length as a result of the consecutive reactions. Therefore, a change in the slope of the ethylene profile can be observed. Near the inlets, ethylene is transported from the catalyst layer by diffusion and convection. Because of the increased mass fraction of ethylene in the reactor tube, its transfer from catalyst layer is only due to convection. Comparing the profiles in both reactors, it can be seen that more ethylene is produced in the reactor with a packed bed because of the better radial mass transfer.

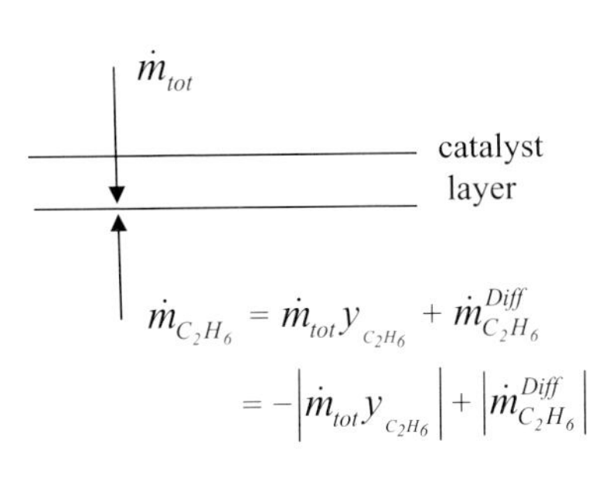

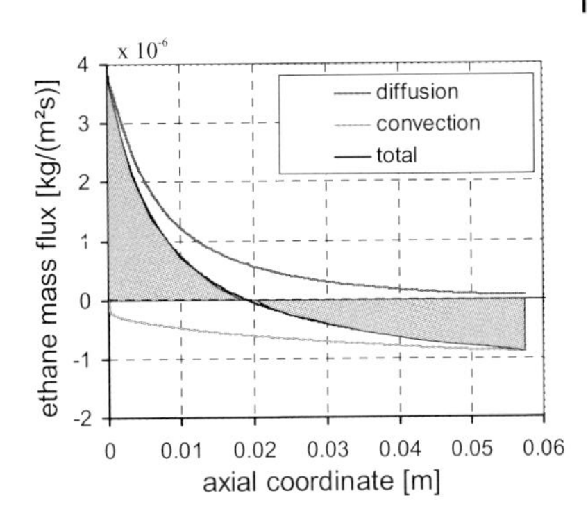

Figure 4.29 Visualization of the diffusion-convection problem.

This result explains the advantage of the CMR with a packed bed, obtained on the basis of conversion of ethane, selectivity and yield of ethylene. The detailed consideration of diffusive species transport in the CMR is especially important for the understanding of the reactor behavior. Therefore, the diffusion-convection problem is discussed in the following.

Locally considered, the mass balance for ethane at the interface between reactor tube and membrane can be expressed as in Figure 4.29.

In order to have transport of ethane into the catalyst layer, the condition $\dot{m}^{\mathrm{Diff}}_{\mathrm{C_2H_6}} > |\dot{m}_{\mathrm{tot}} y_{\mathrm{C2H6}}|$ has to be fulfilled.

This supposes that a maximal total flux or a dosage limit exists, at which the amount transported by diffusion is still bigger than that transported by convection. At the value for which both amounts are equal, the mass flux of ethane is zero and no more ethane can be transported into the catalyst layer.

The local diffusion fluxes can be quantified at the interfaces (not only for the catalyst layer but for all membrane layers) on the basis of Equation 4.24. The total fluxes of ethane are given by Equation 4.25.

$$\dot{m}^{\mathrm{Diff}}_{\mathrm{C_2H_6}} = \rho D_{\mathrm{C_2H_6,eff}} \frac{\partial y_{\mathrm{C_2H_6}}}{\partial r}\bigg| r = R_m \tag{4.24}$$

$$\dot{m}_{\mathrm{C_2H_6}} = \rho D_{\mathrm{C_2H_6,eff}} \frac{\partial y_{\mathrm{C_2H_6}}}{\partial r} + \rho y_{\mathrm{C_2H_6}} v_r | r = R_m \tag{4.25}$$

As an example, the diffusion fluxes are calculated without reaction between the shell side and the membrane and are presented in Figure 4.29. Because the considered reactor has a dead-end configuration, the entire amount of ethane has to be transported back to the tube side and the integral on the total flux of ethane approaches zero with the accuracy of the numeric simulation. This is visualized on the basis of the filled areas under and over the curve of the total flux. They are nearly equal, with an accuracy of $\approx 10^{-13}$. The diffusion flux reduces with the length, because of a decreasing concentration gradient. In contrast, the magnitude of the convective flux increases and the available amount of ethane in the shell side is transported again into the tube.

Because in most of the simulations the SS/TS ratio is varied on the basis of an equal reactor residence time, the integral diffusion streams which are considered in the following are appropriately scaled to the mass flow rate of ethane at the tube inlet. There are many possibilities to influence the back-diffusion phenomenon, from setting optimal operating conditions to choosing appropriate morphological membrane parameters. Therefore, the influence of some selected parameters is discussed in the following.

The parameter ranges used are summarized in Table 4.4 (Section 4.6). At first, the dependence of the diffusion fluxes with variation of the volume flow rate at the shell inlet is considered by keeping constant the tube inlet flow rate and the results are presented in Figure 4.30. Varying the SS/TS ratio in this way corre-

Table 4.4 Influence parameters and their variation bounds.

Temperature, °C	480–620
Total volume flow rate, L/h	15.0–68.36
SS/TS ratio: SS/TS = $\dot{V}_{SS}/\dot{V}_{TS}$	0.4–9.0
Concentration ratio $N = \tilde{y}_{C_2H_6}/\tilde{y}_{O_2}$	2.0–3.5
Tube diameter, mm	1.75–5.0
Number of membrane layers	Support, intermediate layers, separation layer
Pore diameter in catalyst layer, nm	5–60
Catalyst layer thickness, μm	2–10
ε/τ ratio in the catalyst layer	0.05–0.5
Pore diameter in support layer, nm	500–10 000
Support thickness, mm	0.5–3.0
ε/τ ratio in the support layer	0.05–0.5

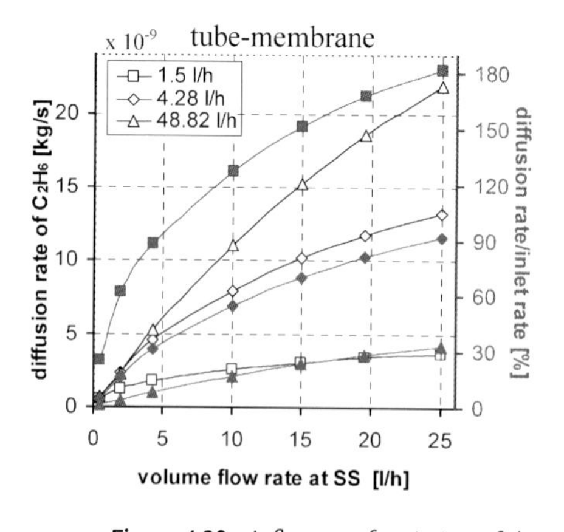

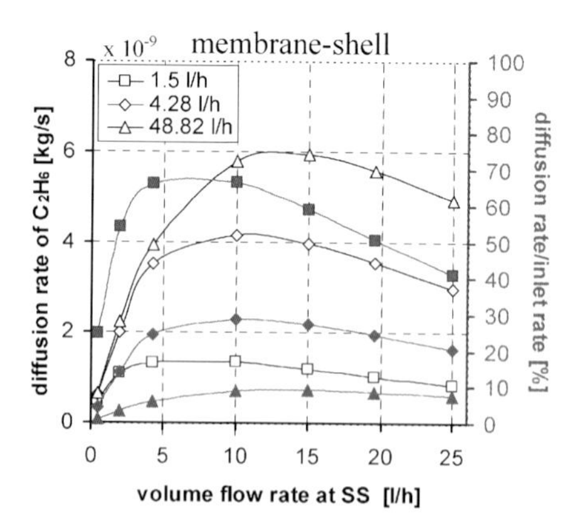

Figure 4.30 Influence of variation of the volume flow rate at the shell side inlet on the diffusion rate at the interfaces tube-membrane and membrane-shell, $T = 600\,°C$, without reaction.

sponds to different total volume flow rates. The simulations are accomplished without consideration of the reactions and for three different constant flow rates at the tube inlet. At the tube–membrane interface, Figure 4.30, a continuous increase of the diffusion stream and the scaled diffusion rate can be seen. By increasing the shell-side stream, the driving force for the diffusion process of ethane grows and the diffusion rate increases at this interface. In contrast, at the membrane-shell side, both diffusion flux and ratio go through a maximum and, for higher shell-side volume flow rates, the ethane loss through the membrane is suppressed despite the present diffusion driving force.

For most of the simulations by parameter variation (see Section 4.6) the SS/TS ratio is changed on the basis of the same residence time. Some selected results for the influence of the morphological parameters as a function of SS/TS ratio on the diffusion rates at the membrane-shell side are presented in Figure 4.31. Among these parameters are the number of membrane layers, the support pore diameter, ε/τ ratio and thickness.

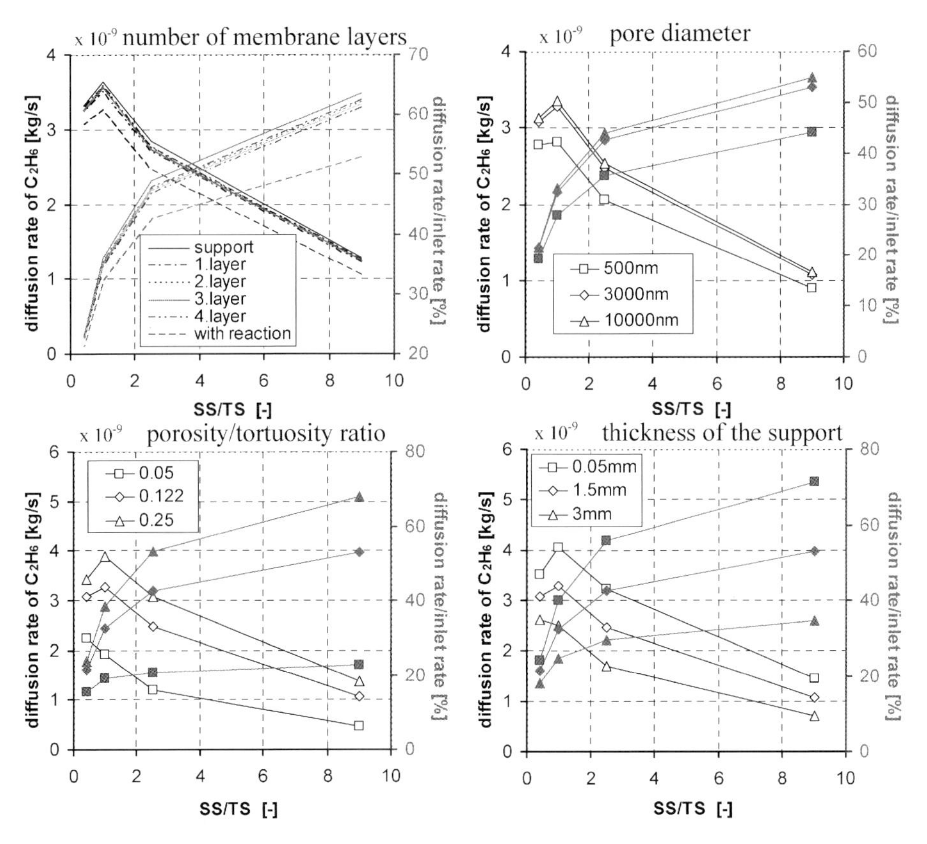

Figure 4.31 Influence of the number of membrane layers, the support pore diameter, ε/τ ratio and thickness on the diffusion fluxes, GHSV = 6086 L/h, T = 600 °C.

By increasing the SS/TS ratio, the inlet mass flow rate of ethane becomes lower and the integral diffusion rate decreases significantly. This tendency is different only for SS/TS = 0.4, whereas the tube-side stream is very high and the influence of external mass transfer resistances can be observed (see Section 4.6). Because of the longer residence time of the ethane molecules at higher SS/TS ratios (lower flow rate at tube inlet), the scaled diffusion rate of ethane increases. By varying the number of the membrane layers at SS/TS = 9.0, 61–64% of the entered ethane mass flow rate diffuses through the membrane. Under the given conditions, the number of layers has a very small influence on the diffusion rates. Taking the reactions into account, the diffusion rates of ethane are reduced in this case by about 10%. For the same total volume flow rate, the influence of the morphological characteristics of the support is also studied with variations of the SS/TS ratio. It can be observed that the pore diameter of the support has only a minor influence on the reduction of the back-diffusion. A sixfold change in the pore diameter leads only to approximately 10% lower diffusion rates. This can be explained by the different dominant transport mechanisms. For higher pore diameters, the molecular diffusion and convective flow are important for mass transfer process. The contribution of the molecular diffusion depends also on the ε/τ ratio. Therefore, the variation of the pore diameter influences the back-diffusion indirectly due to the formation of a larger pressure drop through the membrane. The influence of the pore diameter on the ethane diffusion is stronger in regions where Knudsen diffusion is the dominant mechanism, that is, in the catalyst layer. A very pronounced effect on the scaled diffusion rate of ethane is observed with variation of the ε/τ ratio and the support thickness, whereas the back-diffusion rates are reduced by more than 50% at SS/TS = 9.0. From the presented results can be concluded that the ethane and product loss through the membrane can be significantly decreased with variation of the support thickness and ε/τ ratio, relatively independently from the catalyst properties. This gives the opportunity to choose them in an optimal way. Similar diffusion effects can be seen in the packed-bed membrane reactor, regarding the effect of the oxygen distribution. At low SS/TS ratios a limited distribution of oxygen is obtained (see Figure 5.11). The developing flow behavior and the occurring reactions suppress the oxygen transport in the fixed bed and lead to inefficient use of the catalyst near the inlet. Therefore, the conditions for the reaction and the transport processes have to be adjusted to each other for an optimal operation of the reactor.

4.6
Parametric Study of a CMR

This section investigates the influence of various parameters on the CMR performance for oxidative dehydrogenation of ethane, using the 2-D mathematical model described in Chapter 2. The simulations serve as a basis for the analysis and evaluation of the overall influence of the transport and reaction processes on the integral quantities like the conversion of ethane, selectivity and the yield of ethylene. The limits of variation in the reactor operating and geometrical conditions as well as the morphological membrane parameters are summarized in Table 4.4.

The quantity Gas-Hourly-Space-Velocity ($\text{GHSV} = \dot{V}/V_m = 1/\tau$) is defined after (Ullmann, 1977) and is used by the following analysis. It is inversely proportional to the reactor residence time. Some of these parameters are interrelated and their correlations are given in Equations 4.26 and 4.27:

$$\dot{V}_{\text{tot}} = \dot{V}_{\text{SS}} + \dot{V}_{\text{TS}} \sim \text{GHSV} \tag{4.26}$$

$$\frac{\text{SS}}{\text{TS}} = \frac{\dot{V}_{\text{SS}}}{\dot{V}_{\text{TS}}} = \frac{\dot{V}_{\text{tot}}}{\dot{V}_{\text{TS}}} - 1 = \frac{\text{GHSV}}{V_m \dot{V}_{\text{TS}}} - 1 \tag{4.27}$$

The equations show that the volume flow rate at the tube inlet depends on the total volume flow rate and SS/TS ratio.

The results from the simulation are presented here only for parameters with major influence. The effect of the other parameters, like the influence of the morphological parameters of the support, is discussed by Georgieva-Angelova, 2008.

Before starting the simulation study, it is necessary to validate the model predictions not only for the transport processes, but also for the entire reactor performance. The comparison is carried out with experimental results for the CMR, provided by Klose *et al.* (Klose, 2008) and serves to validate the output of the simulation for the conversion of ethane, the selectivity of ethylene and the pressure drop. The comparison is accomplished for different SS/TS ratios, initial temperature and total volume fluxes. A rather reliable prediction of the reactor tendencies can be obtained (see Figure 4.32). Because the conversions obtained by the experiments are always higher than the predicted ones, the catalytic layer thickness is varied between 2 and 3 μm. Better results are observed by using 3 μm catalyst layer thickness. No other parameters have been fitted.

In the simulations for parameter significance, the morphological parameters after (Thomas, 2003) according to Table 4.2 are used. The usage of these parameters leads to better comparison results between simulation and experiment.

At first, comparisons between simulations with and without considering the energy balance are carried out in a wide range of parameter. Minimal differences of 0.05% under conditions of low conversion and 0.15% at higher conversion levels are obtained for the values of the ethylene yield, as shown in Figure 4.33. This indicates a pseudo-isothermal behavior. Nevertheless, most of the simulations are performed under consideration of the energy balance. A stronger influence of the temperature profiles in the reactor is expected when thicker catalytic layers with higher activity are applied and respectively higher conversions are achieved.

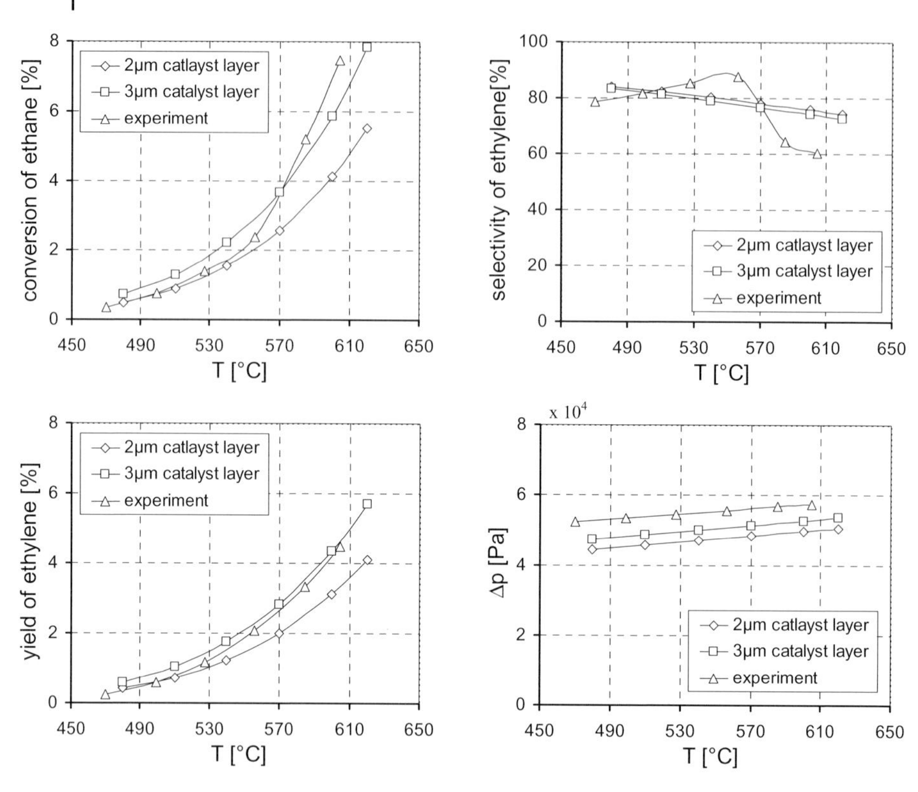

Figure 4.32 Comparison of results between experiment and numerical solution (GHSV = 27 737 L/h, SS/TS = 9, with inert bed).

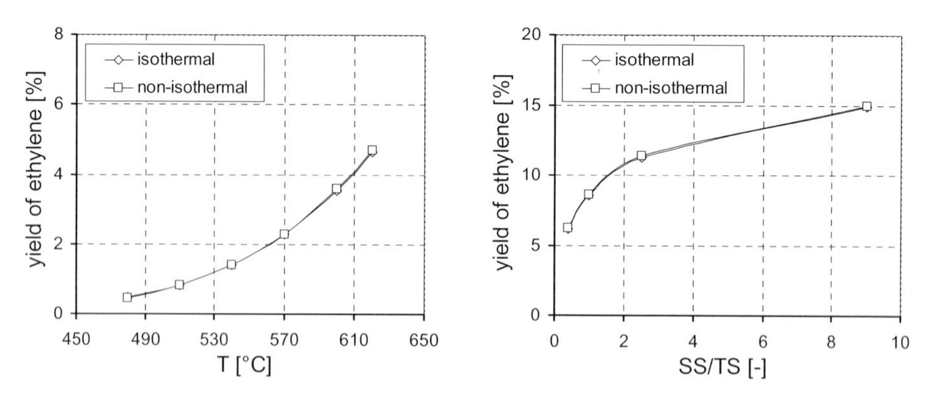

Figure 4.33 Influence on the ethylene yield at GHSV = 27 737 L/h, SS/TS = 9 and GHSV = 6086 L/h, T =620 °C.

4.6.1 Influence of Characteristic Geometrical Parameters

The transport resistances in the fluid phase are often essential under laminar flow conditions and can be a limiting factor for the entire reaction process. Under the given conditions in the CMR, the heat transfer limitations in the boundary layer are significant in comparison with the thermal resistance of the membrane. Higher conversions and temperatures gradients in the reactor lead to a stronger influence on the external heat transfer resistances. Under the conditions of low reactant concentrations, a pseudo-isothermal reactor behavior is observed. Therefore, only the influence of the mass transfer resistances is studied in the following. The diffusion conditions for the reactants and products, for example for ethane, are influenced by varying the tube diameter. The external transport limitations can be estimated at identical gas dynamic conditions for all reactor geometries. The simulations are carried out with the boundary conditions of constant velocity at the tube inlet and constant SS/TS ratio. This means that the mass flow rate at the inlet decreases respectively by lowering the tube diameter from 5.0 mm to 1.75 mm.

The estimation of the external transport resistances is carried out after (Zanfir and Gavriilidis, 2003) and (Cussler, 1997) on the basis of a ratio between the times necessary for convective and diffusive transport. The convection time is given by Equation 4.28 and represents the required time of an element at position z to leave the reactor, that is, the "remaining residence time".

$$\tau_k = \frac{L - z}{v_z(z, r = 0)} \tag{4.28}$$

The diffusive time is defined in Equation 4.29 as the time needed for a component to reach the wall of the catalytic layer from the reactor axis. The diffusion coefficient is calculated according to Equation 2.46.

$$\tau_d = \frac{R^2}{D_{ij}(z)} \tag{4.29}$$

Then, the following ratio F of the two times results in Equation 4.30:

$$F = \frac{\tau_k}{\tau_d} = \frac{L - z}{v_z(z, r = 0)} \frac{D_{ij}(z)}{R^2} \tag{4.30}$$

When $F > 1$ or $\ln(F) > 0$, the molecules of ethane have enough time to reach the wall, because the time needed for the convective transport is higher than that for the diffusion process. When $F < 1$ or $\ln(F) < 0$, the molecules leave the reactor before reaching the catalyst layer.

The simulations are accomplished for GHSV = 6086 L/h, SS/TS from 0.4 to 9.0 and the results are presented in Figure 4.34. The time available for the diffusion of ethane or F decreases significantly with the reactor length. $\ln(F) < 0$ is observed for 55% of the reactor length for the tube radius R_t = 3.5 mm at SS/TS = 9.0. In

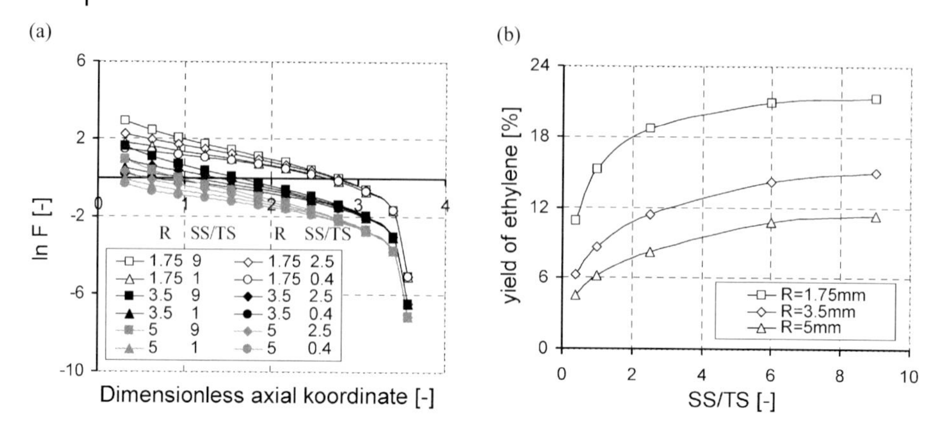

Figure 4.34 Influence of the variation of the tube diameter for GHSV = 6086 L/h, T = 620°C.

comparison at SS/TS = 0.4, $\ln(F) < 0$ is obtained already at 14% of the reactor length. Clearly, the transport limitations of ethane in a radial direction have a significant influence at lower SS/TS ratios.

By increasing the tube radius, the diffusion limitations become more significant. At R_t = 5.0 mm, SS/TS = 1.0 and SS/TS = 0.4, the $\ln(F) < 0$ is observed for the whole reactor. When the tube radius is lowered to R_t = 1.75 mm, 77% of the length is free from diffusion limitations for all SS/TS ratios. This analysis does not consider the limitation of the ethane transport caused by the radial convective transport (discussed in Section 4.5).

These results correspond to the obtained data for conversion of ethane and yield of ethylene. By lowering the tube radius, the conversion of ethane as well as the yield of ethylene increase rapidly. The usage of larger tube diameter leads to lower F ratios and accordingly to enhancement of the external mass transfer limitations. Therefore, the application of 2-D models is recommended for large-scale plants.

4.6.2
Influence of the Morphological Membrane Parameters in the Catalyst Layer

The influence of the pore diameter and the ε/τ ratio of the catalyst layer is investigated on the basis of modification of the permeability constant and the Knudsen coefficient according to Equation 2.51, and Equation 2.54. Their influence on the reaction distribution in the membrane catalyst and on the integral quantities is studied by keeping constant other structure parameters of the membrane. Furthermore, the effect of the catalyst layer thickness on the integral quantities is also of great interest.

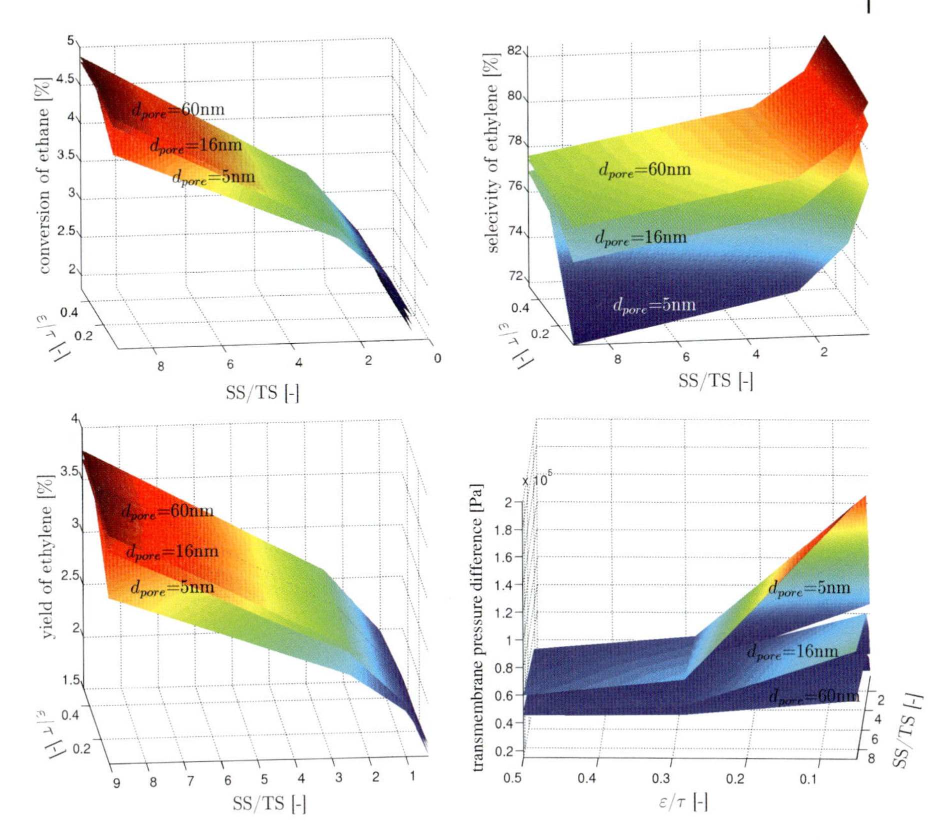

Figure 4.35 Influence of the membrane parameters at GHSV = 27 737 L/h and 3 μm catalyst layer thickness.

An important aspect of the characterization of heterogeneous reactions is the estimation of effectiveness factor η_j, defined in Equation 4.31 as a ratio of heterogeneous to homogeneous reaction rates.

$$\eta_j = \frac{r_{j,\text{heterogeneous}}(\tilde{y}_{i,\text{ed}}, T_{r,z})}{r_{j,\text{homogeneous}}(\tilde{y}_{i,\text{ed,hom}}, T_{r,z})} \tag{4.31}$$

The simulations are carried out at GHSV of 27 737 L/h, temperature of 873.15 K and operating pressure of 120 700 Pa with variation of the ratio SS/TS. Figures 4.35 and 4.36 present the results in terms of ethane conversion, ethylene selectivity and yield for 3 and 10 μm thicknesses, respectively. The latter value was found in the literature (Alfonso *et al.*, 1999), for such catalytic layers. The effect of the pore diameter on the changes of the active surface is not considered. Therefore, the catalytic activity depends only on the layer thickness.

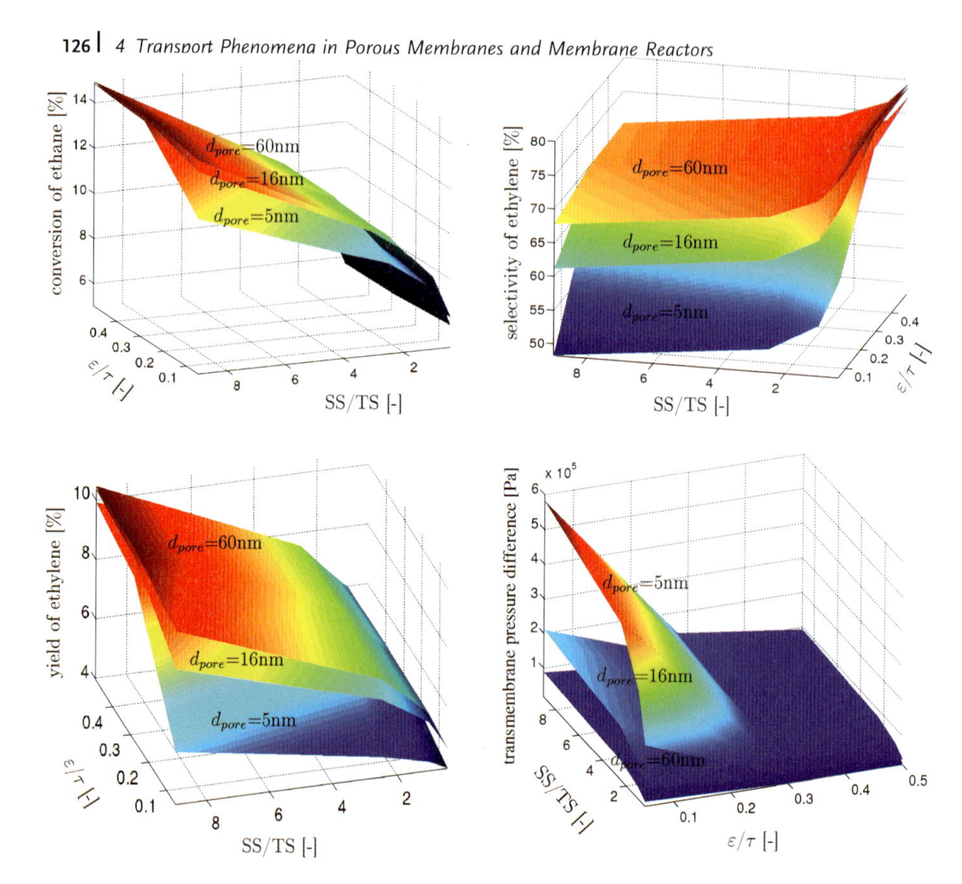

Figure 4.36 Influence of the membrane parameters at GHSV = 27 737 L/h and 10 μm catalyst layer thickness.

The influence of the pore diameter on the ethane conversion is small at high ε/τ ratios for both membranes. Decreasing the pore diameter and ε/τ ratio at higher SS/TS flow ratios leads to lower conversions and higher transmembrane pressure drop. In this case, the high resistance of porous medium limits the ethane diffusion in the catalyst layer. Consequently, the catalyst surface is not completely used. This effect is more pronounced for the 10 μm catalyst layer. The internal mass transfer limitations can be appropriate visualized on the basis of the effectiveness factor, for example, at $\varepsilon/\tau = 0.283$ and SS/TS = 9 (Figure 4.37). The coordinates are dimensionless and are scaled to the reactor diameter.

A good radial reaction distribution is obtained for 60, 16 and even for 5 nm pore diameters in the 3 μm layer. The effect of decreasing the pore diameter in the 10 μm layer leads to drastic reduction of the effectiveness factor up to 40% with the radial coordinate. Not only the conversion of ethane is negatively influenced, but also the selectivity to the intermediate products. Under the above conditions, the removal of ethylene from the reaction zone is limited and respectively lower ethylene selectivity is obtained in the reactor at small ε/τ ratios and pore diameters.

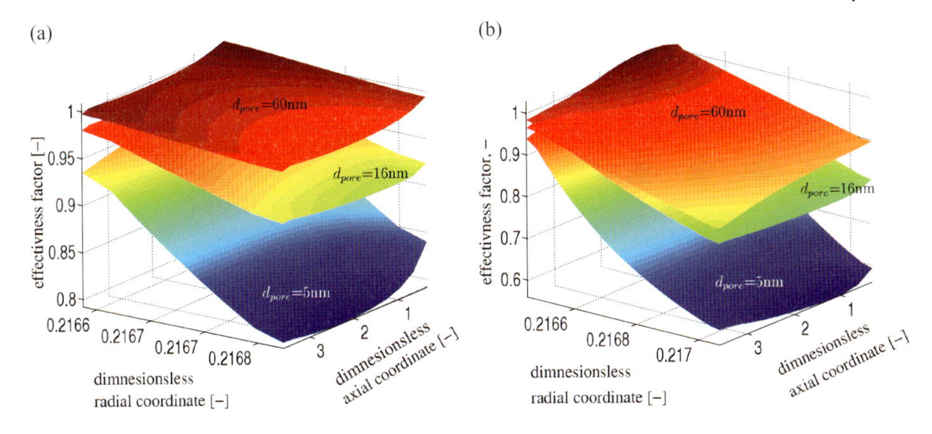

Figure 4.37 Effectiveness factor for GHSV = 27 737 L/h, 3 μm and 10 μm catalyst layer, $\varepsilon/\tau = 0.283$, SS/TS = 9.

Therefore, the lowest yield of ethylene at higher SS/TS ratios is observed at small ε/τ ratio and pore diameter because of the lowest selectivity and conversion.

The inverse tendency for the conversion by variation of the pore diameter and ε/τ ratio is obtained at low SS/TS ratios for both membranes, except 10 μm, $d_{pore} = 5$ nm und $\varepsilon/\tau = 0.05$. A detailed discussion about this phenomenon is presented by (Georgieva-Angelova, 2008).

When no internal transport limitations are observed in the catalyst layer, the maximal ethylene yield is limited by the active membrane surface area. For the example studied increasing the catalyst layer from 3 to 10 μm approximately triples the ethylene yield.

The catalyst layer thickness has an essential impact on the reactor performance and manufacturing membranes with thicker catalyst layers will enhance the results for the conversion of ethane and the yield of ethylene, especially for higher GHSV values.

4.6.3 Influence of the Operating Conditions

This section investigates the influence of the averaged residence time in the reactor, the temperature, the flow rate ratio SS/TS = $\dot{V}_{SS}/\dot{V}_{TS}$ and the concentration ratio SS/TS = $\dot{V}_{SS}/\dot{V}_{TS}$ in the ranges summarized in Table 4.4 and for a catalyst layer thickness of 3 μm. The graphics in Figure 4.38 show the influence of these parameters on the conversion of ethane, the selectivity and yield of ethylene and the transmembrane pressure drop.

The following conclusions can be drawn:

- Increasing the GHSV value (decreasing reactor residence time) leads to lower ethane conversions and ethylene yields. The variation of GHSV from 27 736 to

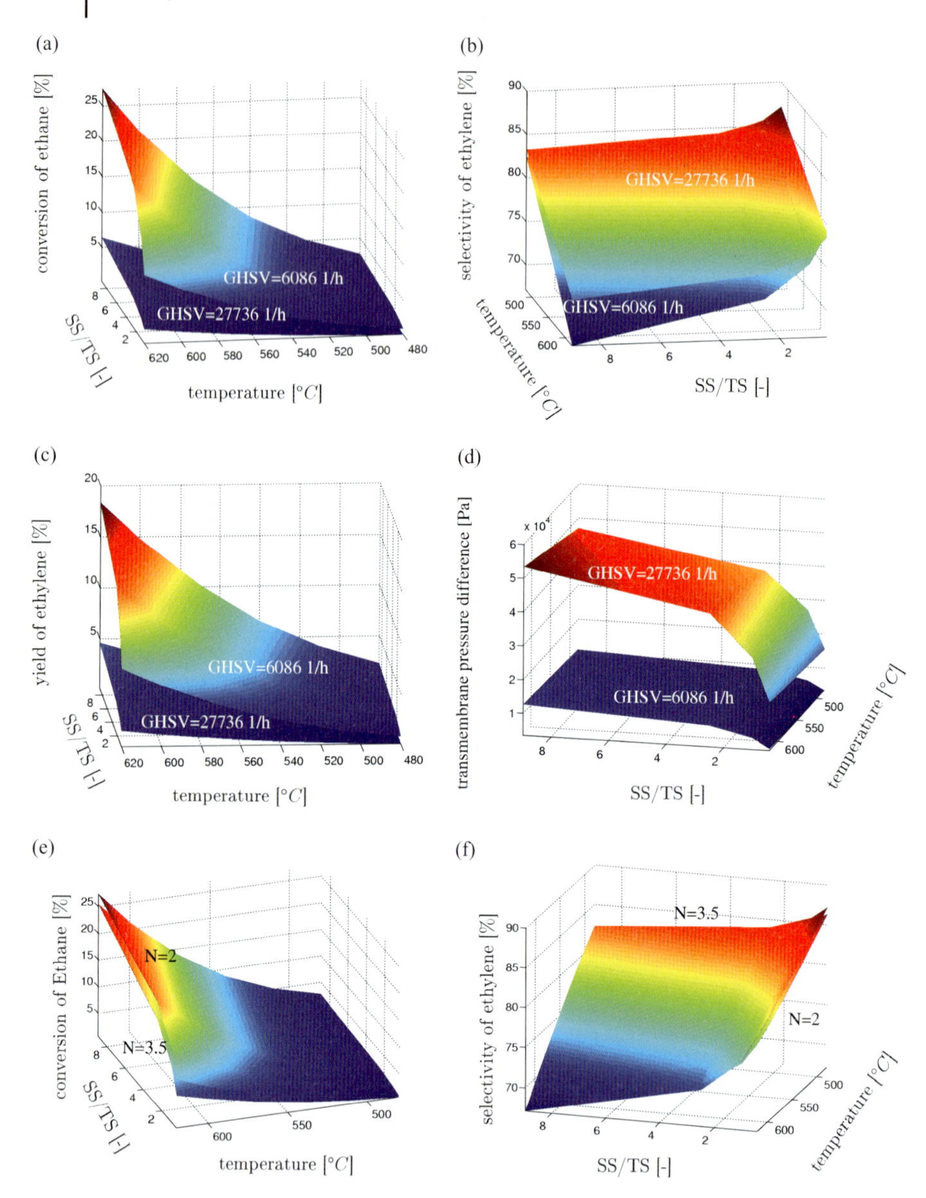

Figure 4.38 Influence of the operating conditions on the integral quantities, 3 μm catalyst layer, GHSV = 27 736 L/h.

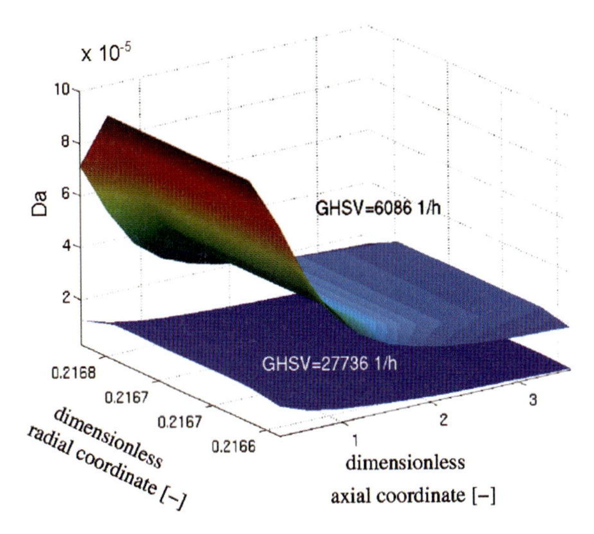

Figure 4.39 Da number as a function of radial and axial coordinate at SS/TS = 9, $T = 600\,°C$ for both residence times.

6086 L/h results in >20% difference in the conversion of ethane. The selectivity values remain high with maximal 5.2% reduction at higher temperatures.

- The ethane conversions and ethylene yields rise significantly with increasing the flow rate ratio SS/TS in the considered range, especially at higher temperatures and higher residence times.
- The temperature has the largest influence on the ethane conversion among all investigated parameters with maximal difference of 25% in the observed range at GHSV = 6086 L/h.

The local reactant concentration determines the local reaction rates and thus determines the conversion and selectivity. Because the simulations are accomplished for low oxygen concentrations, higher selectivity of ethylene can be achieved by varying the others parameters. In contrast to PBMR no reaction limitation is observed in the catalyst layer under lean oxygen conditions even at SS/TS = 0.4 and GHSV = 6086 L/h. Increasing the catalyst layer thickness leads to higher oxygen conversions in the reactor. This phenomenon can be observed especially in the first part of the reactor.

For both GHSV values and SS/TS = 9, the Da number, defined as $\mathrm{Da} = \frac{\tilde{M}_{C_2H_4} V_{cat} r_1}{\dot{M}_{inlet}}$, is calculated and presented in Figure 4.39 as a function of dimensionless axial and radial coordinates. The coordinates are scaled by the reactor diameter. No internal and external transfer limitations are observed for both cases. The obtained tendencies can be explained by the insufficient catalytic surface. By increasing the catalytic surface and the contact time in a thicker catalyst layer the conversion level in the reactor can rise significantly.

Residence time effects are observed in the simulations by varying the SS/TS ratio. The local residence time and the local and average oxygen concentration increase in the reactor tube at higher SS/TS ratios, followed by an increase in the ethane conversion. A longer residence time of the ethane molecules also promotes radial diffusion towards the catalyst layer.

The obtained results show that the operating parameters like residence time, temperature and SS/TS ratio have a significant influence on ethane conversion and ethylene yield. For effective operation of CMR, low total volume flow rates and high SS/TS ratios are needed for higher conversions in the reactor. Increasing the inlet temperature leads to higher conversion and lower selectivity. Because the conversion is the limiting factor under the investigated conditions, its increase causes a rise in the ethylene yields. In the accomplished simulations, the concentration ratio was varied within a relatively small range. Nevertheless, a clear tendency of improving ethylene selectivity at lower oxygen concentration is observed.

Based on this theoretical parameter study, a high potential for the CMR performance can be demonstrated. An application of this type of membrane reactor in mini- or micro-scale modules could be of great interest.

4.7 Conclusion

The presented results contribute to a better understanding of the coupled transport and reaction phenomena in membrane reactors. A significant influence of the transport processes on the yield and the selectivity was observed, especially for the case of the catalytic membrane reactor (CMR). Optimal operation conditions for membrane reactors can only be achieved by selecting appropriate parameters. An analysis of the influence of operating parameters as well as that of the geometry and structural parameters of the membrane was made by numerical experiments, which led to important conclusions regarding this reactor type. More results regarding other membrane reactor configurations are presented in the following chapters.

References

Alfonso, M.J., Julbe, A., Farrusseng, D., Menendez, M., and Santamaria, J. (1999) Oxidative dehydrogenation of propane on $V/\gamma\text{-}Al_2O_3$. *Chem. Eng. Sci.*, **54**, 1265–1272.

Byrd, R.H., Lu, P., Nocedal, J., and Zhu, C. (1995) A limited memory algorithm for bound constrained optimization. *SIAM J. Sci. Comput.*, **16** (5), 1190–1208.

Cussler, L. (1997) *Diffusion, Mass Transfer in Fluid Systems*, Cambridge University Press.

Edreva, V., Zhang, F., Mangold, M., and Tsotsas, E. (2009) Mass transport in multilayer, porous metallic membranes: diagnosis, identification and validation. *Chem. Eng. Technol.*, **32** (4), 632–640.

Georgieva-Angelova, K. (2008) Modellbasierte Analyse der Transportprozesse und des Einflusses der Betriebs- und Geometrieparameter in einem katalytisch beschichteten Membranreaktor, PhD Thesis, University of Magdeburg.

Girault, V., and Raviart, P.-A. (1986) *Finite Element Methods for Navier-Stokes Equations*, Springer, Berlin.

Goering, H. (1977) *Asymptotische Methoden zur Lösung von Differentialgleichun-gen*, Akademie-Verlag, Berlin.

Hamel, Ch. (2008) *Experimentelle und modellbasierte Studien zur Herstellung kurzkettiger Alkene sowie von Synthesegas unter Verwendung poröser und dichter Membranen*, Docupoint Verlag Magdeburg.

Hein, S. (1999) *Modellierung wandgekühlter katalytischer Festtbettreaktoren mit Ein- und Zweiphasenmodellen*, VDI-Verlag, Düsseldorf.

Hussain, A. (2006) Heat and mass transfer in tubular inorganic membranes, PhD Thesis, Otto-von-Guericke-University Magdeburg.

Hussain, A., Seidel-Morgenstern, A., and Tsotsas, E. (2006) Heat and mass transfer in tubular ceramic membranes for membrane reactors, heat and mass transfer in tubular ceramic membranes for membrane reactors. *Int. J. Heat Mass Transf.*, **49**, 2239–2253.

John, V., Knobloch, P., Matthies, G., and Tobiska, L. (2002) Non-nested multi-level solvers for finite element discretizations of mixed problems. *Computing*, **68**, 313–341.

Keil, R. (2007) *Modeling of Process Intensification*, Chapter 5, Packed-bed Membrane Reactors, Wiley-VCH Verlag GmbH, Weinheim.

Klose, F. (2008) *Structure-Activity Relations of Supported Vanadia Catalysts and the Potential of Membrane Reactors for the Oxidative Dehydrogenation of Ethane*, Docupoint Verlag Magdeburg.

Knobloch, P., and Tobiska, L. (2008) *On the Stability of the Finite Element Discretization of Convection-Diffusion-Reaction Equations*, Preprint 08–11, Otto-von-Guericke-University Magdeburg.

Kramers, H.A., and Kistemaker, J. (1943) On the slip of a diffusing gas mixture along the wall. *Physica*, **10**, 699–713.

Krasnyk, M., Bondareva, K., Mikholov, O., Teplinskiy, K., Ginkel, M., and Kienle, A. (2006) The ProMoT/Diana simulation environment, in *Proc. of the 16th European Symposium on Computer Aided Process Engineering* (ed. W. Marquardt and C. Pantelides), Elsevier, pp. 445–450.

Matthies, G., and Tobiska, L. (2007) Mass conservation of finite element methods for coupled flow-transport problems. *Int. J. Comput. Sci. Math.*, **1**, 293–307.

Matthies, G., Skrzypacz, P., and Tobiska, L. (2005) Superconvergence of a 3D finite element method for stationary Stokes and Navier-Stokes problems. *Numer. Methods Part. Diff. Equat.*, **21**, 701–725.

Matthies, G., Skrzypacz, P., and Tobiska, L. (2007) A unified convergence analysis for local projection stabilization applied to the Oseen problem. *ESAIM: M2AN*, **41** (4), 713–742.

Matthies, G., Skrzypacz, P., and Tobiska, L. (2008) Stabilization of local projection type applied to convection-diffusion problems with mixed boundary conditions. *ETNA*, **32**, 90–105.

Rodriguez-Fernandez, M., Mendes, P., and Banga, J.R. (2006) A hybrid approach for efficient and robust parameter estimation in biochemical pathways. *BioSystems*, **83** (2–3), 248–265.

Roos, H.-G., Stynes, M., and Tobiska, L. (2008) *Robust Numerical Methods for Singularly Perturbed Differential Equations. Convection-Diffusion-Reaction and Flow Problems*, Springer Series in Computational Mathematics, vol. **24**, Springer-Verlag, Berlin.

Schlünder, E.U., and Tsotsas, E. (1988) *Wärmeübertragung in Festbetten, durchmischten Schüttgütern und Wirbelschichten*, Georg Thieme Verlag, Stuttgart.

Skrzypacz, P., and Tobiska, L. (2005) Finite element and matched asymptotic expansion methods for chemical reactor flow problems. *Proc. Appl. Math. Mech.*, **5**, 843–844.

Teplinskiy, K., Trubarov, V., and Svjatnyj, V. (2005) Optimization problems in the technological-oriented parallel simulation environment, in *Proc. of the 18th ASIM-Symposium on Simulation Techniques* (eds. F. Hülsemann *et al.*) SCS Publishing House, Erlangen, pp. 582–587.

Thomas, S. (2003) *Kontrollierte Eduktzufuhr in Membranreaktoren zur Optimierung der Ausbeute gewünschter Produkte in Parallel- und Folgereaktionen*. Logos Verlag, Berlin.

Turek, S. (1999) *Efficient Solvers for Incompressible Flow Problems*, Lecture Notes in Computational Science and Engineering, vol. **6**, Springer-Verlag, Berlin.

Uchytil, P., Schramm, O., and Seidel-Morgenstern, A. (2000) Influence of the transport direction on gas permeation in two-layer ceramic membranes. *J. Membr. Sci.*, **170** (2), 215–224.

Ullmann (1977) *Ullmann's Encyclopädie Der Technischen Chemie*, Verlag Chemie, Weinheim.

Young, J.B., and Todd, B. (2005) Modelling of multi-component gas flows in capillaries and porous solids. *Int. J. Heat Mass Transf.*, **48**, 5338–5353.

Zanfir, M., and Gavriilidis, A. (2003) Catalytic combustionassisted methane steam reforming in a catalytic plate reactor. *Chem. Eng. Sci.*, **58**, 3947–3960.

Zhang, F. (2008) Model identification and model based analysis of membrane reactors, PhD thesis, Otto-von-Guericke-Universität, Magdeburg.

5
Packed-Bed Membrane Reactors

Christof Hamel, Ákos Tóta, Frank Klose, Evangelos Tsotsas, and Andreas Seidel-Morgenstern

5.1
Introduction

In the field of chemical reaction engineering intensive research is devoted to developing new processes for many industrially relevant reactions in order to improve the selectivities and yields of intermediate products (e.g., partial oxidations; Hodnett, 2000; Sheldon and van Santen, 1995).

Chapter 1 describes how in reaction networks optimal local reactant concentrations are essential for a high selectivity towards the target product (Levenspiel, 1999). If undesired series reactions can occur, it is usually advantageous to avoid backmixing. This is one of the main reasons why partial hydrogenations or oxidations are preferentially performed in tubular reactors (Westerterp, Swaaij, and Beenackers, 1984). Typically, all reactants enter such reactors together at the reactor inlet (co-feed mode). Thus, in order to influence the reaction rates along the reactor length, essentially temperature is the parameter that could be modulated. In classic papers, summarized by (Edgar and Himmelblau, 1988), the installation of adjusted temperature profiles has been suggested in order to maximize selectivities and yields at the reactor outlet. However, the practical realization of a defined temperature modulation is not trivial. Alternative efforts have been devoted to study the application of distributed catalyst activities (Morbidelli, Gavriilidis, and Varma, 2001). This can be realized, for example, by mixing different catalysts or by local catalyst dilution. An attractive option, which is also capable to influence the course of complex reactions in tubular reactors and which is discussed in this chapter, is to abandon the conventional co-feed mode and to install more complex dosing regimes. The concept is based on the fact that there is a possibility to add one or several reactants to the reactor in a distributed manner. There is obviously a large variety of options differing mainly in the positions and amounts at which components are dosed. Deciding whether a certain concept is useful or not requires a detailed understanding of the dependence of the reaction rates on concentrations. As discussed in Chapter 1 in particular the reaction orders with respect to the dosed component are of essential importance (Lu *et al.*, 1997a, 1997b, 1997c). Besides dosing one or several components at a discrete position into a fixed-bed

Membrane Reactors: Distributing Reactants to Improve Selectivity and Yield.
Edited by Andreas Seidel-Morgenstern

ISBN: 978-3-527-32039-4

reactor there also exists the possibility to realize a distributed reactant feeding over the reactor wall. This can be conveniently realized using tubular membranes. The concept of improving product selectivities in parallel-series reactions by feeding one reactant through a membrane tube into the reaction zone was studied, for example, by (Coronas, Menendez, and Santamaria, 1995; Diakov and Varma, 2004; Hamel *et al.*, 2003; Kürten, Sint Annaland, and Kuipers, 2004; Lafarga, Santamaria, and Menéndez, 1994; Lu *et al.*, 2000; Seidel-Morgenstern, 2005; Tota *et al.*, 2004; Zeng, Lin, and Swartz, 1998).

Theoretical and laboratory-scale studies focusing on the application of various configurations of membrane reactors described in Chapter 1 in order to improve selectivity–conversion relations in complex reaction systems are available. In particular, several industrially relevant partial oxidation reactions were investigated. Relevant studies were performed, for example, for the oxidative coupling of methane (Caro *et al.*, 2006; Diakov, Lafarga, and Varma, 2001; Tonkovich *et al.*, 1996a), for the partial oxidation of ethane (Al-Juaied, Lafarga, and Varma, 2001; Coronas, Menendez, and Santamaria, 1995; Klose *et al.*, 2004a; Klose *et al.*, 2003a; Wang *et al.*, 2006) and propane (Caro, 2006; Grabowski, 2006; Liebner *et al.*, 2003; Ramos, Menendez, and Santamaria, 2000; Schäfer *et al.*, 2003; Ziaka, Minet, and Tsotsis, 1993) and for the oxidative dehydrogenation (ODH) of butane (Alonso *et al.*, 2001; Mallada *et al.*, 2000b; Tellez, Menendez, and Santamaria, 1997).

In most of these works a single tubular membrane was used to distribute one reactant in the inner volume filled with catalyst particles. In this contribution the partial oxidations of ethane to ethylene and propane to propylene as model reactions performed in a pilot-scale membrane reactor were investigated in detail. For such reactions improved integral reactor performance can be achieved further by an optimized stage-wise dosing of one or several reactants (Hamel *et al.*, 2003; Thomas, 2003). Adjusted dosing profiles can be realised, for example, by feeding reactants separately through permeable reactor walls (e.g., through tubular membranes discussed in Chapter 4).

5.2 Principles and Modeling

5.2.1 Reactant Dosing in a Packed-Bed Membrane Reactor Cascade

Figure 5.1 shows schematically the principle of the well established packed-bed reactor (PBR) in comparison to a three-stage membrane reactor cascade (PBMR-3) applied for controlled multi-stage dosing of reactants through the reactor wall, respectively. The three-stage membrane reactor cascade allows the realization of different dosing profiles, distributing the fed amount of reactants in the particular stages, influencing the local concentration and residence time of the reactants.

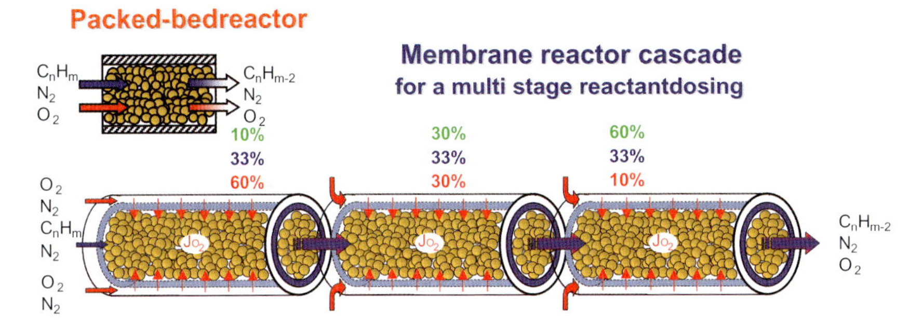

Figure 5.1 Reactant feeding in a packed-bed reactor (PBR) and in a multi-stage membrane reactor (PBMR) cascade.

The reactants are fed separately in the packed-bed membrane reactor (PBMR) on the inner side (tube side, TS) and outer side (shell side, SS) of the membrane. The catalyst is placed in the tube side of the membrane as a packed bed. The applied dead-end reactor configuration allows a feeding of reactants through the membrane in a controlled manner, with defined fluxes. All reactants dosed have to permeate through the membrane, which is considered to be catalytically inert.

The membrane reactor illustrated in Figure 5.1 differs from the conventional PBR with respect to local residence time, local concentration and temperature profiles. Figure 5.2a depicts typical total flow rates for PBR and PBMR for identical overall feed flow and outlet flow rates without chemical reaction, respectively. In this schematic representation the total flow rate in the PBR remains constant along the reactor length. All reactants are fed in a co-feed mode at the reactor entrance and have, therefore, the same average residence time in the reactor. For comparison, PBMR profiles are shown for two different ratios of tube side to shell side flow rate (F_{TS}/F_{SS} = 1:8 and/or 1:1). Assuming a uniformly distributed flux through the membrane there is a linear increase in the total flow rate along the reactor length. The slope decreases as the F_{TS}/F_{SS} flow ratio increases. This causes a decreasing residence time of the reactants at the entrance of the reactor. Reactants fed via the shell side have a different residence time distribution. Due to the fact that reactants entering the reaction zone next to the inlet pass a longer reactor distance than molecules entering close to the outlet, the average residence time of the reactants dosed in a PBMR through the membrane is higher compared to a PBR (Tonkovich *et al.*, 1996b). Thus, a higher conversion can be expected in a PBMR. Additionally to a one-stage membrane reactor a multi-stage cascade allows a feeding of different amounts as well as suitable dosing profiles in each stage. Just like the residence time behavior, the local concentrations of the reactants influences the reactor performance significantly. The local concentration has an influence on local reaction rates as well as on conversion and selectivity.

Demonstrating the differences between the reactor concepts, Figure 5.2b reveals the internal oxygen concentration profiles. Hereby the total amount of oxygen

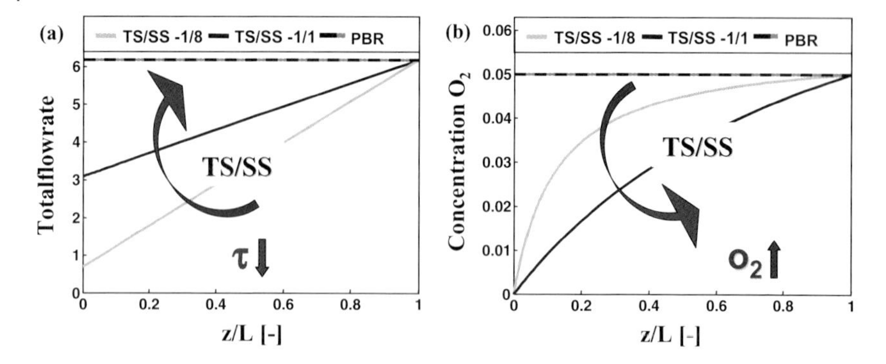

Figure 5.2 Schematic illustration of: a) total flow rates and b) molar fraction of oxygen in a fixed-bed reactor (PBR) and in a membrane reactor (PBMR) for two different tube side/shell side feed flow ratios (TS/SS).

dosed over the membrane in the PBMR is the same as for the PBR. Without the occurrence of chemical reactions in the PBR the molar fraction of oxygen is constant along the reactor length. Considering the ODH of short-chain alkanes, a high oxygen level is undesirable because it favours the consecutive reactions (Hamel *et al.*, 2003; Klose *et al.*, 2003b; Tota *et al.*, 2004). Via distributed feeding the local oxygen concentration can be reduced in a membrane reactor, as shown in Figure 5.2b. Thus, the selectivity of desired intermediate products should be increased by avoiding series reactions. At low F_{TS}/F_{SS} ratios (i.e., for high trans-membrane fluxes) the concentration of oxygen increases rapidly at the entrance. The local and hence also the averaged oxygen concentration is always lower than in the PBR. It is worthwhile to note that, under these flow conditions, the local residence time of the hydrocarbons is relatively high near the PBMR inlet. This behavior will also, together with the rapidly increasing oxygen concentration, significantly influence the temperature profiles in the reactor.

5.2.2
Modeling Single-Stage and Multi-Stage Membrane Reactors

5.2.2.1 Simplified 1-D Model

The main advantage of reduced 1-D pseudo-homogeneous models is that they are computationally inexpensive (see Chapters 1 and 2, respectively). Comprehensive parametric studies can be carried out within a few minutes, which makes this model attractive for optimization calculations.

A reduced reactor model was applied for an efficient calculation and evaluation of PBR and PBMR in a broad range of operation conditions (Caro *et al.*, 2006; Hamel *et al.*, 2006; Hamel *et al.*, 2003; Tota *et al.*, 2004). The simplified "1-D model" allows analysis of a series connection of S equally sized stages of isothermal tubular reactors, as illustrated in Figure 5.1. All reactants can be dosed in a discrete manner in each stage s at the inlets $(\rho_g^{TS} y_i^{TS} F_T^{TS})^{s,mix}$ and/or over the walls ($j_i^{\mathrm{TS},s}$). Every reactor stage is modeled under the following conditions: negligible axial and radial dispersion, plug flow, steady-state and isothermal operation. Based on these assumptions the mass balance for stage s can be written as:

$$\frac{d(\rho_g^{\mathrm{TS}} y_i^{\mathrm{TS}} F_T^{\mathrm{TS}})^s}{d\xi^s} = \frac{d\dot{m}_i^s}{d\xi^s} = V_R \rho_{\mathrm{cat}} \tilde{M}_i \sum_{j=1}^{N_R} \nu_{ij} r_j + A^{\mathrm{TS}} j_i^{\mathrm{TS},s} \text{ with } j = 1, N_R;\ s = 1, S \quad (5.1)$$

In Equation 5.1 $\xi^s = z^s/Z$ is the dimensionless length coordinate and $j_i^{\mathrm{TS},s}$ is, in contrast to the 1+1-D model presented below, the constant convective flux of component i over the wall of stage s. The following boundary conditions were used together with Equation 5.1:

$$\dot{m}_i^1(\xi^1 = 0) = \dot{m}_i^{\mathrm{in}} \quad (5.2)$$

$$\dot{m}_i^s(\xi^s = 0) = \dot{m}_i^{s-1}(\xi^{s-1} = 1) + \dot{m}_i^{s,\mathrm{mix}} \quad (5.3)$$

The total amounts of component i introduced at all inlets (mix, $S \geq 2$) and over all walls (diff) are:

$$\dot{m}_{i,\mathrm{tot}}^{\mathrm{mix}} = \sum_{k=2}^{S} \dot{m}_i^{s,\mathrm{mix}} \quad (5.4)$$

$$\dot{m}_{i,\mathrm{tot}}^{\mathrm{diff}} = \sum_{s=1}^{S} \dot{m}_i^{s,\mathrm{diff}} = A^{\mathrm{TS}} \sum_{s=1}^{S} j_i^{\mathrm{TS},s} \quad (5.5)$$

With the exception of a few cases (simple reaction orders, isothermal condition), the solution of the above system of ordinary differential equations can be performed only numerically. For this there are several numerical software tools available. In this contribution Matlab® was used to solve the initial value problem given in Equations 5.1–5.5.

5.2.2.2 More Detailed 1+1-D Model

The assumption of a purely pressure-driven convective flow through the membrane is usually an adequate approximation, even in laboratory-scale reactors. However, if the barrier effect exerted by the membrane is insufficient, that is, diffusive transport in the pores (e.g., bulk, Knudsen, or surface diffusion) cannot be neglected, transport equations must be defined for the membrane and solved simultaneously with the balance equations for both sides of the membrane. Such a scenario is analyzed in the following.

In this study a simplified transport model, the so-called extended Fick model (EFM) was used to describe the mass transport of the species through the membrane. Compared to the more sophisticated dusty gas model (DGM; Chapters 2 and 4; Mason and Malinauskas, 1983) the interactions between the species are treated on a much simpler way in this modeling approach. The mass flux of a component i is given by:

$$j_i = -D_{i,\mathrm{eff}} \rho_g \frac{\partial y_i}{\partial r} - y_i \rho_g \frac{B_0}{\eta} \frac{\partial p}{\partial r}. \quad (5.6)$$

Table 5.1 Simulation parameters for the 1+1-D reactor models (5.6–5.14).

Reactor length (length of the porous zone) $-L$	60 mm
Inner/outer diameter of the membrane $-d_i/d_o$	7/10 mm
Membrane permeability $-B_0$	$5.4e^{-14}$ m^2
Knudsen coefficient $-K_0$	$9.6e^{-08}$ m
Membrane porosity to tortuosity ratio $-\varepsilon/\tau$	0.096
Membrane pore diameter $-d_{por}$	3 μm
W/F	150–225–400 kg s/m^3
Oxygen concentration (overall)	2 vol%
Ethane concentration (overall)	2 vol%
Tube to shell side flow ratio $-F_{TS}/F_{SS}$	2/1–1/1–1/4–1/9
Reactor temperature	600 °C

The effective diffusion coefficient $D_{i,\mathrm{eff}}$ is calculated by means of Equation 2.57 discussed in Chapter 2. The second term on the right hand side of Equation 5.6 is the convective flux, where according to Darcy's law the superficial velocity is expressed by the pressure gradient. The membrane is characterized by three structure parameters (B_0, ε_p/τ, K_0) which have to be obtained experimentally (Table 5.1 and Chapter 4).

The 1+1-D reactor model has been derived based on the following assumptions:

- Isothermal, steady-state conditions, ideal gas behavior,
- Constant total pressure and plug-flow on TS and SS,
- Absence of mass transport limitations, reactions only occur on the TS,
- Molecular and Knudsen diffusion with superimposed convective flux through the membrane.

The mass balances as well as the corresponding boundary conditions are given below:

- Component mass balance on the "tube side" (TS):

$$\frac{\partial}{\partial z}(\rho_g^{TS} y_i^{TS} F_T^{TS}) = V_R \rho_{cat} \cdot \tilde{M}_i \sum_{j=1}^{N_R} \nu_{ij} r_j + A^{TS} j_i^{TS}(z) \text{ for } i = 1, \ldots, N_C \tag{5.7}$$

- Overall mass balance:

$$\frac{\partial}{\partial z}(\rho_g^{TS} F_T^{TS}) = \sum_{i=1}^{N_C} A^{TS} j_i^{TS}(z) \tag{5.8}$$

- Boundary conditions on TS:

$$y_i^{TS}\big|_{z=0} = y_{i,0}^{TS};\ \rho_g^{TS} F_T^{TS}\big|_{z=0} = \rho_{g,0}^{TS} F_{T,0}^{TS} \tag{5.9}$$

- Component mass balance on the "shell side" (SS):

$$\frac{\partial}{\partial z}(\rho_g^{SS} y_i^{SS} F_T^{SS}) = A^{SS} j_i^{SS}(z) \text{ for } i = 1, \ldots, N_C \tag{5.10}$$

- Overall mass balance:

$$\frac{\partial}{\partial z}(\rho_g^{SS} F_T^{SS}) = \sum_{i=1}^{N_C} A^{SS} j_i^{SS}(z) \tag{5.11}$$

- Boundary conditions on SS:

$$y_i^{SS}\big|_{z=0} = y_{i,0}^{SS};\ \rho_g^{SS} F_T^{SS}\big|_{z=0} = \rho_{g,0}^{SS} F_{T,0}^{SS} \tag{5.12}$$

- Mass balance for the membrane:

$$\frac{1}{r}\frac{\partial}{\partial r}(r \cdot j_i) = 0 \text{ for } i = 1, \ldots, N_C \tag{5.13}$$

- Boundary conditions for the membrane:

$$y_i\big|_{r=r_i} = y_i^{TS} \text{ and } y_i\big|_{r=r_o} = y_i^{SS},\ p\big|_{r=r_o} = p^{SS} \text{ and } p\big|_{r=r_i} = p^{TS} \tag{5.14}$$

These algebraic equations are coupled by the boundary conditions to the system of ordinary differential equations for TS and SS, respectively. The resulting nonlinear differential algebraic equation system was solved using Newton iterations and the direct linear solver UMFPACK in Comsol Multiphysics 3.2 (COMSOL, 2006). For the applied dead-end membrane reactor configuration the SS pressure was found by an iterative procedure.

5.2.2.3 **Detailed 2-D Modeling of a Single-Stage PBMR**

For a theoretical analysis a detailed 2-D reactor model assuming steady-state, ideal gas behavior, no heat and mass transfer limitations between bulk phase and catalyst particle as well as inside the catalyst pellets was developed and implemented in the simulation tool Comsol. This model is based on the Equations 2.7–2.11 presented in Chapter 2 and by (Tota *et al.*, 2007), respectively. Important parameters for packed-bed membrane reactors are the effective radial and axial heat dispersion coefficients, which were calculated according to two different approaches described by (Tsotsas, 2002). The first approach called "α_w model" presumes heat dispersion with different but radially constant coefficients in both directions. The second one, called "$\lambda(r)$ model", was developed by (Cheng and Vortmeyer, 1988) and refined in an extensive comparison with experimental data by (Winterberg and Tsotsas, 2000b; Winterberg *et al.*, 2000). The authors argued that especially in reactive flow problems at low Reynolds numbers the assumption of an inhibiting laminar sublayer at the wall is not appropriate. Therefore, the use of a wall heat transfer coefficient, α_w, can only be artificial. The correlations applied for the calculation of the radial/axial heat dispersion coefficient are given by (Tota *et al.*, 2007; Winterberg and Tsotsas, 2000a).

To calculate the flow field under reactive conditions the extended Navier–Stokes equation and the mass continuity equation defined in Chapter 2 (2.7–2.8) were solved. Regarding the boundary conditions for the membrane reactor, there is one considerable difference compared to the PBR: the radial component of the velocity at the membrane wall is not zero. The latter is proportional to the trans-membrane flux. The friction force arising due to the flow through the packed bed can be calculated according to (Ergun, 1952) by defining the friction coefficient. For low tube-to-particle diameter ratios a significant part of the flow shifts from the core of the bed towards the reactor wall. This so-called flow maldistribution is caused by the non-uniform porosity profile of the packed bed. The effect on the local velocity is pronounced especially at low-particle Reynolds numbers (Winterberg and Tsotsas, 2000b). To describe the experimentally found damped oscillation of the radial porosity profiles, the correlation given by (Hunt and Tien, 1990) was applied.

5.3 Model-Based Analysis of a Distributed Dosing via Membranes

5.3.1 Model Reactions

As model reactions the oxidative dehydrogenation (ODH) of ethane to ethylene and the ODH of propane to propylene on a $VO_x/\gamma\text{-}Al_2O_3$ catalyst were chosen. A detailed analysis of the networks and the five main reactions taking place is given by (Klose *et al.*, 2004b), (Liebner, 2003), and in Chapter 3 of this book respectively.

According to the scheme given in Figure 5.3a, ethane is converted in a parallel reaction to CO_2 and the desired product ethylene. This parallel reaction limits the maximal achievable ethylene selectivity in the reaction (Klose *et al.*, 2004a). Additionally, the consecutive reaction to CO and the total oxidation can further decrease the olefin selectivity. The authors suggested a kinetic model based on a Mars–van Krevelen type redox mechanism for the ethylene production and Langmuir–Hinshelwood–Hougen–Watson kinetics for the deep oxidation reactions (Joshi, 2007; Klose *et al.*, 2004a). The derived kinetic model for the ODH of ethane was found to describe a large set of experimental laboratory-scale data with good accuracy (see Chapter 3).

In principle the reaction network of propane, given in Figure 5.3b, is similar. Also propane is converted in a parallel reaction to CO_2 and the desired product propylene. According to the studies of (Liebner, 2003), propylene can be formed,

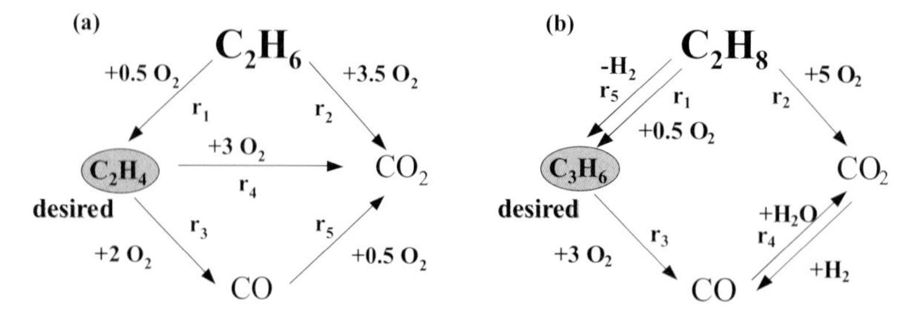

Figure 5.3 Reaction network of the oxidative dehydrogenations of ethane and propane.

respectively, by ODH (r_1) and by the thermal dehydrogenation of propane (r_5) in the absence of oxygen. In contrast to the ethane network and considering the lower temperature range the ratio of CO and CO_2 is determined by the water gas shift reaction.

5.3.2
Simulation Study for ODH of Ethane Using the 1-D Model

This section investigates the performance of a conventional PBR and a single-stage membrane reactor in a simulation study and compares the results for the ODH of ethane to illustrate the reactor behavior in a broad range of operating conditions. Therefore the reduced membrane reactor model given by Equations 5.1–5.3 for one-stage, as well as the reaction rates given by (Klose *et al.*, 2004b) were applied.

As shown in Figure 5.4a, the single-stage PBMR shows a higher ethylene selectivity than the PBR in the oxygen-controlled region for a high mass of catalyst/total flow rate (W/F) corresponding to a high conversion. Increasing the oxygen level decreases the ethylene selectivity. Because of unavoidable uncertainties of the

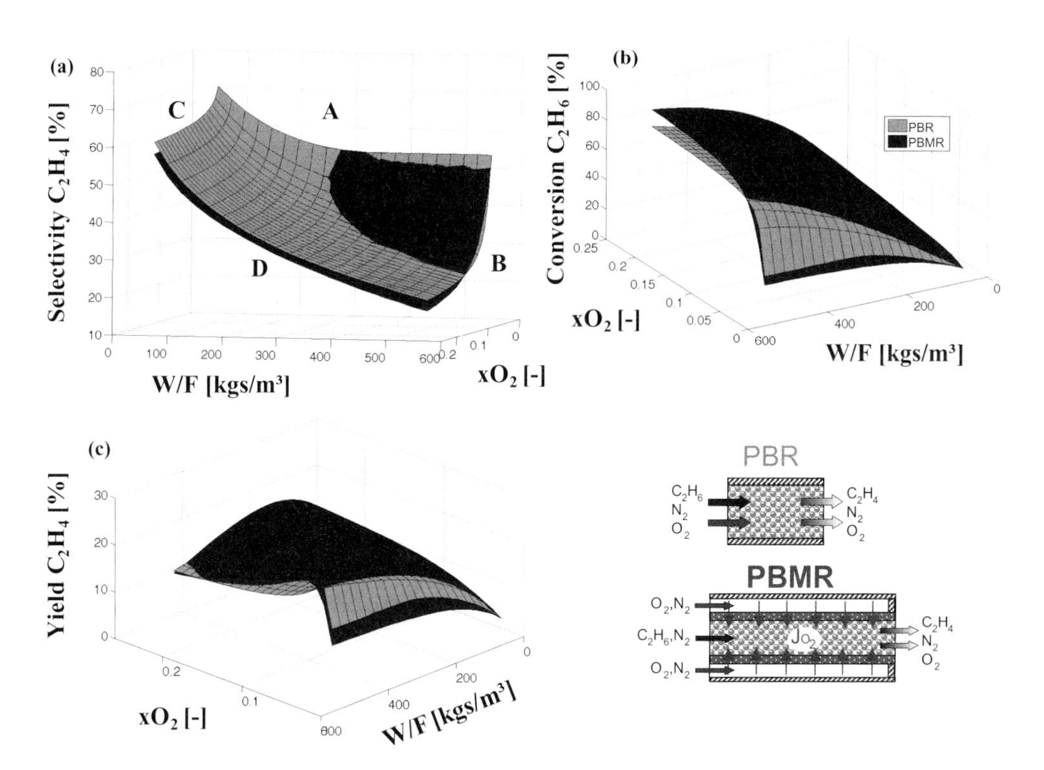

Figure 5.4 Comparison between PBR and PBMR: a) selectivity of ethylene, b) conversion of ethane, c) yield of ethylene. $x_{C2H6}{}^{in} = 1.5\%$, $x_{O2}{}^{in} = 0.5–21.0\%$, $W/F = 50–550\,kg\,s/m^3$, $T = 600\,°C$, $F_{TS}/F_{SS} = 1/8$.

kinetic model the simulation results below an oxygen level of 1 vol% should be evaluated with care. In the conversion plot given in Figure 5.4b the benefits of the PBMR can be recognized. Conversion differences of up to 15% can be obtained for a broad range of oxygen concentration and residence time. Below an oxygen concentration of 1.5 vol% the PBR shows slightly higher conversions, which can be explained by the higher local oxygen concentration in the PBR. This region is extended towards higher oxygen concentrations with increasing residence time. Differences in the mean residence time of the reactants between PBR and PBMR are still present, but not so significant as below $W/F = 250\,kg\,s/m^3$. With increasing contact time the oxygen level is the only conversion-determining factor. The resulting ethylene yields (Figure 5.4c) reveal that there are two preferable regions for the application of a PBMR. The first one is in the residence time-controlled region (in this case below approximately $250\,kg\,s/m^3$). In this region the oxygen concentration can be relatively high. The higher ethylene yields achievable here result from the high reactant conversion. The second region, where simulation postulates the highest PBMR yields, is at longer contact times for oxygen concentrations below 1.5 vol%.

Finally, a comparison between the dosing concepts will be given by means of a selectivity–conversion plot. To get an insight into the selectivity behavior of the PBMR for the whole parameter range studied, one can extract all the simulation results for ethylene selectivity and ethane conversion and plot the corresponding values together in one diagram (Figure 5.5). The resulting envelope describes the attainable region for the reactor at 600 °C, for the flow rates and reactant concentrations considered. In this special case the borders of the envelope are the edges indicated in Figure 5.4a as lines A, B, C and D. The results given in Figure 5.5 show that for low residence times (line C) and at high oxygen concentration (line D) the PBMR outperforms the PBR. However, in this region the yields are

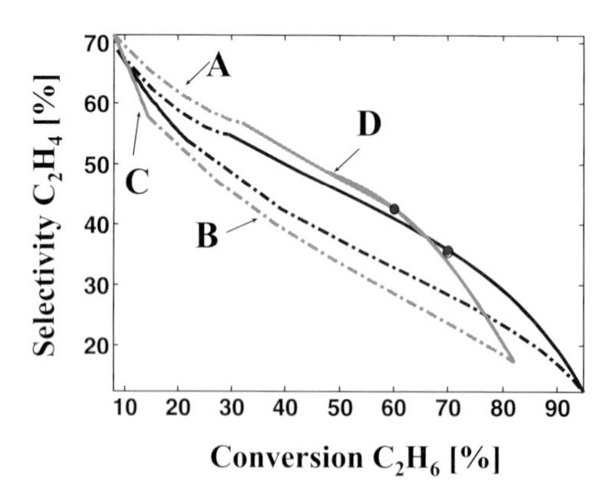

Figure 5.5 Ethylene selectivity versus ethane conversion in a fixed-bed und one-stage membrane reactor; $x_{C2H6}^{in} = 1.5\%$, $x_{O2}^{in} = 0.5$–21.0%, $W/F = 50$–$550\,kg\,s/m^3$, $T = 600\,°C$, $F_{TS}/F_{SS} = 1/8$. A, B, C, and D are related to lower and upper limits of parameter ranges with respect to oxygen concentration and residence time.

quite low in both reactors. Further, the curves of the PBMR are located within the operating window of the PBR; thus, one can easily find operating parameters where the PBR is superior to the PBMR. For ethane per pass conversion below approximately 65% the PBR allows higher ethylene selectivity and therefore higher yields. If higher per pass conversions are needed, the reaction should be carried out in a PBMR, because the ethylene selectivity does not decrease as rapidly as in the reference concept (PBR). The maximal ethylene yields, marked by dots, are hardly different in the two reactors. According to the simulation result, 25.2% of the fed ethane is converted to ethylene for the chosen simulation conditions. This is achieved in the membrane reactor for around 10% higher conversion, while the ethylene selectivity is approximately 7% below that of the PBR.

The following section presents simulation results of the 1+1-D reactor model, describing the laboratory-scale membrane reactor used (see chapter 5.4.2). For the sake of simplicity, the calculations are carried out for a homogeneous membrane with a pore diameter of 3 μm, and a typical membrane support with structure parameters is summarized in Table 5.1 (Hussain, 2006; Thomas, 2003). Figure 5.6a,b

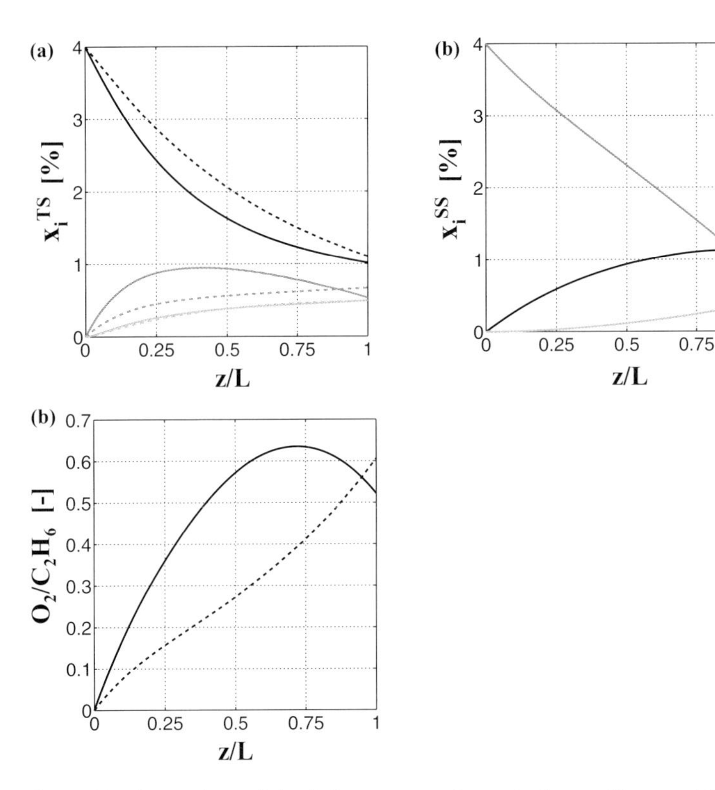

Figure 5.6 Comparison of simulation results for the 1-D (dashed) and the 1+1-D (solid) reactor model. Axial ethane (black), O_2 (dark gray), ethylene (light gray). Concentration profiles on: a) TS and b) SS. c) Oxygen to ethane concentration ratio on: $W/F = 400\,kg\,s/m^3$, $T = 600\,°C$, $x_{C2H6}^{in} = 2\%$, $x_{O2}^{in} = 2\%$, $F_{TS}/F_{SS} = 1/1$.

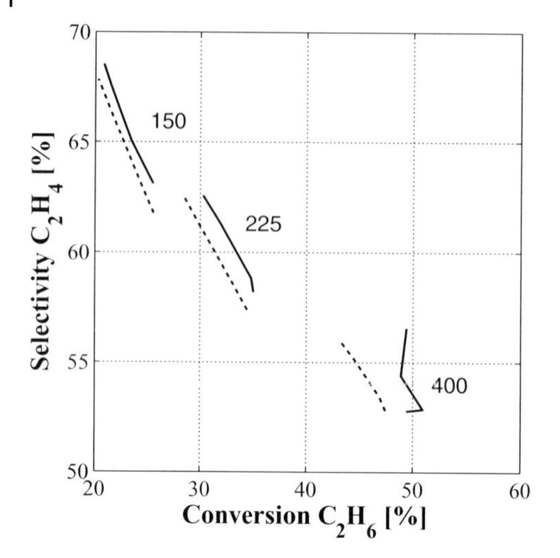

Figure 5.7 Comparison of simulation results for the 1-D (dashed) and the 1+1-D (solid) reactor model. Conditions: $W/F = 150/225/400\,\text{kg}\,\text{s/m}^3$, $T = 600\,°\text{C}$, $x_{C2H6}{}^{in} = 2\%$, $x_{O2}{}^{in} = 2\%$, $F_{TS}/F_{SS} = 2/1–1/9$.

illustrates axial concentration profiles of the reactants and ethylene on both membrane sides. These results correspond to extremely low trans-membrane pressure differences of approximately 50 mbar. It can be seen that the simple model given by Equations 5.7–5.14 (dashed line) gives higher local ethane concentrations than the more realistic 1+1-D reactor model (solid line). According to this model relatively high ethane and even ethylene concentration can be expected on the SS. At the same time the diffusive flux considerably enhances the transport of oxygen through the membrane. This results in a high local oxygen to ethane concentration ratio (Figure 5.6c) and is responsible for the higher ethane conversions. The results suggest that the simpler model (5.1–5.3) underestimates the achievable ethylene yields in the membrane reactor. As a check simulations were carried out for higher trans-membrane pressure differences by changing the total flow rate and F_{TS}/F_{SS} flow ratio. As expected, the results show (Figure 5.7) that, at short contact times ($W/F = 150\,\text{kg}\,\text{s/m}^3$), both models give an almost identical reactor performance. Small discrepancies can be seen only by dominating tube side flow conditions ($F_{TS}/F_{SS} > 0.5$). In these cases a slightly higher ethylene selectivity is to be expected. With increased contact time ($W/F = 225$ and $400\,\text{kg}\,\text{s/m}^3$) the difference between the models increases. Ethane conversion increases, while a slight increase of the calculated ethylene selectivity is obtained. These results reveal that for the calculated laboratory-scale PBMR using a thin membrane thickness of 1.5 mm above approx. $200\,\text{kg}\,\text{s/m}^3$ space time the more complex model should be used to compare reactor concepts. In cases where the membrane transport is dominated by the viscous flow, the simple 1-D model can give sufficiently accurate results for a first comparison.

5.4 Experimental

5.4.1 Catalyst and Used Membrane Materials

As described in Chapter 3 the VO_x/γ-Al_2O_3 catalyst used in this study for the ODH of ethane and propane in all reactors was prepared by soaking and impregnation of γ-Al_2O_3 with a solution of vanadyl acetylcetonate in acetone followed by a calcination step at 700 °C. The vanadium content of the catalyst was 1.4% V. The surface of the calcinated catalyst was 158 m^2/g, measured by the single-point BET method. The color of the fresh catalyst is yellow, indicating that vanadium was mainly in the +5 oxidation state (Oppermann and Brückner, 1983). After the measurements, the catalyst color had changed to a light blue-green. This can be attributed to a significant reduction of V(V) to V(IV). Ceramic (HITK e.v.) and sintered metal (GKN) membranes for oxygen dosing were investigated experimentally. The tubular ceramic composite membranes in a pilot-scale used for feeding air as the oxidant were provided by the Hermsdorfer Institut für Technische Keramik. They were already characterized in Chapter 4. They consisted of a mechanically stable α-Al_2O_3 support (average pore diameter: 3 μm; thickness: 5.5 mm) on which two more α-Al_2O_3 layers (pore diameters: 1 μm and 60 nm; thickness 25 μm) and finally one γ-Al_2O_3 layer were deposited (pore diameter: 10 nm; thickness: 2 μm). The whole membrane tube had a length of 350 mm and an inner/outer diameter of 21/32 mm. Both ends of the membrane were vitrified, leaving in the center of the membrane a permeable zone of 104 mm (see Figure 5.8a). The mass transport properties of this membrane type were characterized by the authors in a previous study (Hussain, Seidel-Morgenstern, and Tsotsas, 2006). The sintered metal membranes were provided by GKN Sintered Metals Filters GmbH. These mechanically stable membranes are made of Inconel 600, and had a geometry similar to ceramic membranes (average pore diameter: 0.3 μm; thickness: 3 mm). During assembly of the membrane reactor, the membrane tubes were filled with inert material in the non-permeable zones and with the above-described VO_x/γ-Al_2O_3 catalyst in the porous reaction zone.

5.4.2 Single-Stage Packed-Bed Membrane Reactor in a Pilot-Scale

In order to study a PBMR with radial dimensions close to an industrial scale, a pilot-plant was realized allowing total feed flow rates up to 4500 L/h. Beside the possibility to evaluate the development of radial and axial gradients the large scale had the advantage that the impact of measuring installations, especially thermocouples within the catalyst bed, which disturb fluid dynamics, was significantly reduced. The single-stage PBMR consisted of a stainless steel tube (inner

(a)

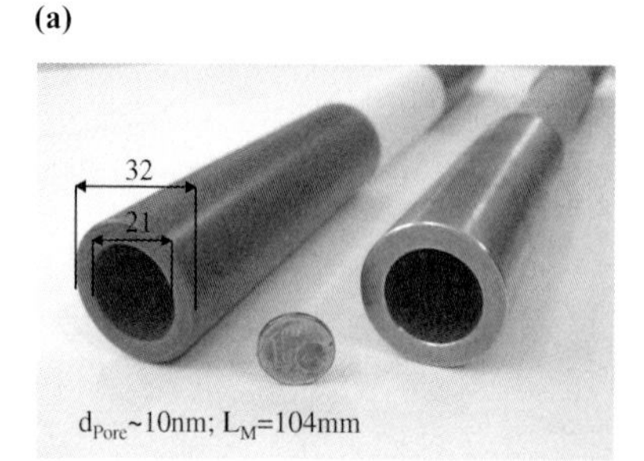

(b)

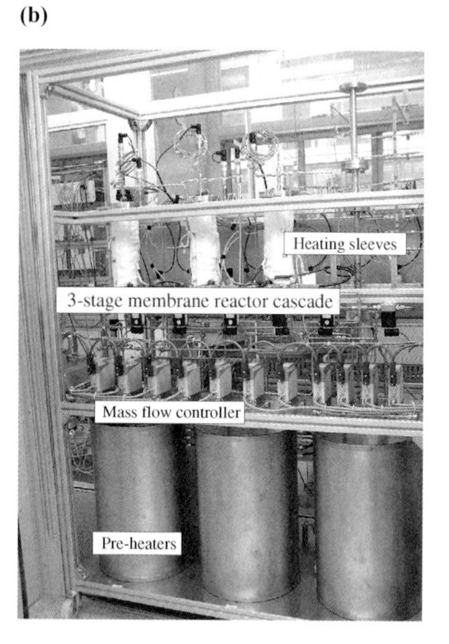

Figure 5.8 a) Porous asymmetric Al_2O_3 membrane (left) and sintered metal membrane (right). b) Multi-stage membrane reactor cascade for the oxidation of short-chain hydrocarbons on a pilot scale.

diameter: 38.4 mm) into which the membrane tube filled with catalyst was inserted. The reactor was heated by an electric heating sleeve outside the steel tube. The hydrocarbons, for safety reasons diluted 1 : 10 in nitrogen, were fed on the tube side of the membrane. Air was dosed from the shell side over the membrane into the reactor. The shell side outlet was closed so that all dosed air was pressed through the membrane into the catalyst bed (see Figure 5.1). This configuration is similar to that reported most often in the recent literature, for example, by (Mallada, Menéndez, and Santamaría, 2000a; Ramos, Menendez, and Santamaria, 2000; Tonkovich *et al.*, 1996b). However, it differs from those studies, where the shell side outlet was open, in that oxidant transfer over the membrane is influenced by diffusion (Farrusseng, Julbe, and Guizard, 2001; Kölsch *et al.*, 2002). Pressing all air over the membrane has the advantage of an easy control of the amount of air inserted. In this way the membrane was used as a non-perm selective oxidant distributor in a "dead-end" configuration. During assembly of the reactor the membrane tube was filled with inert material in the vitrified zones and with $VO_x/\gamma\text{-}Al_2O_3$ catalyst (17 g) in the permeable section. Five thermocouples were placed in the center of the tube: at the reactor inlet/outlet, at inlet/outlet of the porous zone and in the middle of the catalyst bed (see Figure 5.10a). Additionally, thermocouples were installed on the surface of the membrane on shell and tube side. Gas samples were taken at the tube side inlet and directly after the reactor outlet. The membrane tube was fixed in housing with fixed flanges. Both membrane sides were sealed gas-proof by pressing the applied graphite seals (Novafit) during the assembling of the slip-on flanges at the ends. These flanges had a distance of 125 mm from the hot reaction zone. Because of the material of the seals the upper head segment temperature was limited to 550 °C. It was possible to operate the reactor at catalyst temperatures up to 650 °C without any damage to the seals or the membrane itself.

5.4.3
Reference Concept – Conventional Fixed-Bed Reactor

For the fixed-bed experiments the same reactor equipment was used as for the PBMR. A completely vitrified alumina membrane with the same geometry was installed in the membrane reactor housing. All reactants were fed in a co-feed mode (Figure 5.1). The application of a vitrified membrane avoids wall reactions compared to a simple fixed-bed reactor made from stainless steel. Further it allows experiments to be performed under similar heat transfer conditions.

5.4.4
Multi-Stage Membrane Reactor Cascade

The membrane reactor cascade (PBMR-3) applied for studying the local concentration distribution of oxygen and the residence time behavior consists of a series connection of three identical membrane reactors, as described in the previous section and illustrated in Figure 5.1. The catalyst masses placed in every stage of the PBMR-3 were 17 g. The tube side outlet of each reactor was connected with the tube side inlet of the following reactor using a heated transfer line. The hydrocarbons were inserted in the tube side inlet of the first reactor. Air was fed via a membrane from the shell sides of all reactors. To realise as different dosing profiles of air, for example, increasing (10/30/60%), uniform (33/33/33%) and decreasing (60/30/10%), the distribution of the overall air amount between the reactors was modified systematically. Further, each reactor could be heated separately by an electrical heating sleeve. However, for the experiments the temperature was kept equal in all the reactors. Gas sampling was possible at the tube side inlet of the first reactor and at each tube side outlet of the three reactor stages.

5.4.5
Analytics

The set-up used consists of several units: reactor modules, catalytic afterburner and a gas chromatograph (Agilent 6890 GC/TCD with 5973 MSD) with a 12-port multiposition valve for reactant and product stream analysis. The catalytic afterburner had to prevent hazardous emissions to the environment. A SIMATIC S7-based process control system was implemented to run the unit automatically and recover all process data. Feed mixtures and flow configurations were realized by using electronic mass flow controllers (Bürkert, type 8712). Gas samples were taken from different positions of the reactors as described below by switching the multiposition valve. They were transferred via a heated line to the sample valve of the GC-MSD to prevent condensation. All gas samples were analyzed with a GC-TCD/MSD system equipped with a four-column configuration. This configuration included a HP Plot Q column for the quantification of CO_2, ethane/propane and ethylene/propylene, a HP Molsieve 5A column for the separation of permanent gases and CO and a FFAP column to analyze oxygenates.

5.4.6 Experimental Conditions

Based on the preliminary theoretical analysis shown above, a large set of experimental studies was carried out in a temperature range between 520–650 °C (ethane) and 350–500 °C (propane). The molar O_2/C_nH_{2n+2} ratio was varied between 0.5–5.0 and especially near the stoichiometric ratio (hydrocarbon/oxygen = 2 : 1) of the ODH of ethane. In the investigations, high space velocities were applied, closer to the requirements set by industry. The inverse weight hourly space velocity was varied between W/F = 100–400 kg s/m^3 (catalyst mass/total volumetric flow rate). Additionally, overall oxygen hydrocarbon ratios were kept below the lower explosion limit, which favored safety of operation. In all PBMR measurements the shell side/tube side feed ratio was adjusted at 8, meaning 88% of total feed flow was permeated over the membrane. Hereby this ratio was kept constant independently on the dosed oxygen amount by mixing air with nitrogen for the permeating feed. The reproducibility was checked by conducting product stream analysis more than three times on every set of experimental parameters and additionally by twice repeating every complete run to examine additionally changes of the catalyst itself. The basic reference used to compare the performance of the reactors was the weight hourly space velocity, which was the same for all considered dosing strategies. Thus, the overall feed flow rates of the PBMR-3 were related to the feed flow rates of the PBR and PBMR, by applying a corresponding mass of catalyst.

5.5 Results for the Oxidative Dehydrogenation of Ethane to Ethylene

5.5.1 Comparison Between PBR and PBMR Using Ceramic Membranes in a Single-Stage Operation Mode

Figure 5.9a,c,e compares the performance of a single-stage PBMR using a ceramic membrane and a conventional PBR on a pilot scale for lean oxygen conditions (O_2/C_2H_6 = 1), and Figure 5.9b,d,f does the same for excess oxygen conditions (O_2/C_2H_6 = 5).

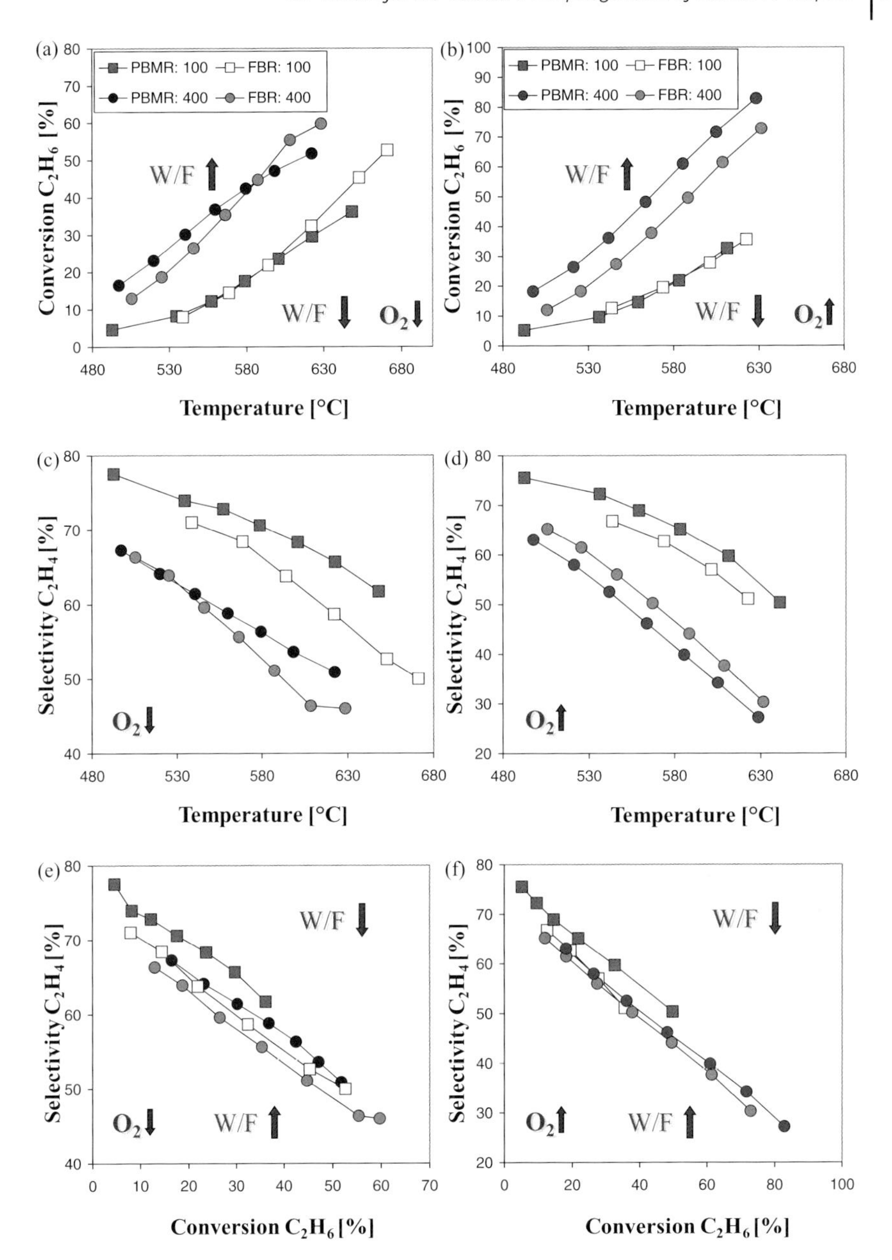

Figure 5.9 a), b) Conversion of ethane versus temperature. c), d) selectivity of ethylene versus temperature. e), f) Selectivity of ethylene versus conversion, ethane molar fraction $xC_2H_6^{in} = 1.5\%$, $O_2/C_2H_6 = 1O_2\downarrow a.5O_2\uparrow$, $W/F = 100/400\,kg\,s/m^3$, catalyst: VO_X/γ-Al_2O_3, (1.4%), BET: $157\,m^2/g$.

According to the simulations presented in the theoretical analysis above, a higher ethane conversion can be expected in the PBMR as long as the oxygen availability does not become rate-determining. This is due to the prolonged contact time of ethane over the catalyst bed compared to the PBR. Further, the parameter study revealed that, under oxygen excess conditions, the sensitivity of ethylene selectivity to oxygen partial pressure is low and for this reason, at moderate excess of oxygen, the PBMR should provide higher ethane conversion as well as higher ethylene yields.

Figure 5.9a,b gives the impact of membrane-assisted oxidant dosing on ethane conversion under lean oxygen and excess oxygen conditions, respectively. For an excess of oxygen (Figure 5.9b) and high contact times ($W/F > 100\,kg\,s/m^3$) the PBMR outperforms the PBR significantly. At short contact times ($W/F = 100$ $kg\,s/m^3$) and low temperatures (below 600 °C) the conversions obtained in the PBMR and PBR are hardly different, as predicted by simulations. The enhanced conversion in the PBMR can be explained by the higher residence time induced by the distributed feeding of O_2 and N_2 over the membrane (see Section 5.2.1). In contrast, under lean oxygen conditions and for temperatures over 600 °C the PBR shows a better performance concerning conversion. For these operation parameters the PBMR seems to be limited with respect to oxygen. Thus, the dosed amount of O_2 cannot be distributed radially over the catalyst bed. Latter aspect will be investigated in detail using 2-D reactor models. Higher conversions of ethane in the PBR can be obtained also at short contact times and at high temperatures.

The selectivity of the desired intermediate product ethylene could be significantly increased in the PBMR, especially for lean oxygen conditions, but also for conditions of oxygen excess and short residence times. Thus, the concept of decreasing the local oxygen concentration by distributed dosing to avoid series reactions forming CO and CO_2 has proven to be successful applying ceramic membranes. For further information see Hamel *et al.*, 2003; Thomas, Klose, and Seidel-Morgenstern, 2001.

Under lean oxygen conditions ethylene selectivity can be significantly enhanced using the PBMR at similar levels of ethane conversion (Figure 5.9e); in contrast under conditions of oxygen excess the selectivity–conversion plots fall together (Figure 5.9f) with those of the corresponding PBR reference experiments. Thus, a benefit regarding ethylene selectivity is given especially at low oxygen-to-hydrocarbon ratios.

A further intention of investigating membranes on a pilot scale was to study the developing radial temperature profiles. Thus, seven thermocouples were placed in the PBMR to measure the temperature at different positions of the catalyst bed and the membrane surface, respectively. The position of the thermocouples illustrated in Figure 5.10a was already described in Section 5.4. A more detailed analysis of the distributed dosing of reactants on the axial temperature profiles was alreadly given by (Tota *et al.*, 2007). The study revealed that, for the PBMR, the heat transfer is characterized by heat conduction and additionally a high convective

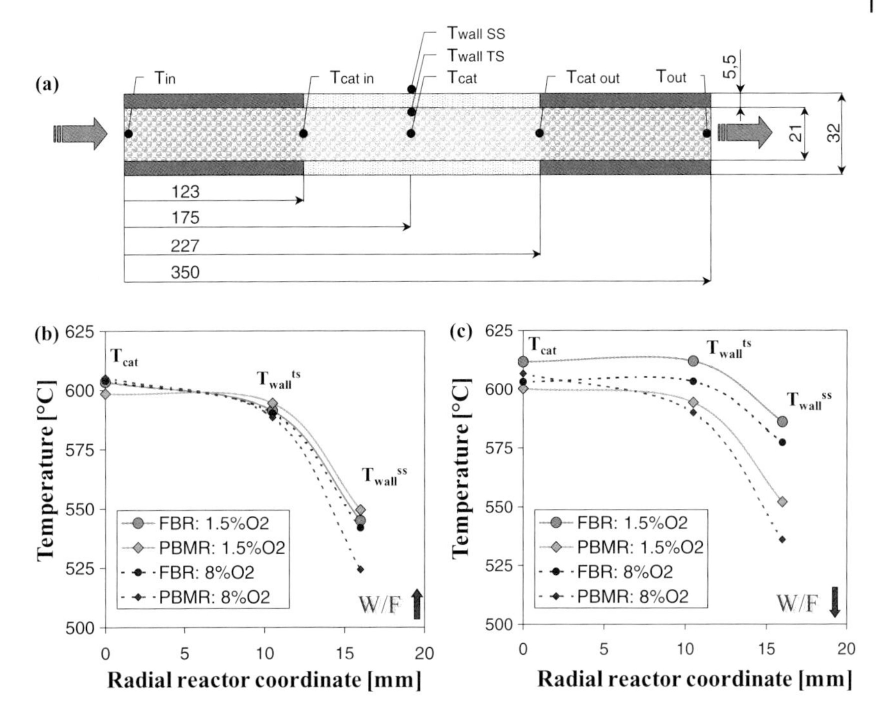

Figure 5.10 a) Placement of thermocouples, b) temperature versus radial reactor coordinate ($W/F = 400\,\text{kg s/m}^3$), c) temperature versus radial reactor coordinate ($W/F = 200\,\text{kg s/m}^3$), $xC_2H_6^{in} = 1.5\%$.

radial flow due to the distributed feeding of O_2 and N_2. Thus, the feed of the PBMR can be heated up more rapidly and/or the temperature can be controlled significantly better than in co-feed mode in the PBR, where heat transfer takes place only by heat conduction over the reactor walls.

The temperature regime with respect to the radial distribution is given in Figure 5.10b,c. As can be seen in Figure 5.10b at high contact times ($W/F = 400\,\text{kg s/m}^3$, low feed rates) the differences between PBMR and PBR are small, approximately 5 K. For a higher concentration of oxygen (8%) the temperatures at the shell side of the membrane wall (T_{wall}^{ss}) are lower than in lean oxygen conditions in both reactor concepts. This is because the catalyst bed temperature T_{cat} is used as the control variable for the heating sleeves. A higher feed concentration of oxygen leads to an increase of the reaction rate, hence to enhanced heat generation. The latter is compensated by the controller, which is reducing the heating power of the heating jacket over the reactor length.

In the case of higher feed flow rates ($W/F = 200\,\text{kg s/m}^3$) the radial temperature gradients are small and can be neglected for the investigated membrane diameter (Figure 5.10c).

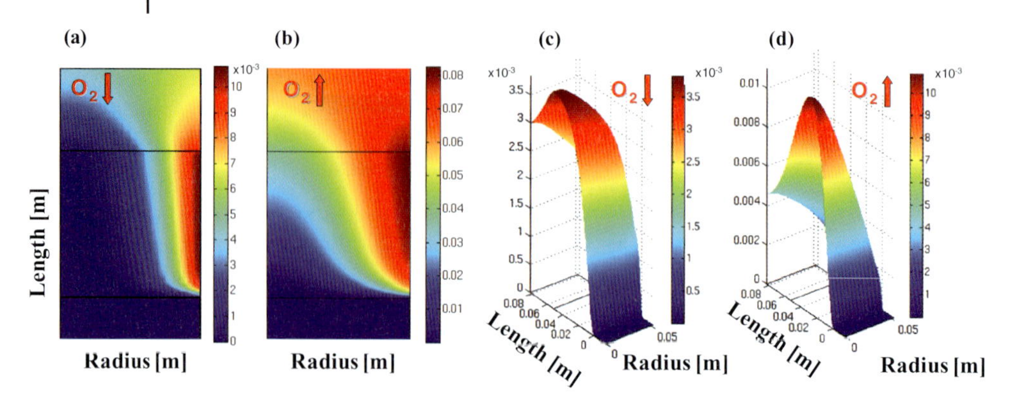

Figure 5.11 a), b) Simulated oxygen concentration fields in the PBMR, c), d) ethylene concentration using the $\lambda(r)$ model. $W/F = 400\,\text{kg s/m}^3$, ethane molar fraction $xC_2H_6^{in} = 1.5\%$, $O_2/C_2H_6 = 1O_2\downarrow$ and $5O_2\uparrow$, $T = 610\,°C$.

5.5.2
2-D Simulation Results – Comparison Between PBR and PBMR

Based on the detailed model presented in Chapter 2 the concentration, temperature and velocity fields in the PBMR were calculated. As selected results the calculated concentration fields of oxygen and the desired intermediate product ethylene using the "$\lambda(r)$ model" are illustrated in Figure 5.11. The simulated concentration fields of oxygen (Figure 5.11a,b) illustrate clearly the PBMR principle: lowering the local oxygen concentration to avoid series reactions by membrane-assisted distributed dosing. As shown in Figure 5.11a, under lean oxygen conditions and for temperatures over 600 °C the PBMR is limited with respect to oxygen. Thus, the dosed amount of O_2 cannot be distributed radially in the whole catalyst provided. Consequently, the PBMR reveals under this condition a better performance concerning conversion. In contrast, for higher concentrations of oxygen (Figure 5.11b) the catalyst bed is used more efficiently, but the impact of series reactions increases, indicated by decreasing ethylene concentration (Figure 5.11c).

Table 5.2 gives an overview with respect to the results of the reduced 1-D model (5.1–5.3) and the more detailed "α_w model" and "$\lambda(r)$-model", respectively, in comparison to the experimental data. The oxygen concentrations calculated with the 1-D model are considerably lower than those of the 2-D model. The predicted ethane conversion values (Table 5.2) are 3–7% higher. Thus, the lower oxygen concentration of the 1-D model results from higher oxygen consumption. Especially for high contact times ($W/F = 400\,\text{kg s/m}^3$), that is, low volumetric flow rates as well as low Reynolds numbers, ethane conversion as well as ethylene selectivity could be described in a relatively good agreement with the detailed "$\lambda(r)$ model", taking into account the radial oxygen profile and porosity distribution. This conclusion was also published by (Hein, 1999; Winterberg *et al.*, 2000) especially for reactive conditions. For short contact times ($W/F = 100\,\text{kg s/m}^3$), that is, high

Table 5.2 Evaluation of the detailed 2-D $\alpha_w/\lambda(r)$ models and the reduced 1-D reactor model by means of experimental data: $xC_2H_6^{in} = 1.5\%$, $x_{O2}^{in} = 8\%$, $T = 610\,°C$.

	Experiment		1-D model		α_w model		$\lambda(r)$ model	
	[%]		[%]		[%]		[%]	
W/F [kg s/m^3]	100	400	100	400	100	400	100	400
Conversion C_2H_6	33	72	36	79	32	66	32	72
Selectivity C_2H_4	60	34	49	30	50	32	51	33
Yield C_2H_4	20	25	18	24	16	21	16	24

Reynolds numbers, the applied "α_w model" and "$\lambda(r)$ model" are close together, as confirmed by (Bey, 1998; Tsotsas and Schlünder, 1990)

5.5.3 Application of Sintered Metal Membranes for the ODH of Ethane

In spite of the performance improvement by membrane-assisted oxidant dosing, the drawback of a higher construction effort has to be addressed, too, and temperature resistant sealings of the membranes have to be found which should be safe and cost-effective. In this study graphite was used for sealing the ceramic membrane. One alternative can be the use of sintered metal membranes characterized in Chapter 4. Unfortunately, sintered metal membranes have much larger pores than ceramic ones. Thus, the trans-membrane pressure drop can be expected to be much lower than for mesoporous ceramic membranes. This is problematic on two counts. First, when the trans-membrane pressure drop is of the same order of magnitude as the pressure drop over the catalyst bed, the distribution of the reactants along the length is no longer uniform, which can significantly affect the reactor performance (see also Section 5.5.4). Second, under the conditions of laboratory experiments, according to the theoretical analysis shown above, it seems impossible to compensate uncontrolled diffusive transport for the sintered metal membranes. Only at higher feed rates, when the dominating convective transport can be anticipated, does the application of sintered metal membranes seem to be attractive in the pilot plant.

The sintered metal membrane (PBMR-SM) described in Section 5.4.1 was investigated under the same condition as the PBR and the PBMR using ceramic composite membranes. A comparison is given in Figure 5.12. It has to be noted that the pressure drop over the metal membrane is approximately one-third that of the ceramic membrane (Figure 5.12c). For this reason, the focus is set on those experiments with varying contact times because they demonstrate most sensitively the impact of trans-membrane pressure drop on reactor performance.

In comparison to the PBMR experiments with the ceramic membrane for short contact times, that means at high feed flow rates ($W/F = 100\,kg\,s/m^3$, see

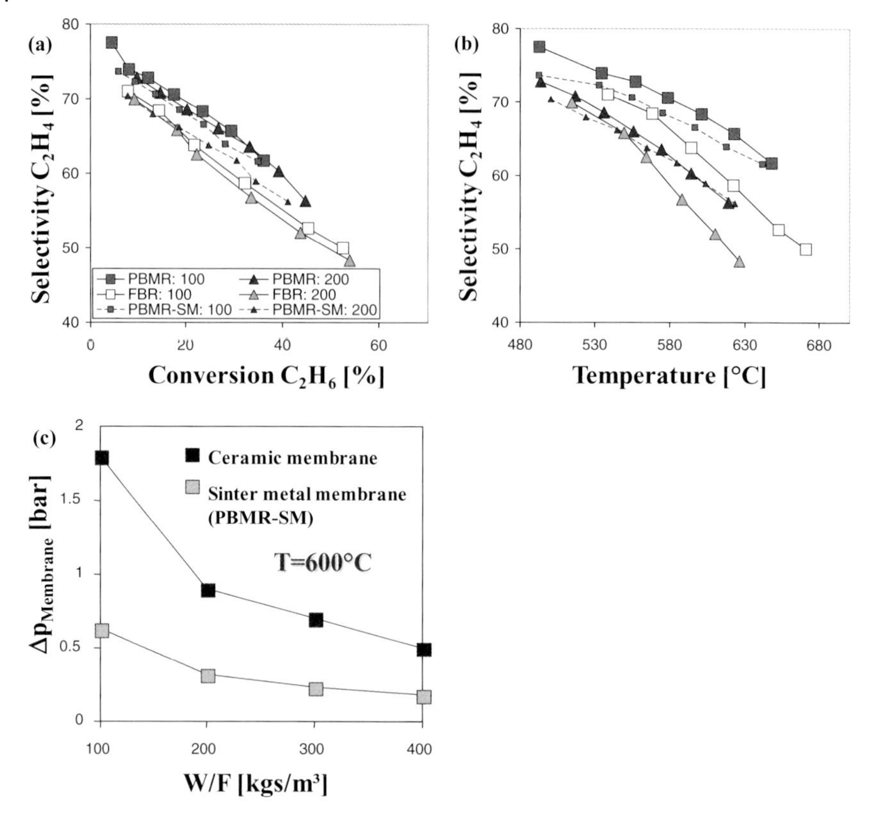

Figure 5.12 Comparison of fixed-bed (PBR) and one-stage membrane reactors using ceramic (PBMR) and sintered metal membranes (PBMR-SM). a) Ethylene selectivity versus conversion. b) Ethylene selectivity versus temperature. $xC_2H_6^{in} = 1.5\%$, $O_2/C_nH_m = 1$, $W/F = 100$–$200\,kg\,s/m^3$. c) Pressure drop over the ceramic and sintered metal membrane at 600 °C.

Figure 5.12a,b), the performance of the PBMR with the sintered metal membrane is slightly lower. Plotting ethylene selectivity versus ethane conversion also for higher contact times ($W/F = 200\,kg\,s/m^3$) as illustrated in Figure 5.12a, the pattern of PBMR-SM is between those of PBMR using ceramic membranes and PBR. This indicates that, due to the lower trans-membrane pressure drop shown in Figure 5.12c for the metal membrane, back-diffusion is expected to play a larger role than for the ceramic membrane. The problem of back-diffusion was investigated in detail and already discussed by (Tota *et al.*, 2006, 2007). However, despite this limitation, the usage of metal membranes with their better characteristics regarding construction and mechanical stability is possible in principle and can be considered as an attractive alternative to overcome the problems with ceramic membranes if a sufficient level of trans-membrane pressure drop can be reached.

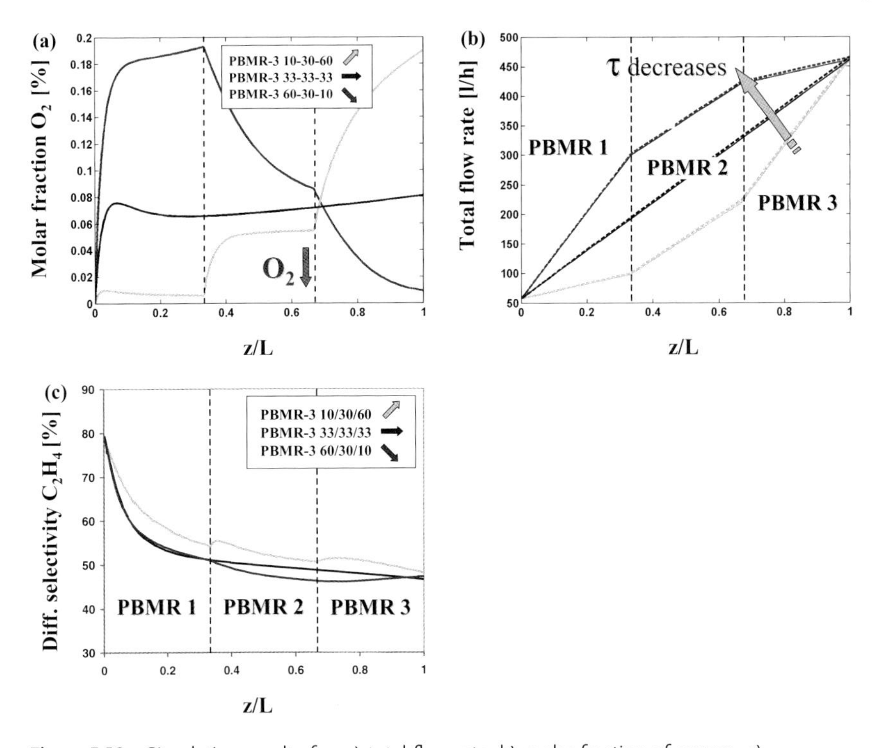

Figure 5.13 Simulation results for: a) total flow rate, b) molar fraction of oxygen, c) differential ethylene selectivity in a three-stage membrane reactor cascade. $W/F = 400\,\mathrm{kg\,s/m^3}$, $xC_2H_6^{in} = 1.5\%$, $x_{O2} = 0.75\%$, $T = 600\,°C$.

5.5.4 Investigation of a Membrane Reactor Cascade – Impact of Dosing Profiles

A detailed proposition concerning the local reaction rates including undesired parallel–series reactions is given by the differential selectivity. For the postulated reaction network of the ODH of ethane to ethylene (Figure 5.3) the differential selectivity introduced in Chapter 1 (Equation 1.30) is defined for the ODH of ethane as follows:

$$\text{Differential selectivity } C_2H_4 = \frac{r_1 - r_3 - r_4}{r_1 + r_2} \tag{5.15}$$

Especially for the investigation of suitable dosing profiles influencing the local oxygen concentration and contact time in the three-stage membrane reactor cascade (PBMR-3) the differential selectivity is a useful parameter. In the PBMR-3 three different dosing profiles were investigated: increasing (10/30/60%), uniform (33/33/33%) and decreasing (60/30/10%). The resulting axial total flow rates and axial molar fractions of oxygen are depicted in Figure 5.13a,b, respectively. The increasing dosing profile is characterized by a very low oxygen concentration, especially in the first and second stages (Figure 5.13a). Based on the obtained local oxygen concentration the highest selectivity of the intermediate ethylene can be expected. Corresponding to the lowest total flow rate as well as the highest residence times of the reactants can be obtained by realising an increasing dosing profile. So a high ethane conversion can be expected as long as oxygen is available. Figure 5.13c illustrates the differential selectivity of ethylene, calculated by means of the reduced 1-D model, as a function of the dimensionless cascade length, $\xi = Z/L$.

In general the differential selectivity decreases over the reactor length, based on the increasing intermediate concentration of ethylene and the resulting increasing reaction rates r_3 and r_4, respectively. For the chosen simulation conditions of long contact times and low oxygen concentration ($W/F = 400\,\mathrm{kg\,s/m^3}$, $x_{O2} = 0.75\%$) the increasing dosing profile (10/30/60%) reveals the highest differential selectivity of ethylene.

The increasing dosing profile is characterized by a very low oxygen level, especially in the first PBMR, corresponding to a long contact time. Thus, the desired reaction r_1 forming ethylene can take place. Simultaneously, the ethylene-consuming consecutive reactions r_3 and r_4 are suppressed due to the low oxygen level (see Figure 5.13a). In contrast, the uniform (33/33/63%) and decreasing (60/30/10%) dosing profiles are characterized by a higher local oxygen level.

The obtained performance parameters of the experimentally investigated PBMR-3 are given in Figure 5.14. Based on the kinetics and the reaction network the ethane conversion is increasing and selectivity of the desired intermediate product ethylene is decreasing with increasing temperature as discussed for the single-stage membrane reactor PBMR. The highest conversion (Figure 5.14a) can be obtained for the uniform and decreasing dosing profiles. The temperature-dependence of the ethane conversion is very similar for both. The increasing dosing profile (10/30/60%) reveals a slightly lower conversion, due to oxygen limitation, especially for high temperatures. In general the membrane reactor cascade demonstrates a better performance with respect to conversion of ethane compared to the single-stage PBMR and the conventional fixed-bed reactor. An explanation is the higher residence time of the reactants by means of distributed dosing, which is more significant in the cascade. The integral ethylene selectivity of the cascade is illustrated in Figure 5.14b. As predicted in the calculations, the increasing dosing profile shows the highest ethylene selectivity followed by the constant and decreasing profiles. Thus, the sequence of ethylene selectivity for a comparable conversion is shown in Figure 5.14c. Additionally, the selectivity of ethylene at the outlet of the particular stages (axial concentration profile) for $T = 600\,°\mathrm{C}$ is given in Figure 5.14d. It can be recognised that, at the outlet of the first stage, the increasing dosing profile shows a significant enhancement of the intermediate selectivity. The corresponding low local concentration of oxygen is responsible for the suppression of further oxidation to CO and CO_2, respectively. Due to the non-optimal condition that the total amount of dosed oxygen was taken the same in all reactor configurations, the rapid increase of local oxygen concentration, especially in the

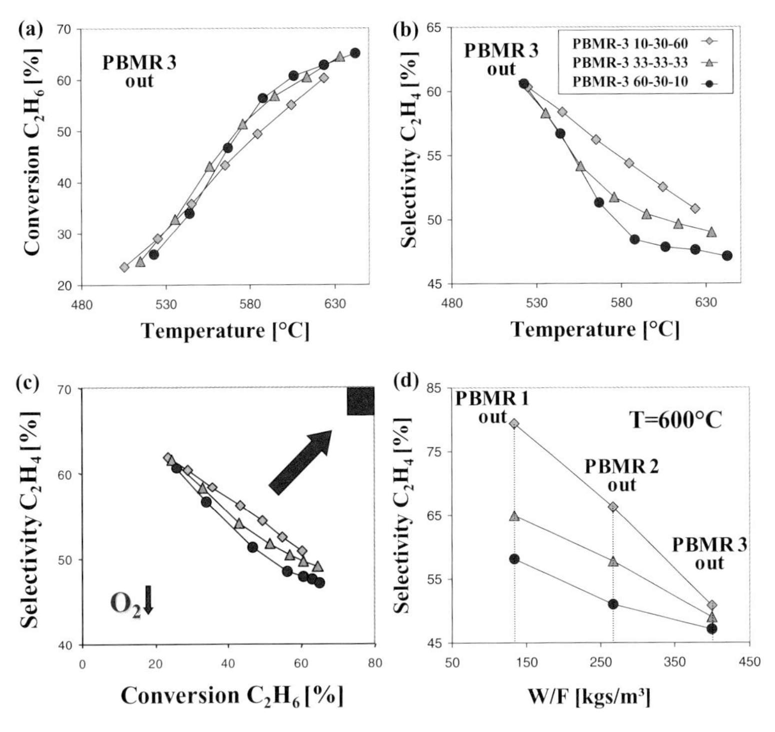

Figure 5.14 a) Conversion of ethane versus temperature, b) selectivity of ethylene versus temperature, c) selectivity of ethylene versus ethane conversion, d) selectivity of ethylene at the outlet of every single membrane reactor in the cascade. $xC_2H_6^{in} = 1.5\%$, $O_2/C_2H_6 = 1$, $W/F = 400\,kg\,s/m^3$.

third stage, impairs the ethylene selectivity due to undesired consecutive reactions. Nevertheless, the trend of an enhanced selectivity obtained by the increasing dosing profile is obvious.

5.5.5
Quantitative Comparison of the Investigated Reactor Configurations

Table 5.3 gives a short summary with respect to performance parameters obtained from the experimental investigation of the ODH of ethane to ethylene using ceramic membranes in single-stage and multi-stage operation.

It can be seen that a distributed dosing of oxygen via membranes leads to a significant increase of the ethylene selectivity from 46.3% (PBR) to 52.3% (PBMR) applying a single-stage operation. Further improvement of reactor performance is possible using a multi-stage reactant dosing and applying an increasing dosing profile. In particular the investigated increasing dosing profile (PBMR-3; 10/30/60%) revealed an enhanced selectivity of ethylene and simultaneously a

Table 5.3 Performance parameter of all investigated reactor configurations: T = 610 °C, $xC_2H_6^{in}$ = 1.5%, O_2/C_2H_6 = 1, W/F = 400 kg s/m^3.

	Conversion C_2H_6	Selectivity C_2H_4	Yield C_2H_4	Yield CO	Yield CO_2
Configuration	[%]	[%]	[%]	[%]	[%]
PBR	55.4	46.3	25.6	21.2	5.6
PBMR[a)]	50.1	52.3	26.2	17.6	5.1
PBMR-3 10/30/60%	60.2	52.5	31.6	20.0	6.1
PBMR-3 33/33/33%	60.5	50.1	30.3	22.8	6.7
PBMR-3 60/30/10%	61.8	47.8	29.5	24.7	7.0

a) Ceramic membrane, γ Al_2O_3 10 nm.

higher ethane conversion. Thus, a maximal ethylene yield of 31.6% was observed in the PBMR-3 for the considered conditions (PBR 25.6%).

The improved reactor performance is due to the favorable change of the local concentration and residence time profiles in the PBMR, which is more pronounced in a membrane reactor cascade.

5.6 Results for the Oxidative Dehydrogenation of Propane

5.6.1 Comparison Between PBR and a Single-Stage PBMR Using Ceramic Membranes

As a second and from the industrial point of view more interesting reaction, the oxidative dehydrogenation (ODH) of propane to propylene on the same VO_x/γ-Al_2O_3 catalyst was investigated experimentally. For the propane system a detailed analysis of the reaction network was recently given by Liebner, 2003. Figure 5.15a,b compares the performance of a single-stage PBMR using a ceramic membrane and the conventional PBR in a pilot-scale for lean oxygen conditions (O_2/C_3H_8 = 1), and Figure 5.15c,d does the same for excess oxygen conditions (O_2/C_3H_8 = 4).

In comparison to the ODH of ethane a similar trend with respect to the selectivity can be obtained (Figure 5.15). However, propylene selectivity depends stronger on temperature. Thus, the selectivity of propylene is decreasing strongly with increasing temperature as well as conversion, nearly independently from the investigated oxygen concentration. Nevertheless, under lean oxygen conditions (Figure 5.15a,b) and at high contact times, significantly higher propylene selectivity was obtained. This trend is more pronounced for temperatures above 450 °C, where the influence of series reactions is stronger. The latter can be avoided by a distributed dosing via membranes. In contrast to the operation window at lean oxygen conditions the performance of PBMR and PBR is comparable for high

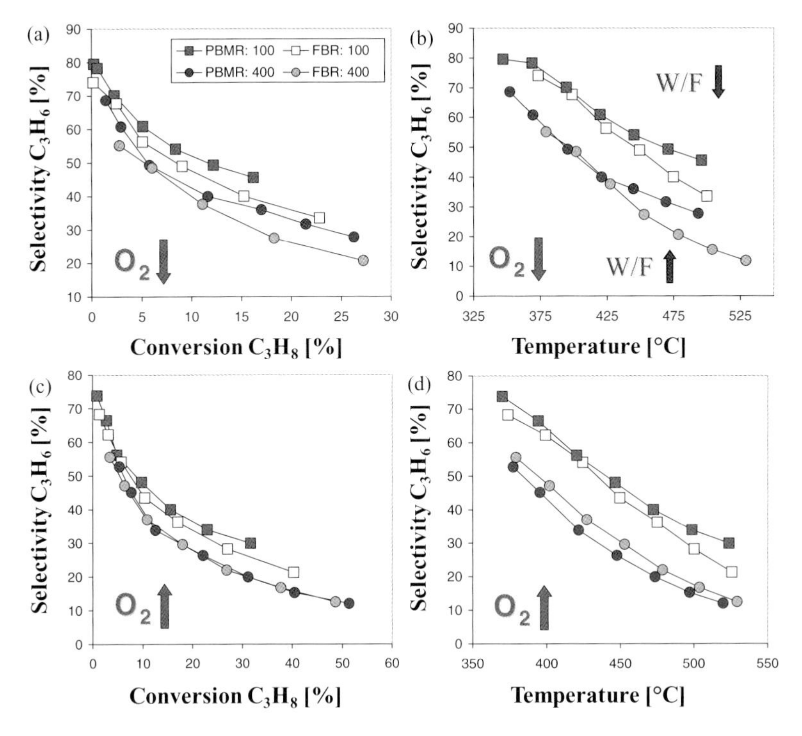

Figure 5.15 a), c) Selectivity of propylene versus conversion. b), d) Selectivity of propylene versus temperature. $xC_3H_8^{in} = 1\%$, $O_2/C_3H_8 = 1O_2\downarrow$ and $4O_2\uparrow$, $W/F = 100\text{–}400\,\text{kg s/m}^3$.

contact times ($W/F = 400\,\text{kg s/m}^3$) and excess oxygen conditions. In this operation region the application of a PBMR is not favorable.

In general the dosing concept via a PBMR is less efficient than for the ODH of ethane discussed above.

5.6.2
Investigation of a Three-Stage Membrane Reactor Cascade

Figure 5.16 summarizes selected results of the experimental investigations of the cascade for the ODH of propane. Regarding propane conversion, depicted in Figure 5.16a, the results of all considered dosing profiles are close together. The dependency with respect to the local oxygen concentration influenced by the dosing profile is even less pronounced than for the ODH of ethane. Nevertheless, an enhancement of the propylene selectivity of approximately 9% was obtained for the increasing dosing profile (10/30/60%) at higher reaction temperatures (Figure 5.16b).

Figure 5.16c gives the propylene selectivity along the three-stage membrane reactor cascade. In this case the increasing dosing profile shows the highest selectivity along the whole cascade length. Due to the less pronounced sensitivity with respect to oxygen, compared to the ODH of ethane the strong increase of

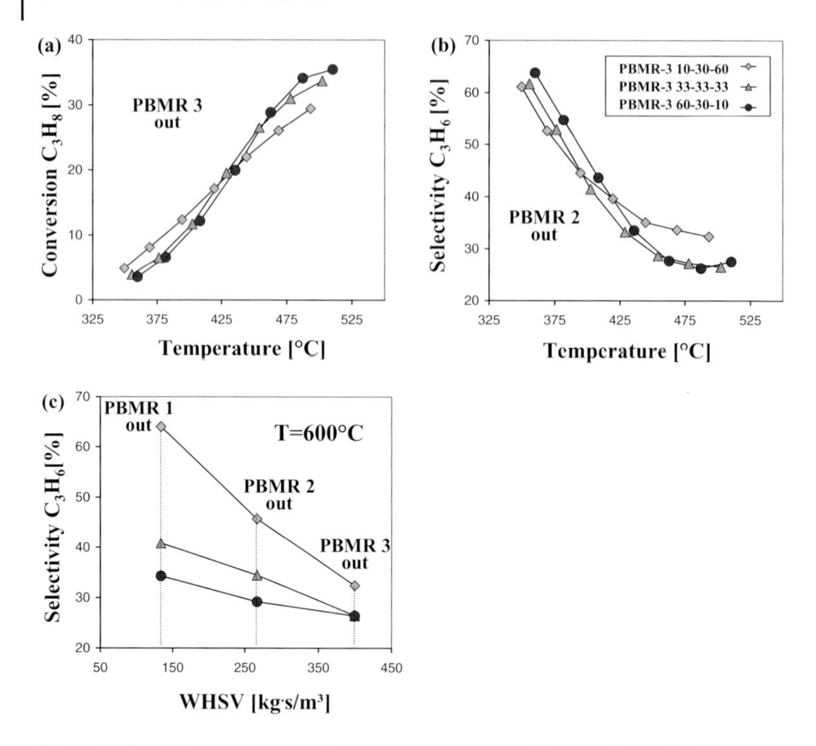

Figure 5.16 a) Propane conversion versus temperature, b) propylene selectivity versus temperature, c) propylene selectivity at the outlet of every single membrane reactor in the cascade. $xC_3H_8^{in} = 1\%$, $O_2/C_3H_8 = 1$, $W/F = 400\,kg\,s/m^3$.

local concentration in the third stage does not affect overall propylene selectivity significantly.

5.7
Summary and Conclusions

Based on an optimal distributed dosing of reactants and the resulting concentration and residence time effects, the product spectrum can differ significantly in a packed-bed membrane reactor compared to a conventional fixed-bed reactor.

A detailed experimental investigation of single- and multi-stage membrane reactors as well as a conventional fixed-bed reactor concept was carried out in a pilot-scale reactor set-up, allowing the experimental validation of theoretical results.

At low oxygen concentrations the experiments using ceramic membranes in a single-stage PBMR indicate that the selectivity of the desired product ethylene can be increased significantly for simultaneously high ethane conversions compared to a conventional fixed-bed reactor (PBR: 46.3%; PBMR: 52.3%). Experiments carried out using sintered metal membranes revealed a comparable performance for high trans-membrane fluxes. Due to the back-diffusion of reactants in the study, the more promising operating region, at higher contact times and lean oxygen conditions, could not be realized with these membranes. In this case a more favorable relation between catalyst activity and membrane permeability (i.e., more active catalyst) would be beneficial.

The application of a multi-stage dosing concept, by means of an increasing oxygen dosing profile, reveals a pronounced increase of ethane conversion as a result of the higher residence time of the reactants for the conditions investigated. Thus, a maximal ethylene yield of 31.6% was observed in the membrane reactor cascade, in comparison to the yield of 25.6% in the reference concept (PBR). The improved reactor performance is due to the favorable change of the local concentration and residence time profiles by a distributed dosing, which is significantly more pronounced in a membrane reactor cascade.

The obtained results for the ODH of propane are comparable to those of the ODH of ethane, even though the increase in propylene selectivity is not so pronounced as in the case of ethylene. Propylene yield could be further enhanced using a multi-stage reactant feeding with an increasing dosing profile.

For the mathematical description of a packed-bed membrane reactor, models of different complexity were applied. Even a relatively simple 1-D pseudo-homogeneous model was able to reflect the main tendencies in reactor behavior. However, resulting from the simplifying assumptions made, more accurate predictions can be achieved only at high trans-membrane flow rates and oxygen excess conditions. The main difference compared to the detailed 2-D model is neglected radial local oxygen distribution applying the 1-D model. Thus, the predicted ethane conversions overestimate the obtained experimental data. Especially for high contact times, botheth-ane conversion and ethylene selectivity could be described in a very good agreement with the detailed "$\lambda(r)$ model" taking account of the radial oxygen concentration profile and radial porosity distribution of the catalyst bed. The results of the performed theoretical analysis of PBMR and PBR revealed optimal operation conditions for the PBMR, that is, a maximum ethane conversion and ethylene selectivity, respectively, at higher residence times and at lean oxygen concentration.

Future work should be devoted to further improve the compatibility between membrane and reaction as well as catalyst to exploit even better the potential of a distributed dosing of reactants in order to compensate the higher investment costs of the membrane reactor. Hereby potential is seen in particular in the application of sintered metal membranes.

Special Notation not Mentioned in Chapter 2

Latin Notation

A	m^2	cross-sectional area
BC		boundary condition
F_T	m^3/s	total volumetric flow rate
N_R		number of reactions

s		stage number in the cascade
S		total number of stages
$S_{diff}^{C_2H_4}$		differential selectivity of ethylene
TS/SS		ratio of tube side to shell side
V_R	m^3	reactor volume
W/F	$kg\,s/m^3$	weight of catalyst/total flow rate
Z	m	total length of the cascade

Greek Notation

ξ	Dimensionless length

References

Al-Juaied, M.A., Lafarga, D., and Varma, A. (2001) Ethylene epoxidation in a catalytic packed-bed membrane reactor: experiments and model. *Chem. Eng. Sci.*, **56**, 2.

Alonso, M., Lorences, M.J., Pina, M.P., and Patience, G.S. (2001) Butane partial oxidation in an externally fluidized bed-membrane reactor. *Catal. Today*, **67**, 151–157.

Bey, O. (1998) Strömungsverteilung und Wärmetransport, Universität Stuttgart, Dissertation.

Caro, J. (2006) Membranreaktoren für die katalytische Oxidation. *Chem. Ing. Tech.*, **78**, 899–912.

Caro, J., Schiestel, T., Werth, S., Wang, H., Kleinert, A., and Kölsch, P. (2006) Perovskite hollow fibre membranes in the partial oxidation of methane to synthesis gas in a membrane reactor. *Desalination*, **199**, 415–417.

Cheng, P., and Vortmeyer, D. (1988) Transverse thermal dispersion and wall channeling in a packed-bed with forced convective flow. *Chem. Eng. Sci.*, **43**, 9.

COMSOL (2006) *Handbook, Comsol Multiphysics 3.2*, 1997–2007 COMSOL, Inc.

Coronas, J., Menendez, M., and Santamaria, J. (1995) Use of a ceramic membrane reactor for the oxidative dehydrogenation of ethane to ethylene and higher hydrocarbons. *Ind. Eng. Chem. Res.*, **34**, 12.

Diakov, V., Lafarga, D., and Varma, A. (2001) Methanol oxidative dehydrogenation in a catalytic packed-bed membrane reactor. *Catal. Today*, **67**, 1–3.

Diakov, V., and Varma, A. (2004) Optimal feed distribution in a packed-bed membrane reactor: the case of methanol oxidative dehydrogenation. *Ind. Eng. Chem. Res.*, **43**, 309–314.

Edgar, T.F., and Himmelblau, D.M. (1988) *Optimization of Chemical Processes*, McGraw-Hill Education, ISBN: 0070189927.

Ergun, S. (1952) Fluid flow through packed columns. *Chem. Eng. Prog.*, **35**, 89–94.

Farrusseng, D., Julbe, A., and Guizard, C. (2001) Evaluation of porous ceramic membranes as O2 distributors for the partial oxidation of alkanes in inert membrane reactors. *Sep. Purif. Technol.*, **25**, 137–149.

Grabowski, R. (2006) Kinetics of oxidative dehydrogenation of C2-C3 alkanes on oxide catalysts. *Catal. Rev.*, **48**, 199–268.

Hamel, C., Thomas, S., Schädlich, K., and Seidel-Morgenstern, A. (2003) Theoretical analysis of reactant dosing concepts to perform parallel-series reactions. *Chem. Eng. Sci.*, **58**, 4483–4492.

Hamel, C., Seidel-Morgenstern, A., Schiestel, T., Werth, S., Wang, H., Tablet, C., and Caro, J. (2006) Experimental and modeling study of the O_2-enrichment by perovskite fibers. *AIChE J.*, **52**, 3118–3125.

Hein, S. (1999) *Modellierung wandgekühlter katalytischer Festbettreaktoren mit Ein- und Zweiphasenmodellen*, VDI-Verlag, Düsseldorf, ISBN: 3-18-359303-3, Dissertation, München.

Hodnett, B.K. (2000) *Heterogeneous Catalytic Oxidation : Fundamental and Technological Aspects of the Selective and Total Oxidation of Organic Compounds*, John Wiley and Sons (Asia) Pte Ltd, ISBN: 0-471-48994-8.

Hunt, M.L., and Tien, C.L. (1990) Non-darcian flow, heat and mass-transfer in catalytic packed-bed reactors. *Chem. Eng. Sci.*, **45**, 1.

Hussain, A. (2006) Heat and mass transfer in tubular inorganic membranes, Magdeburg, Dissertation.

Hussain, A., Seidel-Morgenstern, A., and Tsotsas, E. (2006) Heat and mass transfer in tubular ceramic membranes for membrane reactors. *Int. J. Heat Mass Transf.*, **49**, 2239–2253.

Joshi, M. (2007) Statistical analysis of models and parameters in chemical and biochemical reaction networks, Otto von Guericke University Magdeburg, Dissertation.

Klose, F., Wolff, T., Hamel, C., Alandjiyska, M., Weiß, H., Joschi, M., Tota, A., and Seidel-Morgenstern, A. (2003a) Partial oxidation of ethane- the potential of membrane reactors. *J. Univ. Chem. Technol. Metall.*, **38**, 25–40.

Klose, F., Wolff, T., Thomas, S., and Seidel-Morgenstern, A. (2003b) Concentration and residence time effects in packed bed membrane reactors. *Catal. Today*, **82**, 1–4.

Klose, F., Joshi, M., Hamel, C., and Seidel-Morgenstern, A. (2004a) Selective oxidation of ethane over a VOx/Al_2O_3 catalyst investigation of the reaction network. *Appl. Catal. A Gen.*, **260**, 101–110.

Klose, F., Joshi, M., Hamel, C., and Seidel-Morgenstern, A. (2004b) Selective oxidation of ethane over a VOx/gamma-Al2O3 catalyst—investigation of the reaction network. *Appl. Catal. A Gen.*, **260**, 1.

Kölsch, P., Noack, M., Schäfer, R., Georgi, G., Omorjan, R., and Caro, J. (2002) Development of a membrane reactor for the partial oxidation of hydrocarbons: direct oxidation of propane to acrolein. *J. Membr. Sci.*, **198**, 119–128.

Kürten, U., van Sint Annaland, M., and Kuipers, J.A.M. (2004) Oxygen distribution in packed-bed membrane reactors for partial oxidations: effect of the radial porosity profiles on the product selectivity. *Ind. Eng. Chem. Res.*, **43**, 1823–1835.

Lafarga, D., Santamaria, J., and Menéndez, M. (1994) Methane oxidative coupling using porous ceramic membrane reactors – I. reactor development. *Chem. Eng. Sci.*, **49**, 2005–2013.

Levenspiel, O. (1999) *Chemical Reaction Engineering*, John Wiley & Sons Inc.

Liebner, C. (2003) Einführung der Polythermen Temperatur Rampen Methode für die Ermittlung kinetischer Daten, Universität Berlin, Dissertation.

Liebner, C., Wolf, D., Baerns, M., Kolkowski, M., and Keil, F.J. (2003) A high-speed method for obtaining kinetic data for exothermic or endothermic catalytic reactions under non-isothermal conditions illustrated for the ammonia synthesis. *Appl. Catal. A Gen.*, **240**, 95–110.

Lu, Y.P., Dixon, A.G., Moser, W.R., and Ma, Y.H. (1997a) Analysis and optimization of cross-flow reactors for oxidative coupling of methane. *Ind. Eng. Chem. Res.*, **36**, 3.

Lu, Y.P., Dixon, A.G., Moser, W.R., and Ma, Y.H. (1997b) Analysis and optimization of cross-flow reactors with distributed reactant feed and product removal. *Catal. Today*, **35**, 4.

Lu, Y.P., Dixon, A.G., Moser, W.R., and Ma, Y.H. (1997c) Analysis and optimization of cross-flow reactors with staged feed policies = isothermal operation with parallel series, irreversible reaction systems. *Chem. Eng. Sci.*, **52**, 8.

Lu, Y., Dixon, A.G., Moser, W.R., and Ma, Y.H. (2000) Oxidative coupling of methane in a modified g-alumina membrane reactor. *Chem. Eng. Sci.*, **55**, 4901–4912.

Mallada, R., Menéndez, M., and Santamaría, J. (2000a) Use of membrane reactors for the oxidation of butane to maleic anhydride under high butane concentrations. *Catal. Today*, **56**, 191–197.

Mallada, R., Pedernera, M., Menendez, N., and Santamaria, J. (2000b) Synthesis of maleic anhydride in an inert membrane reactor. Effect of reactor configuration. *Ind. Eng. Chem. Res.*, **39**, 3.

Mason, E.A., and Malinauskas, A.P. (1983) *Gas Transport in Porous Media: The Dusty-Gas Model*, Elsevier, Amsterdam, ISBN: 0-444-42190-4.

Morbidelli, M., Gavriilidis, A., and Varma, A. (2001) *Catalyst Design: Optimal Distribution of Catalyst in Pellets, Reactors and Membranes*, Cambridge University Press Cambridge.

Ramos, R., Menendez, M., and Santamaria, J. (2000) Oxidative dehydrogenation of propane in an inert membrane reactor. *Catal. Today*, **56**, 1–3.

Schäfer, R., Noack, M., Kolsch, P., Stohr, M., and Caro, J. (2003) Comparison of different catalysts in the membrane-supported dehydrogenation of propane. *Catal. Today*, **82**, 1–4.

Seidel-Morgenstern, A. (2005) *Analysis and Experimental Investigation of Catalytic Membrane Reactors, in Integrated Chemical Processes : Synthesis, Operation, Analysis, and Control*, Wiley-VCH Verlag GmbH, Weinheim, ISBN: 3-527-30831-8.

Sheldon, R.A., and van Santen, R.A. (1995) *Catalytic Oxidation: Principles and Applications -A Course of the Netherlands Institute for Catalysis Research*, World Scientific Publishing, Catalysis Research.

Tellez, C., Menendez, M., and Santamaria, J. (1997) Oxidative dehydrogenation of butane using membrane reactors. *AIChE J.*, **43**, 3.

Thomas, S. (2003) Kontrollierte Eduktzufuhr in Membranreaktoren zur Optimierung der Ausbeute gewünschter Produkte in Parallel- und Folgereaktionen, Dissertation.

Thomas, S., Klose, F., and Seidel-Morgenstern, A. (2001) Investigation of mass transfer through inorganic membranes with several layers. *Catal. Today*, **67**, 205–216.

Tonkovich, A.L.Y., Jimenez, D.M., Zilka, J.L., and Roberts, G.L. (1996a) Inorganic membrane reactors for the oxidative coupling of methane. *Chem. Eng. Sci.*, **51**, 11.

Tonkovich, A.L.Y., Zilka, J.L., Jimenez, D.M., Roberts, G.L., and Cox, J.L. (1996b) Experimental investigations of inorganic membrane reactors: a distributed feed approach for partial oxidation reactions. *Chem. Eng. Sci.*, **51**, 789–806.

Tota, A., Hamel, C., Thomas, S., Joshi, S., Klose, F., and Seidel-Morgenstern, A. (2004) Theoretical and experimental investigation of concentration and contact time in membrane reactors. *Chem. Eng. Res. Des.*, **82**, 236–244.

Tota, A., Hamel, C., Klose, F., Tsotsas, E., and Seidel-Morgenstern, A. (2006) Enhancement of intermediate product selectivity in multi-stage reactors-Potential and Pitfalls, ISCRE 19, Potsdam, Poster 275.

Tota, A., Hlushkou, D., Tsotsas, E., and Seidel-Morgenstern, A. (2007) Packed Bed Membrane Reactors, in *Modelling of Process Intensification* (Herausgeber: ed. F. Keil), Wiley-VCH Verlag GmbH, Weinheim, ISBN: 3-527-31143-2 99, pp. 99–149.

Tsotsas, E. (2002) *VDI-Wärmeatlas*, Springer, Berlin, p. 1366.

Tsotsas, E., and Schlünder, E.-U. (1990) Heat transfer in packed beds with fluid flow: remarks on the meaning and the calculation of a heat transfer coefficient at the wall. *Chem. Eng. Sci.*, **45**, 4.

Wang, H., Tablet, C., Schiestel, T., and Caro, J. (2006) Hollow fiber membrane reactors for the oxidative activation of ethane. *Catal. Today*, **118**, 98–103.

Westerterp, K.R., van Swaaij, W.P.M., and Beenackers, A.A.C.M. (1984) *Chemical Reactor Design and Operation*, John Wiley & Sons, Ltd, ISBN: 0471901830.

Winterberg, M., and Tsotsas, E. (2000a) Correlations for effective heat transport coefficients in beds packed with cylindrical particles. *Chem. Eng. Sci.*, **55**, 23.

Winterberg, M., and Tsotsas, E. (2000b) Impact of tube-to-particle-diameter ratio on pressure drop in packed beds. *AIChE J.*, **46**, 5.

Winterberg, M., Tsotsas, E., Krischke, A., and Vortmeyer, D. (2000) A simple and coherent set of coefficients for modelling of heat and mass transport with and without chemical reaction in tubes filled with spheres. *Chem. Eng. Sci.*, **55**, 5.

Zeng, Y., Lin, Y.S., and Swartz, S.L. (1998) Perovskite-type ceramic membrane: synthesis, oxygen permeation and membrane reactor performance for oxidative coupling of methane. *J. Membr. Sci.*, **150**, 87–98.

Ziaka, Z.D., Minet, R.G., and Tsotsis, T.T. (1993) Propane dehydrogenation in a packed-bed membrane reactor. *AIChE J.*, **39**, 3.

6
Fluidized-Bed Membrane Reactors

Desislava Tóta, Ákos Tóta, Stefan Heinrich, and Lothar Mörl

6.1
Introduction

Many membrane reactor applications, which lead to higher educt conversion and yields of intermediates, are related to packed-bed configurations (Marcano and Tsotsis, 2002; Saracco *et al.*, 1999; Dixon, 2003). However, packed-bed membrane reactors (PBMR) also possess the disadvantage of large temperature gradients for highly exothermic reactions, which may lead to catalyst coking, further undesirable side reactions and reactor instability.

These disadvantages can be largely overcome by integrating the membranes inside a fluidized bed, referred to as a fluidized-bed membrane reactor (FLBMR). A FLBMR is a special type of reactor that combines the advantages of a fluidized bed and a membrane reactor. One of the main advantages of the fluidized-bed reactor (FLBR) is the excellent tube-to-bed heat transfer, which allows a safe and efficient reactor operation even for highly exothermic reactions. The intense macro-scale solids mixing induced by the rising bubbles results in a remarkable temperature uniformity, even in beds as large as 10 m (Miracca and Capone, 2001). Other advantages of FLBRs are low pressure drop and the potential to control the catalyst activity due to continuous processing, for example, catalyst regeneration in circulating fluidized beds. There are a few drawbacks compared to the fixed-bed reactors – gas back-mixing, bypassing and erosion of reactor internals and catalyst attrition – which can hinder the development of the fluid-bed technology.

By the insertion of membranes in a fluidized bed a synergistic effect can be accomplished. First, optimal concentration profiles can be created via controlled dosing or withdrawal; and second, the fluidization behavior can be improved via the presence of inserts and permeation of the gas through the membranes, so that large improvements in conversion and selectivity might be achieved.

Some first concepts of fluidized-bed reactors with controlled reactant distribution used nozzles for secondary gas injection. Mleczko *et al.* (1994) investigated the oxidative coupling of methane to C_{2+} hydrocarbons. To improve the mass transfer by reduced bubble size, the oxygen was injected via secondary distributor

Membrane Reactors: Distributing Reactants to Improve Selectivity and Yield.
Edited by Andreas Seidel-Morgenstern

ISBN: 978-3-527-32039-4

in the fluidized bed. The product selectivity and yield were increased in part of the operation range. A significant influence of the secondary gas injections was approved.

A concept for separate reactant dosing was proven by Soler *et al.* (1999) for the oxidative dehydrogenation of *n*-butane to butadiene by separation of the oxidation and the reduction zones in the same catalytic fluidized bed. The butadiene yield was about 200% higher than in the conventional reactor.

Al-Sherehy, Grace, and Adris (2005) investigated the effect of the discrete distributing gaseous feed along a bubbling fluidized bed on the ethane partial oxidation to ethylene and acetic acid. An improvement of the product selectivity and the reactor performance was achieved by using multiple nozzles for the secondary gas injection.

Recent developments in the material sciences, and especially the development of suitable inorganic membranes which are chemically resistant and stable at high temperatures, offer the integration of membranes into catalytic reactors and a new reactor design. Only a limited number of applications of the membrane-assisted FLBR for the distributive feeding of one of the reactants have been investigated and most of these applications involve the controlled dosing of reactant via porous membranes. Alonso *et al.* (2001) studied the butane partial oxidation in an externally FLBMR. Air from a pre-heated fluidized bed (catalytic inert) flows across a catalyst-filled membrane tube. A detailed reactor model demonstrated a minimized hot spot effect and maleic anhydride yields were predicted to be 50% higher compared to a conventional fixed-bed reactor. However, later experimental data (Alonso *et al.*, 2005) show that, on average, the overall performance in the conventional fixed bed is superior to the membrane configuration because selectivity is higher and the reactor may be operated at higher temperatures resulting in superior MA yields.

Ramos *et al.* (2001) proposed a similar concept for the partial oxidation of propane to propene. Air fluidized the shell side where catalyst-filled membrane tubes and cooling coils were immersed. Oxygen transport through the membrane was controlled by the pressure drop. The controlled oxygen addition along the axis improved the propene selectivity and broadened the operating range with respect to the hydrocarbon and oxygen feed rates.

Recently, Deshmukh *et al.* (2005a, 2005b) constructed a small laboratory-scale FLBMR for the partial oxidation of methanol to formaldehyde. High methanol conversion and high selectivity to formaldehyde were achieved with safe reactor operation (isothermal reactor conditions) at higher methanol inlet concentrations than that currently employed in industrial processes.

An area of much current interest is the production of hydrogen via methane steam reforming or autothermal reforming. Fluidized-bed concepts were proposed to solve the problems of thermal control encountered in fixed-bed reactors. These studies showed that, for steam reforming of natural gas, the thermodynamic equilibrium restrictions can be overcome by *in situ* separation and removal of hydrogen via perm-selective thin-walled palladium-based membranes, leading to

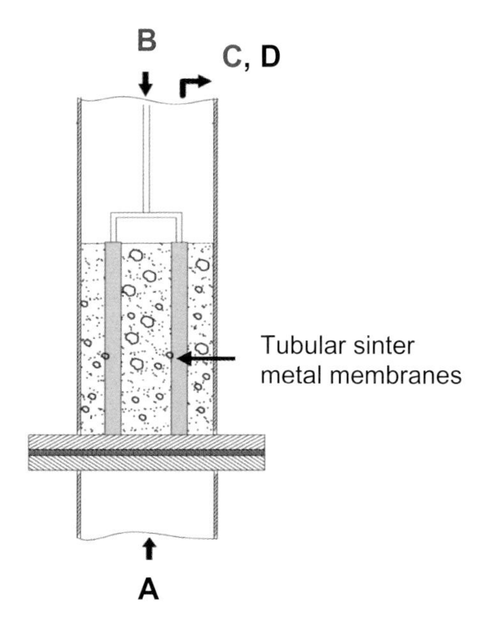

Figure 6.1 Schematic of the dosing concept of the FLBMR.

increased synthesis gas yields in comparison to the industrial fixed-bed steam reformer (Chen, Yan, and Elnashaie, 2003; Patil, Annaland, and Kuipers, 2005; Adris, Lim, and Grace, 1997).

This short overview demonstrates the wide field of applications and the crucial role of the reaction system. A more comprehensive review on the potential and hurdles of FLBMRs is given by Deshmukh *et al.* (2007).

The present chapter studies experimental and theoretical analysis of the potential of the FLBMR concept (continuous distributed reactant dosing) for the same reaction as introduced in Chapter 3, that is, the oxidative dehydrogenation of ethane to ethylene on a $VO_x/\gamma\text{-}Al_2O_3$ catalyst in comparison to a FLBR with co-feed reactant. In the FLBMR concept one of the reactant (oxygen) was distributed via vertically inserted tubular membranes in a bubbling FLBR of catalysts (Figure 6.1).

The performance of the reactor was investigated in comparison to the conventional FLBR under various experimental conditions.

This chapter gives an overview of the FLBMR by theoretical and experimental study. The next section presents a developed phenomenological reactor model and simulation results which illustrate the influence of some model parameters and attractive ranges for the new reactor concept. In the second part, experimental results demonstrate the benefits and limitations of this reactor concept compared to co-feed reactant dosing in a conventional FLBR.

6.2 Modeling of the Distributed Reactant Dosage in Fluidized Beds

6.2.1 Theory

A theoretical study has been carried out to describe the effects of segregated reactant dosage on the reactor performance in fluidized beds and to identify attractive parameter ranges for reactor operation.

Due to the complex fluid dynamics of the fluidized bed, in this work a phenomenological model was applied. It is well known for a fluidized bed that bubbles, particle cluster and clouds greatly influences the reactor performance. Their characteristics can be changed by internals (Jin, Wei, and Wang, 2003) or by secondary gas injection (Al-Sherehy, Grace, and Adris, 2004; Deshmukh, 2004; Christensen, Coppens, and Nijenhuis, 2005). Therefore, this simulation study should show the uncertainties and limits of phenomenological models of the FLBMR.

A heterogeneous two-phase model has been derived and numerically solved for the simulation of the co-feed dosage and of the distributed reactant dosage in fluidized beds via vertically installed membranes. The model describes the advective-dispersive mass transfer in the suspension and bubble phases, using effective dispersion coefficients (dispersion model). By means of a schematic representation, the main features, notation and assumptions of the model are depicted in Figure 6.2. Important assumptions and features of the model are:

- Steady-state one-dimensional reactor model.
- Ideal gas behavior.
- Distinction between a particle-free bubble phase and a suspension phase with particles; consideration of mass transfer at the phase interface.
- Convective–dispersive mass transfer in the suspension phase.
- Plug flow in the bubble phase.
- The chemical reaction takes place in the suspension phase. Concentration gradients inside the catalyst as well as between catalyst surface and suspension gas are neglected.
- The change of the superficial gas velocity due to the gas supply versus the bed height and the change of the mole fraction due to the reaction are considered.
- The secondary gas can be fed to the bubble phase or distributed between the suspension gas and the bubble phase according to their fractions.
- The suspension gas velocity is constant. Therefore, a unilateral convective mass flow rate is defined.
- Reactions in the freeboard region are neglected.

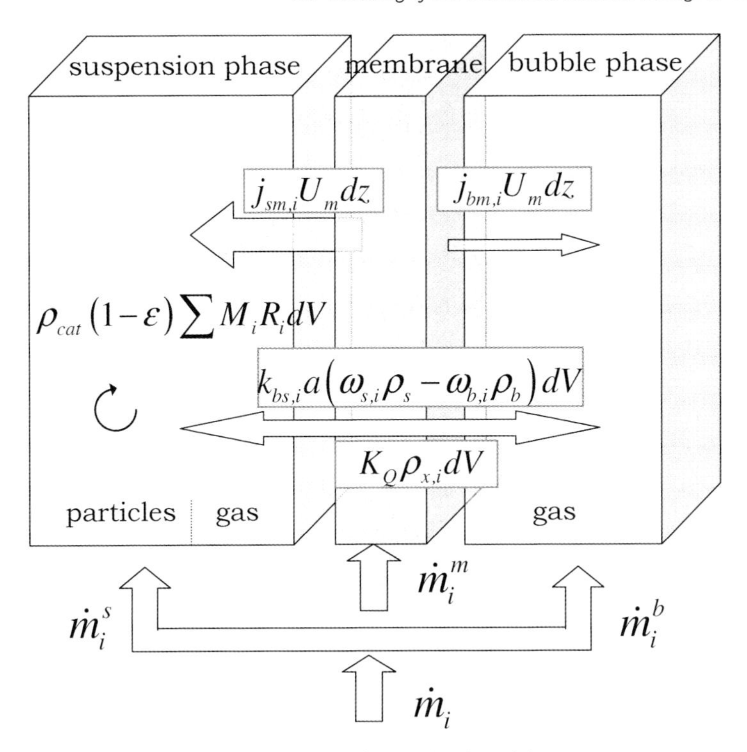

Figure 6.2 Scheme of the two-phase fluidized-bed model.

The resulting mass balances for the suspension and bubble phase are given in Equations 6.1–6.10. In order to describe the mass balances of both phases by means of second-order differential equations an axial dispersion coefficient has been defined for the balances of the bubble phase. The introduction of this "artificial" diffusion term serves solely numerical purposes. Therefore, the axial dispersion coefficient has been set to a very low value of 10^{-5}–$10^{-6}\,m^2/s$. The rate of the reactions considered leading to the component specific source terms R_i are described in detail in Chapter 3.

- Mass balance of a component i in the suspension phase:

$$\frac{\partial \omega_{s,i}\rho_s u_s(1-\varepsilon_b)}{\partial z} = -\frac{\partial}{\partial z}\left[(1-\varepsilon_b)D_{ax,i}^{s}\frac{d\omega_{s,i}\rho_s}{dz}\right] + \rho_{Cat}(1-\varepsilon_s)(1-\varepsilon_b)M_i R_i + \frac{(1-\varepsilon_b)\dot{m}_m}{L_m A_{app}} - k_{bs,i}a(\omega_{s,i}\rho_s - \omega_{b,i}\rho_b) - K_Q\rho_{Q,i} \tag{6.1}$$

- Mass balance of a component i in the bubble phase:

$$\frac{\partial \omega_{b,i}\rho_b u_b \varepsilon_b}{\partial z} = -\frac{\partial}{\partial z}\left[(1-\varepsilon_b)D_{ax,i}^{b}\frac{d\omega_{b,i}\rho_b}{dz}\right] + \frac{\varepsilon_b \dot{m}_m}{L_m A_{app}} + k_{bs,i}a(\omega_{s,i}\rho_s - \omega_{b,i}\rho_b) + K_Q\rho_{Q,i} \tag{6.2}$$

- Total mass balance of the bubble phase:

$$\frac{\partial(\rho_b u_b \varepsilon_b)}{\partial z} = \sum_i [k_{bs,i} a(\omega_{s,i}\rho_s - \omega_{b,i}\rho_b)] + \frac{\varepsilon_b \dot{m}_m}{L_m A_{app}} + K_Q \rho_Q \tag{6.3}$$

The superficial gas velocity u is obtained solving the continuity Equation 6.4.

- Total mass balance:

$$\rho u A_{app} = \rho_s u_s (1-\varepsilon_b) A_{app} + \rho_b u_b \varepsilon_b A_{app} \tag{6.4}$$

The superficial gas velocity changes due to: (a) the gas supply over the bed height and (b) the chemical reaction. Therefore, a convective mass flow is considered to keep the suspension gas velocity constant. In order to estimate the mass transfer coefficient, the total mass flow rate in the differential volume element dV of the suspension phase must be balanced:

$$\frac{d\dot{m}_s}{dz} dz = -K_Q \rho_Q dV - \sum_i [k_{bs,i} a(\omega_{s,i}\rho_s - \omega_{b,i}\rho_b)] dV + \frac{(1-\varepsilon_b)\dot{m}_m}{L_m} dz, \tag{6.5}$$

with $\dot{V}_Q = K_Q \rho_Q dV$ as the volume flow, which is transferred into the bubble or suspension phase. From the ideal gas law follows:

$$p\dot{V}_s = \frac{\dot{m}_s}{M_s} RT \tag{6.6}$$

$$\frac{d\dot{m}_s}{dz} = \rho_s \frac{d\dot{V}_s}{dz} + \dot{V}_s \frac{d\rho_s}{dz} \tag{6.7}$$

The gas volume flow in the suspension phase is:

$$\dot{V}_s = u_s (1-\varepsilon_b) A_{app} \tag{6.8}$$

or rather:

$$\frac{d\dot{V}_s}{dz} = -A_{app} u_s \frac{d\varepsilon_b}{dz} \tag{6.9}$$

By inserting Equations 6.9 and 6.7 in Equation 6.5, the mass transfer coefficient can be calculated from:

$$K_Q \rho_Q = \rho_s u_s \frac{d\varepsilon_b}{dz} - u_s(1-\varepsilon_b)\frac{d\rho_s}{dz} - \sum_i [k_{bs,i} a(\omega_{s,i}\rho_s - \omega_{b,i}\rho_b)] + \frac{(1-\varepsilon_b)\dot{m}_m}{L_m A_{app}} \tag{6.10}$$

The first two terms in Equation 6.10 describe mass flow rate induced by the change of the local gradients of the bubble volume fraction and the suspension gas density. The convective mass transfer between the phases is expressed by the summand, while the last term describes the mass flow of the component, which is distributed via the membranes.

The composition of the above-defined convective stream is affected by the suspension phase. When the mass transfer coefficient K_Q is positive, the density ρ_Q is equal to the density of the suspension phase ρ_s. If the calculated mass transfer

Table 6.1 Boundary conditions for the two-phase fluidized-bed reactor model.

At z = 0 flux density	At z = H_{bed} convective flux
$(\omega_{b,i}u_b)_0 = -\mathcal{D}_{b,i}\left(\frac{\partial \omega_{b,i}}{\partial z}\right)_{z=0} + \omega_{b,i0}u_b$	$\frac{\partial \omega_{b,i}}{\partial z} = 0$
$(\omega_{s,i}u_s)_0 = -\mathcal{D}_{s,i}\left(\frac{\partial \omega_{s,i}}{\partial z}\right)_{z=0} + \omega_{s,io}u_s$	$\frac{\partial \omega_{S,i}}{\partial z} = 0$

coefficient K_Q is negative, the convective mass transfer occurs from the bubble phase to the suspension phase and has the composition of the bubble phase.

For the solution of aforementioned second-order differential equation system, two boundary conditions must be defined. Modeling both of the reactors as a closed system, Dankwert's boundary condition at inlet and outlet are defined (Table 6.1). The overall mass continuity equation for calculation of the local gas velocities can be obtained by summing up the individual component balances in the several phases. The resulting system of ordinary differential equations is solved simultaneously to the component balance equations.

For the solution of the coupled differential equation system, the commercial simulation tool Femlab® 3.1 (Comsol AB, Stockholm) was used. With this package, the differential equation system was discretized by means of the finite element method. The resulting ordinary nonlinear equation system was solved using a UMFPACK algorithmus at each Newton iteration step.

In phenomenological models of fluidized-bed reactors, parameters like bubble diameter and bed porosity have to be considered, which again depend on the fluid dynamics at the particular operation point. The hydrodynamic quantities needed were calculated using empirical or semi-empirical approximations (Table 6.2).

6.2.2 Parametric Sensitivity of the Model

6.2.2.1 Bubble Size

By the use of membranes in a fluidized bed optimal concentration profiles can be realized via controlled dosing; further the fluidization behavior can be improved via internals and secondary gas permeation through the membranes. Internals in a fluidized bed reduce the reactor equivalent diameter and the bubble size; therefore, conversion can be increased by better gas–solid mass transfer (Volk, Johnson, and Stotler, 1962; Kunii and Levenspiel, 1992). Kleijn van Willigen *et al.* (2005) found a significant decrease in the bubble size and bubble hold-up due to the distributed secondary gas supply. Different investigations in fluidized beds show a lower effective axial dispersion by means of the installation of tubular membranes (Deshmukh, 2004) or by the injection of secondary gas (Al-Sherehy, Grace, and Adris, 2004; Deshmukh, 2004; Christensen, Coppens, and Nijenhuis, 2005). This results in fluid dynamics with plug flow characteristics.

Table 6.2 Parameters used in the model.

Mass transfer coefficient between bubble and suspension phase (Werther and Schössler, 1999)	$k_{sb}a_b = \frac{6\varepsilon_b}{d_v}\left[\frac{u_{mf}}{3} + 0.95(\mathcal{D}_{mi}\varepsilon_{mf})^{0.5}\left(\frac{g}{d_v}\right)^{0.25}\right]$	(6.11)
Bubble size equation (Hilligardt, 1986)	$\frac{d(d_v)}{dz} = \left(\frac{2\varepsilon_b}{9\pi}\right)^{1/3} - \frac{d_v}{3\left(280\frac{u_{mf}}{g}\right)u_b}$	(6.12)
Bubble phase fraction	$\varepsilon_b = \frac{v_{rel}(u-u_{mf})}{u_b}$, where: $v_{rel} = 0.67$ for $\frac{H_{bed}}{d_{app}} \leq 1.7$, or: $v_{rel} = 0.51\sqrt{\frac{H_{bed}}{d_{app}}}$ for $1.7 \leq \frac{H_{bed}}{d_{app}} \leq 4.0$	(6.13)
Bubble rise velocity (Hilligardt and Werther, 1986)	$u_b = v_{rel}(u-u_{mf}) + 0.71\cdot\delta\sqrt{g\cdot d_v}$, where $\delta = 3.2D_a^{0.33}$	(6.14)
Axial dispersion coefficient (Lee and Kim, 1989)	$Pe_a = \frac{u_0 D_a}{\mathcal{D}} = 0.02566\left(\frac{u_0}{u_{mf}}\right)^{-0.54}\left(\frac{\rho_p}{\rho_g}\right)^{0.067}\left(\frac{d_p}{\mathcal{D}}\right)^{-0.588}$	(6.15)
Binary diffusion coefficient by Fuller *et al.* (Poling, Prausnitz, and O'Connell, 2001)	$\mathcal{D}_{ij} = \frac{0.01013T^{1.75}\left(\frac{1}{M_i}+\frac{1}{M_j}\right)^{0.5}}{P\left[\left(\sum \upsilon_i\right)^{1/3}+\left(\sum \upsilon_j\right)^{1/3}\right]^2}$	(6.16)
Multi-component diffusion coefficient by Wilke (Poling *et al.*, 2001)	$\mathcal{D}_{im} = \frac{1-y_i}{\sum_{j, j\neq 1} y_i/\mathcal{D}_{ij}}$	(6.17)

Despite intensive research, there were no correlations to describe back-mixing and bubble growth in fluidized beds with internals and secondary gas injection. Thus, in our model, the calculation of the size and rise velocity of the bubbles is based on empirical correlations obtained in fluidized beds without internals and without secondary gas injection. In order to evaluate the sensitivity of the simulation results regarding bubble size, the empirical correlation for the bubble growth rate (6.12) was modified by multiplying the calculated values with a factor K_{dv}. Since the aforementioned studies report small bubbles in fluidized beds with internals and secondary gas injection, the factor K_{dv} was varied between 0.1 and 1.2. All hydrodynamic quantities were calculated according to the correlations given in Table 6.2. Figure 6.3 shows the results for oxygen concentrations at 1 and 5 vol% at a secondary to primary gas flow ratio of $F_s/F_p = 1$.

The simulation results can be interpreted as mass transfer limitations between the bubble and the suspension phase due to the decrease of the total bubble surface area with increasing bubble size. Additionally, the larger bubbles rise faster over

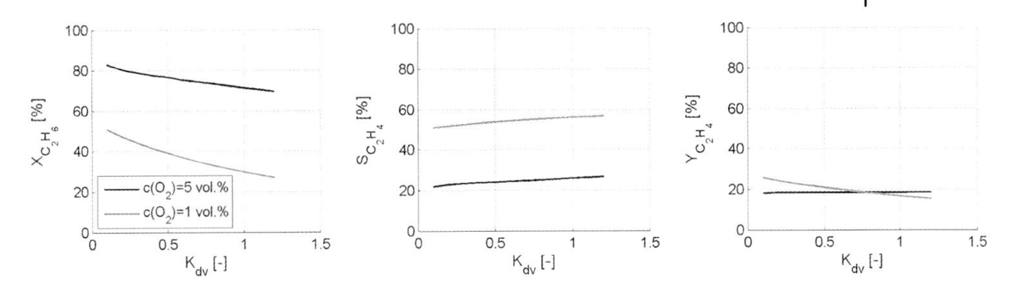

Figure 6.3 Variation of the bubble diameter in the FLBMR: $c(C_2H_6) = 1$ vol%, $u = 0.3$ m/s, $m_{Cat}/F = 1000$ kg·s/m^3, $F_s/F_p = 1.0$, $T = 590$ °C.

the bed and have a shorter contact time. Assuming that the reaction takes place only in the suspension phase, the ethane conversion decreases, because a large fraction of the reactants is captured in the bubble phase. An important fact is that ethane conversion is more sensitive than ethylene selectivity concerning this factor. In particular, this is pronounced in the range of low O_2 concentrations, which is characterized by low driving forces in the suspension phase, resulting in limitations of the reaction rate. It must be pointed out that the ethane is fed at the reactor inlet (i.e., in both phases), whereas the oxygen is transferred into the suspension phase only via the bubbles. A small amount of co-fed oxygen could therefore improve the ethane conversion.

For a detailed analysis of bubble growth in the FLBMR, further experimental studies are necessary in order to describe the effects of reactor geometry and gas supply on the hydrodynamics. Recently, *in silico* measurements, based on more sophisticated modeling approaches like, for example, the discrete particle method or the Lattice–Boltzmann method are considered to fill the lack of reliable correlations (Bokkers *et al.*, 2006).

6.2.2.2 **Secondary Gas Distribution**

There are different assumptions about the mechanism of the secondary gas distribution in a fluidized bed. Christensen, Coppens, and Nijenhuis (2005) found that the secondary injection reduces bubble size and bubble fraction. Deshmukh *et al.* (2005a, 2005b) consider that the secondary gas is distributed between the bubble and suspension gases, although they could not find clear evidence for this hypothesis in their experiments. A hydrodynamic investigation of the distribution of increased gas volume due to reaction (Adris, Lim, and Grace, 1993) showed that at least some of the additional flow ends up in the bubble phase. However, it is not clear how long the additional moles generated stay in the dense phase or whether they transfer very quickly to the bubble phase.

In our simulations, shown in Figure 6.4, two cases are compared at different oxygen concentrations, in order to evaluate the magnitude of uncertenity of the different assumptions. In "case a" a secondary gas is transferred only into the bubble phase. Through mass transfer between both phases the oxygen is

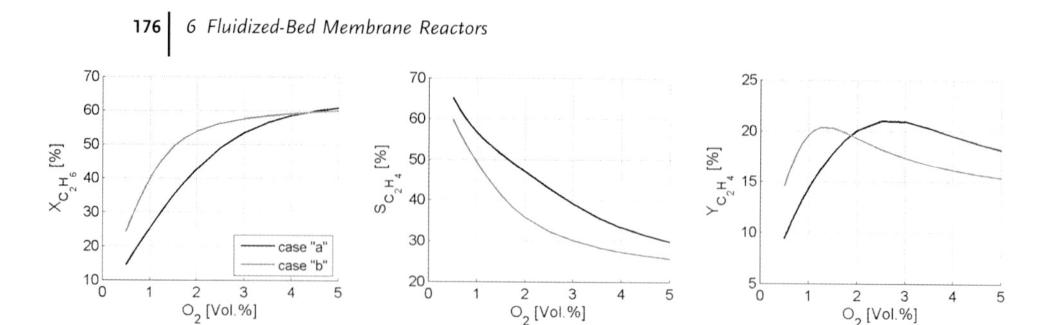

Figure 6.4 Influence of the distribution of the secondary gas between bubble and suspension phase. Case "a": secondary gas is transferred only into the bubble phase. Case "b": secondary gas is distributed into both phases. $c(C_2H_6) = 1$ vol%, $u = 0.45$ m/s, $m_{Cat}/F = 700$ kg·s/m^3, $F_s/F_p = 1.0$, $T = 590$ °C.

transported into the suspension phase. "Case b" assumes that the secondary gas is injected into both phases according to their fractions. Both scenarios assume that the gas velocity through the suspension phase is constant. The excess gas is transferred by convection into the bubbles.

The simulated ethane conversion and ethylene selectivity in Figure 6.4 demonstrate a large difference (up to 15%) between the two scenarios. In case "a", where the oxygen level in the suspension phase is significantly influenced by mass transfer limitations between the bubble and suspension phases, ethane conversion increases more slowly with increasing oxygen amount in the feed than in "case b" with direct oxygen distribution to bubbles and suspension.

The higher oxygen to ethane ratio in case "b" explains the significant enhancement of ethane conversion and loss of ethylene because of total oxidation to carbon oxides. These tendencies disagree with our experimental observations (see Section 6.3.4), in which the ethane conversion rises slow with increasing oxygen partial pressure. Therefore we assumed in the next simulation that the secondary gas is fed into the bubbles and subsequently the oxygen is transfered to the suspension phase. The empirical correlations used were neither fitted to the experimetal data nor modified. Nevertheless, this dosing concept needs further investigation regarding the fluidized-bed hydrodynamics.

6.2.3
Comparison Between Co-Feed and Distributed Oxygen Dosage

In the following simulation study the reactor concepts of co-feed and distributed oxygen dosage are compared with regard to achievable ethane conversions, ethylene selectivity and ethylene yield. The goal of this study was to identify attractive operating ranges for the application of FBMRs and the estimation of reactor performance in comparison to a conventional FLBR. The calculations were carried out with both reactor models at a constant temperature of 590 °C, as in Chapter 5. In order to cover a wide operation range, the oxygen concentration was varied between 0.5 and 5.0 vol% and the contact time was adjusted between 700 and 1875 kg s/m^3. The ethane concentration was kept constant at 1 vol% in all calculations.

Figure 6.5 shows the simulated ethane conversion, ethylene selectivity and yield for both reactor concepts. Below 2 vol% O_2 the conversion strongly depends on the concentration. In this range, the model predicts a slow rise of ethane conversion in the FLBMR. This can be explained with the lower local oxygen concentration.

The ethane and the oxygen profiles of both dosing concepts are depicted in Figure 6.6.

The chemical reaction in the suspension and the mass transfer limitations between the phases define the lower mole fractions in the suspension. In the FLBMR the low oxygen level results from feed distribution via the membranes; however a large amount of oxygen is captured in the bubbles and does not reach the reaction zone. This hinders the ethane conversion and enhances the ethylene selectivity due to the low oxygen to ethane ratio.

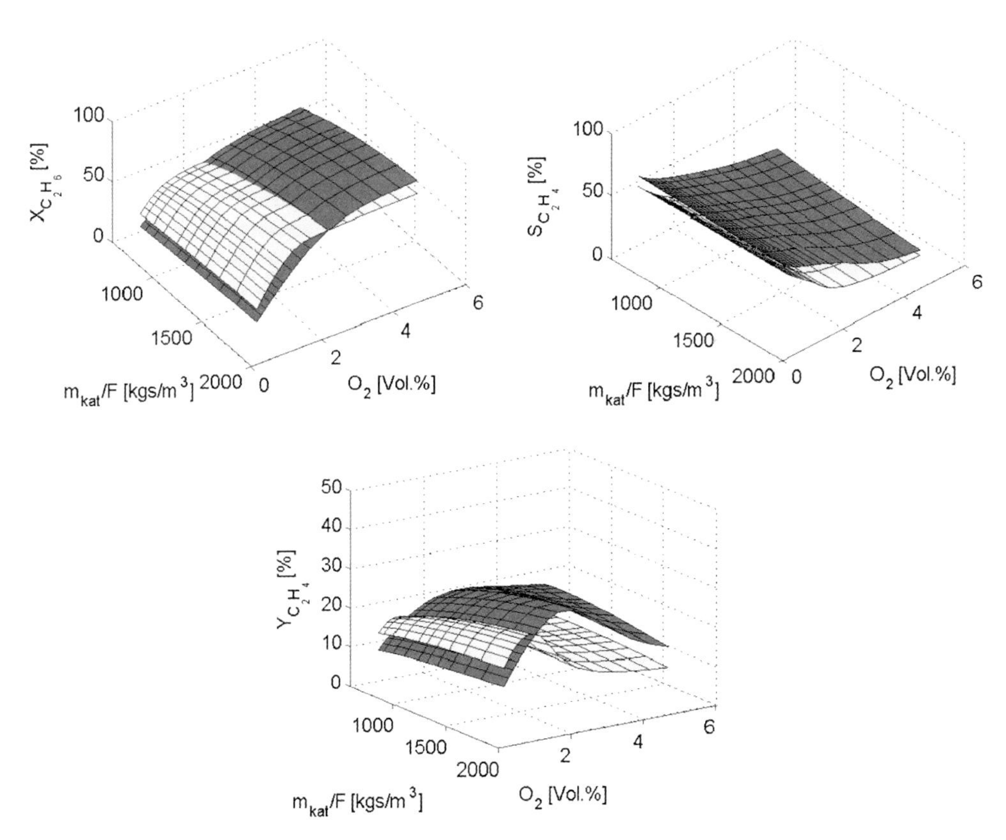

Figure 6.5 Simulation study for comparison of FLBR and FLBMR (gray: FLBR; black: FLBMR). $c(C_2H_6)$ = 1 vol%, $c(O_2)$ = 0.5–5.0 vol%, T = 590 °C, m_{Cat}/F = 700–1875 kg·s/m^3, F_s/F_p = 1.0.

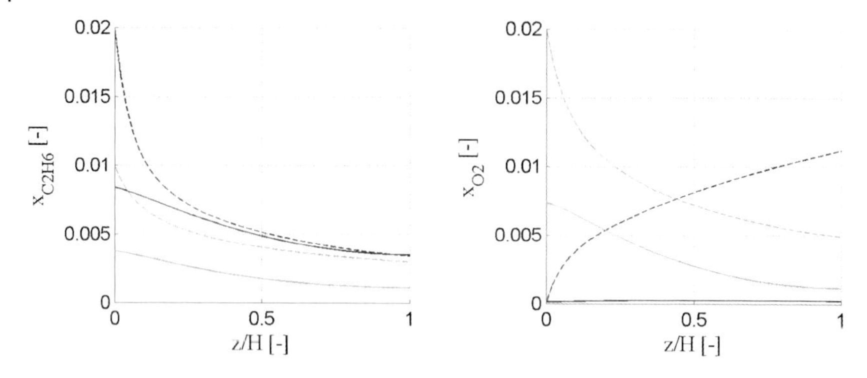

Figure 6.6 Simulated profiles of oxygen and ethane in the FLBR (gray) and in the FLBMR (black). Solid line: suspension phase; dashed line: bubble phase. $c(O_2)$ = 2 vol%, $c(C_2H_6)$ = 1 vol%, T = 590 °C, m_{Cat}/F = 1875 kg·s/m^3, F_s/F_p = 1.0.

The reactor comparison was made at identical volumetric flow rate, thus the local gas velocity in the FLBR is always higher than in the FLBMR. Because mass transfer limitations between bubble and suspension phase increase with higher fluidisation velocity due to increasing bubble size and bubble rise velocity, the dependency of ethane conversion on contact time is slightly higher for the FLBR.

As expected, the ethylene selectivity decreases with increasing oxygen concentration. The lower oxygen partial pressure in the FLBMR results in much higher ethylene selectivity compared to co-feeding. Considering the ethylene yield, the FLBMR is less sensitive regarding the oxygen concentration, which results in a broader yield maximum.

The simulation results demonstrate a better performance of the distributed reactant feeding in a wide parameter range, compared to the conventional co-feed reactor. However, at very low oxygen concentrations a higher ethylene yield should be expected in the FLBR.

6.3 Experimental

6.3.1 Experimental Set-Up

The experiments for both reactor configurations, FLBMR and FLBR, were performed in the same pilot plant, depicted in Figures 6.7 and 6.8.

It consists of a gas-dosing unit providing the fluidization gas (1) and the permeation gas through the membranes (2), the pilot-scale reactor (3) with the catalyst and the membrane (4), a cyclone (5) and a washer (6) at the reactor outlet to collect particles from catalyst attrition and to avoid a carryover of fines into the

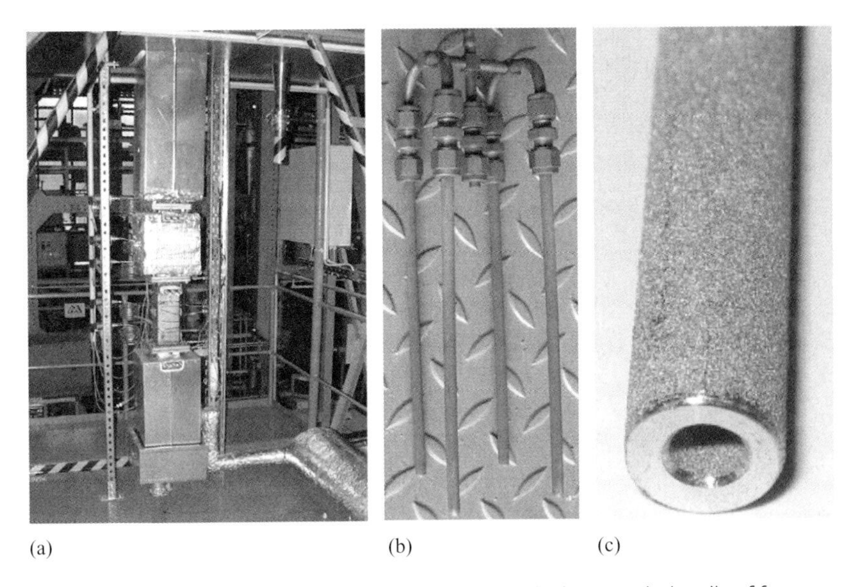

Figure 6.7 Photo of the experimental set-up: a) fluidized-bed reactor, b) bundle of four tubular sinter metal membranes, c) single tubular sinter metal membrane.

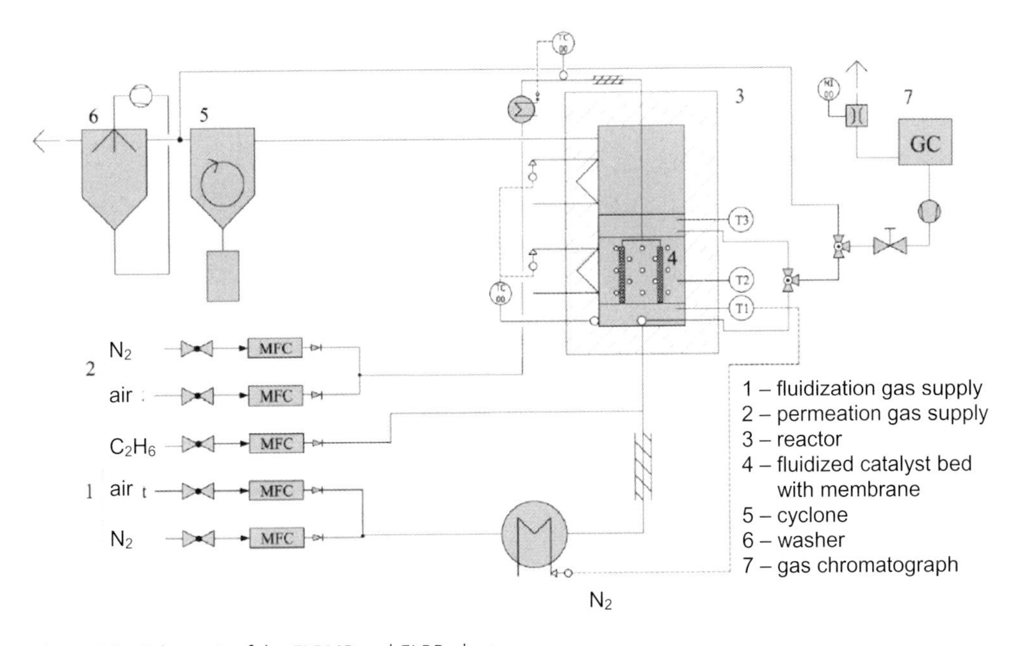

Figure 6.8 Schematic of the FLBMR and FLBR plant.

environment (see Figure 6.8). Reactant and product gas streams are analyzed by gas chromatography (7) The gas-dosing unit is equipped with electronic mass flow controllers allowing preparation of the desired compositions of reactant gases in the different experiments. Nitrogen and ethane had a purity of 99.998 and 99.95%, respectively. Air (dried) was provided by a compressor. The mixed fluidization gas could be heated by a separate electrical pre-heater. In the case of FLBMR measurements C_2H_6/N_2 mixtures were used for fluidization and the oxidant (air or an air/nitrogen mixture) was dosed in a distributed manner over the installed sintered metal membranes. During the FLBR experiments no membrane was installed inside the reactor and the complete reactant mixture was fed as fluidization gas.

The reactor was made of stainless steel (AlCr, Sicromal) and had an inner cross-section of 100×100 mm and a total height of 2.1 m. The preheated fluidization gas entered the reaction zone through a sintered metal plate (Inconel) with an average pore diameter of 20 μm and 43% porosity. A spiral wound electrical heater was used to heat the reactor walls. Additionally, reactor walls were insulated with a ceramic fiber banket. The membranes (sintered porous tubular stainless steel membrane with dead-end configuration, GKN Sinter Metals Filters), used as an inert oxidant distributor in the FLBMR, had a length of 165 mm, inner and outer diameters of 3 and 6 mm, respectively, and an average pore diameter of 8 μm. The metal membrane was selected because of its better thermal stability and mechanical strength, compared to ceramic membranes. The four membrane tubes were vertically mounted in the reaction zone directly above the gas distributor. The oxidant as permeation gas was fed from the top of the tube in the opposite direction to the fluidization gas. Temperature was measured by NiCr/Ni thermocouples at three different vertical positions within the reactor: at the inlet, at the center of fluidized catalyst bed and at the outlet of the reaction zone. Gas samples were taken at the reactor inlet and downstream of the cyclone, using a small membrane pump and were analyzed using a gas chromatograph. The gas chromatograph (HP6890, Agilent Technologies) operated with helium as carrier gas and was equipped with an TCD and an FID; two capillary columns (PoraPlot Q and Plot-Molesieve) allowed the separation of ethane, ethylene, CO, CO_2, O_2 and N_2. The carbon balance was accurate typically within ±3%.

The γ-alumina supported vanadium oxide catalyst was applied as bed particles with a diameter of 0.44 mm and a particle density of 1316 kg/m^3. Details about the preparation process are given by Klose *et al.* (2004b).

6.3.2
Experimental Procedure

The experiments in both operation modes were performed at atmospheric pressure. The reactor was preheated with air as a fluidizing gas until the desired temperature was reached. Subsequently, the air was replaced by nitrogen and the appropriate reactant mixture. Measurement of the gas samples was carried out after 20–60 min, after a steady state was reached. Every measurement was repeated three times. The absence of wall and gas phase reactions was tested with the empty reactor up to 630 °C and with inert α-Al_2O_3 particles up to 600 °C; and less than 2% ethane conversion was found. The temperature distribution in the catalyst bed was measured in axial and radial directions under reaction conditions in the FLBR and in the FLBMR. The temperature differences in the reaction zone were below 2–3 K for both reactor types and showed isothermal conditions in the catalyst bed.

6.3.3
Results and Discussions

The performance of both dosing strategies – co-feed and segregated reactant distribution – was studied by varying the oxygen concentration, the temperature and the contact time (superficial gas velocity). The ratio of the secondary gas to primary gas was a additional operation parameter in the FLBMR, which influenced the local concentration as well as the residence time profiles and was studied by feeding different gas amounts through the membranes. Separate experiments were carried out applying a mixed dosing strategy with simultaneous supply of oxygen at the reactor inlet and through the membranes.

6.3.4
Influence of the Oxygen Concentration

To show the influence of the local distribution of the reactants the oxygen concentration was varied for both configurations (FLBR, FLBMR) in the range of 0.6–3.7 vol%, while the ethane concentration was kept constant (1 vol%). The experiments were performed at a temperature of 590 °C and a contact time of 1750 kg s/m^3. To increase the oxygen concentration in the FLBR, nitrogen was stepwise replaced by air at a constant total gas volume flow. In the FLBMR only the composition of the secondary gas was varied, to keep a constant secondary to primary gas flow ratio. Both ethane conversion and the selectivity and yield of ethylene are compared in Figure 6.9.

For the FLBR one can distinguish between two oxygen concentration ranges: The first one with low oxygen concentration (0.6–1.5 vol%) is characterized by a distinct ethane conversion dependency. A small increase in the O_2 concentration leads to a significant conversion enhancement. Close to the stoichiometric ratio of the oxidative dehydrogenation of ethane (ODHE), the reaction rate depends strongly on the oxygen concentration. Above 1.5 vol% the influence of oxygen concentration on conversion decreases. Compared to the FLBR, the FLBMR shows no conversion plateau. Over the wide range of studied oxygen concentrations, conversion in the FLBR is higher than that in the FLBMR. To reach higher conversions by distributed dosing, a larger oxygen concentration or a higher O_2/C_2H_6 ratio compared to the FLBR is needed. In contrast, ethylene selectivity is significantly higher in the FLBMR and the loss of ethylene selectivity with increasing oxygen supply (due to higher CO_x formation rates) in the FLBMR is less pronounced than that in the FLBR. Under oxygen-limiting conditions, the main product is ethylene. Above 1.5 vol% the ethylene selectivity in the FLBR is nearly constant. The better

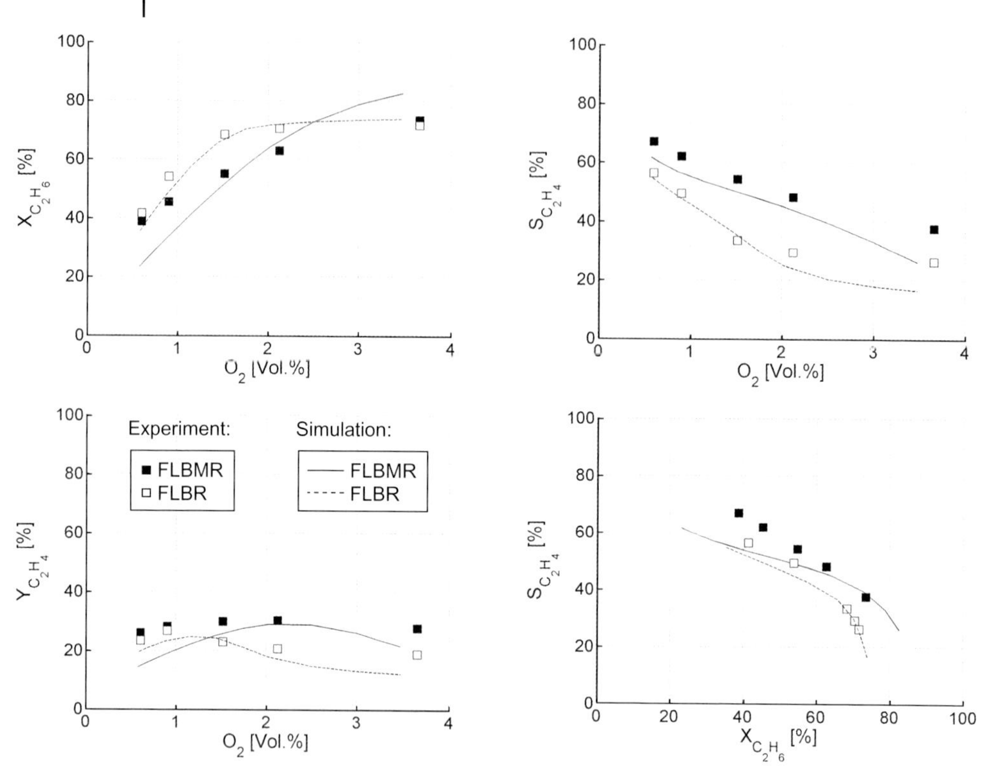

Figure 6.9 Variation of the oxygen concentration. $c(C_2H_6) = 1\,\text{vol}\%$, $T = 590\,°\text{C}$, $m_{\text{Cat}}/F = 1750\,\text{kg·s/m}^3$, $F_s/F_p = 1.0$.

performance of the FLBMR concept is more obvious from the selectivity–conversion diagram. Applying a distributed dosing startegy, 7–8% higher ethylene selectivity was reached at an identical ethane conversion. Considering the ethylene yield a maximum can be observed in both reactor configurations. For the FLBR this maximum appears at 1 vol% O_2, while for the FLBMR the yield maximum is broader between 1.5 and 2.5 vol% O_2 and is less sensitive regarding the oxygen concentration. The higher oxygen amount is needed to compensate for the shorter residence time of the oxygen molecules in the reaction zone. Under oxygen excess conditions the ethylene yield is up to 9% higher in the FLBMR compared to the FLBR. This behavior can be clearly attributed to the decreased average axial oxygen concentration in the FLBMR caused by the distributed oxidant and was also partially observed for packed-bed membrane reactors (Klose *et al.*, 2004a).

These results show a very good agreement between the simulations and the experiments for the FLBR and reproduce the main tendencies for the FLBMR. As described in Section 6.2.1, the discrepancies are the consequence of the model simplifications and the coarse empirical correlation for the fluid dynamic parameters of the fluidized bed.

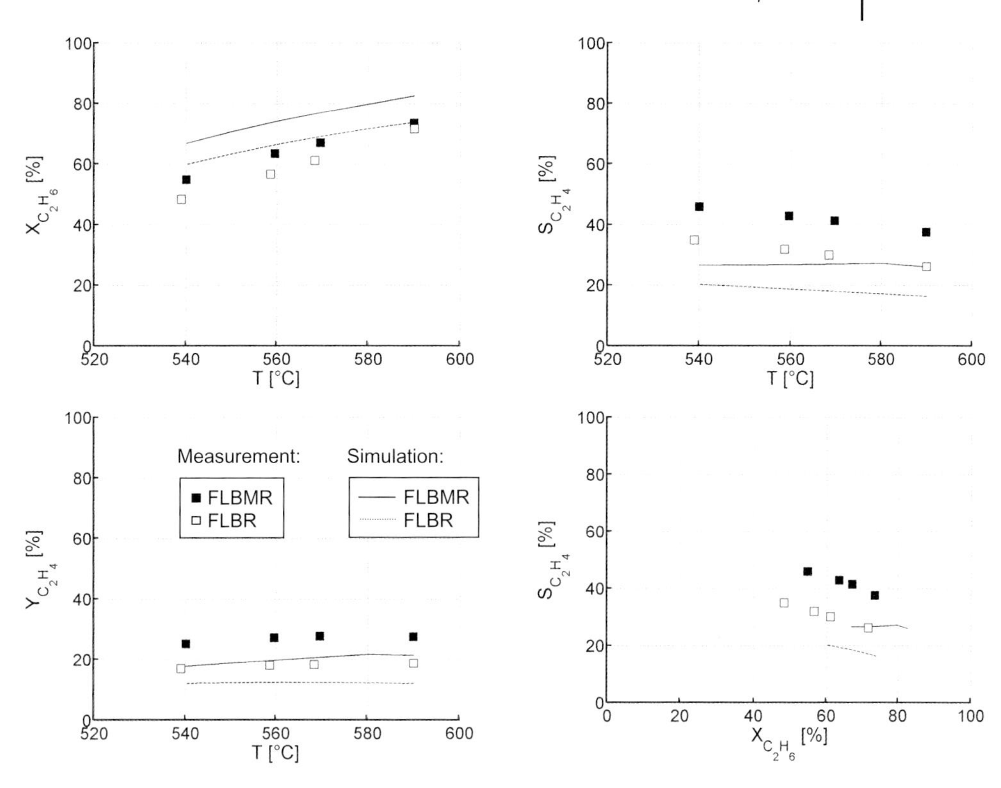

Figure 6.10 Variation of the temperature under oxygen excess conditions. $c(C_2H_6) = 1\,vol\%$, $c(O_2) = 3.7\,vol\%$, $m_{Cat}/F = 1750\,kg{\cdot}s/m^3$, $F_s/F_p = 1.1$.

6.3.5
Influence of the Temperature

Both reactor concepts were compared under oxygen excess conditions in the temperature range between 540 °C and 590 °C. The ethane concentration was 1 vol% and the oxygen concentration was set to 3.7 vol%. The course of the curves in Figure 6.10 show, in contrast to oxygen limiting conditions, a distinct dependency of the reactor performance on the bed temperature.

Comparing the two reactors, ethane conversion is slightly higher in the FLBMR compared to the FLBR (especially at T = 540 °C). Co-feeding (FLBR) shows an increase in ethane conversion from 49 to 72 vol%, while the FLBMR delivers a stronger conversion increase from 55 to 74 vol%. The lower ethane conversion may have different reasons. One reason can be the lower average residence time of the ethane molecules in the FLBR (Tonkovich *et al.*, 1996; Pena *et al.*, 1998, Klose *et al.*, 2004b), but this probably plays no role due to the long contact times. More reasonable is that the higher back-mixing affects and the larger bypass fraction in the FLBR leads to a lower conversion in comparison to the FLBMR, because the higher primary gas flow promotes the formation of larger bubbles. This results in a larger bypass and a reduced mass transfer rate between the suspension and bubble phases in the FLBR. The difference in ethane conversion between the two reactor concepts becomes smaller with increasing temperature.

Higher temperature leads to lower ethylene selectivities in both reactors due to increased carbon oxide formation. However, by distributing the oxygen along the reactor height, total oxidation can be suppressed more efficinetly in the FLBMR for identical overall oxygen amounts in the feed. Furthermore, the lower backmixing in the FLBMR enhances ethylene selectivity. Comparing the concepts at same ethane conversion the ethylene selectivity was 9% higher in the FLBMR over the whole range studied. Remarkable again is the superiority of the membrane reactor regarding ethylene yield. The maximum ethylene yield is 8% higher in the FLBMR (27.8%) in comparison to the FLBR (18.9%).

6.3.6 Influence of the Superficial Gas Velocity

The performance of the fluidized bed depends on the fluid dynamics, which are influenced by the particle properties, bed heights, fluidization medium and fluidization velocity. In order to investigate the influence of the fluid dynamics on reactor performance, varying the fluidization gas velocity for both dosing strategies were carried out between $4u_{mf}$ and $10u_{mf}$ ($u_{mf,500°C} = 0.05\,m/s$). The gas velocities are related to the total gas volume flow. To rule out the influence of other parameters, the primary to secondary gas flow ratio was kept constant. The results of this study are depicted in Figure 6.11.

The tendencies correspond to the results from the literature reported on bubbling fluidized beds (Levenspiel, 1999; Kunii and Levenspiel, 1992). An increase in the gas velocity leads to a decrease in the conversion in both reactors. The FLBR shows higher ethane conversion but has a more pronounced sensitivity regarding the fluidization velocity. Increasing the primary gas flow by a factor of two, ethane conversion decreases by 14% in the FLBR. The conversion loss is only 6% in the FLBMR, which is attributed to the 50% lower primary gas flow and, thus, to the smaller bubble fraction compared to the co-feed FLBR. The latter was also reported by Deshmukh *et al.* (2005a, 2005b). The bubble phase fraction and the bubble size increases with increasing gas velocity and hence reduces the contact time of the reactants as well as the mass transfer rate between the bubble and suspension phases. At higher gas velocities a larger fraction of ethane flows into the bubbles and has therefore no contact with the catalyst surface. Due to the lower oxygen concentration in the feed the effect of the contact time on the ethylene selectivity is negligible. Altough the ethane conversion is lower, the higher selectivity in the FLBMR leads to 4% higher ethylene yield. The more relevant comparison made at identical ethane conversion demonstrates once again the main advantage of the membrane-dosing concept.

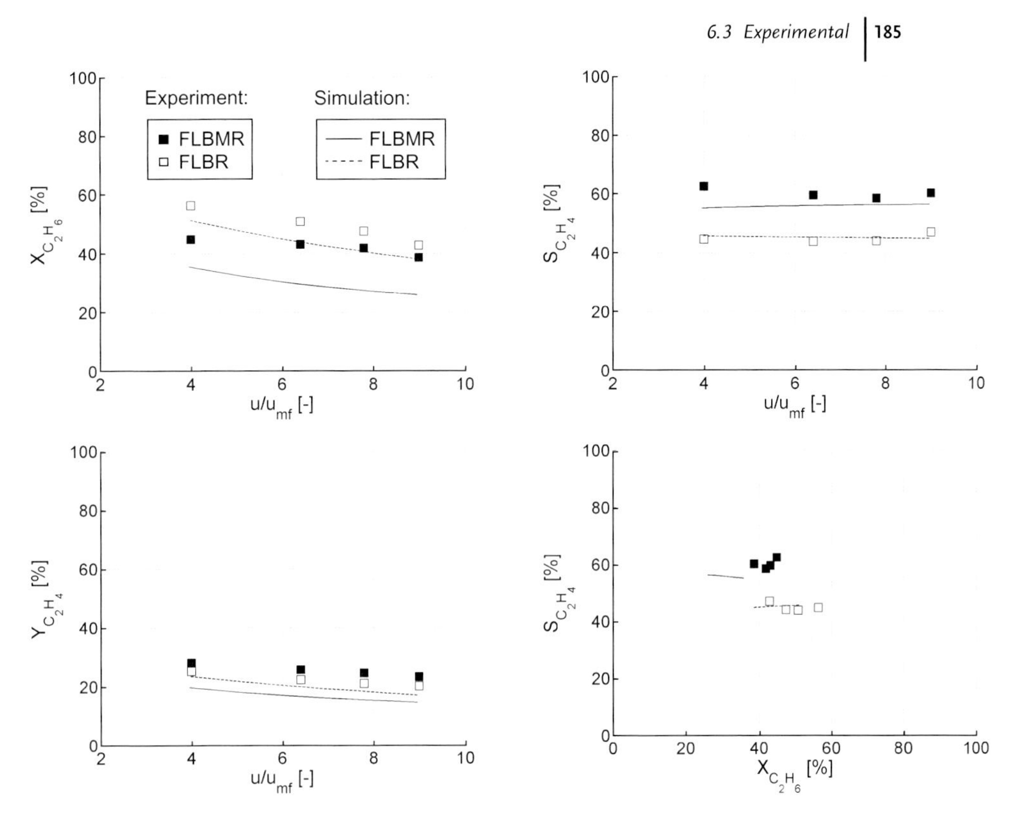

Figure 6.11 Variation of the superficial gas velocity: $c(C_2H_6) = 1$ vol%, $c(O_2) = 1$ vol%, $T = 590\,°C$, $F_s/F_p = 1.1$.

A good agreement between model and experiments could be achieved for the conversion in the FLBR. The differences for the FLBMR could be the result of the correlations used by the calculation of the bubble fractions, which were obtained for fluidized beds without secondary gas injection.

6.3.7
Influence of the Secondary to Primary Gas Flow Ratio

The experimental results above show that remarkable improvements in the ethylene selectivity can be achieved in both reactor concepts by decreasing the oxygen amount in the feed. However, in most cases this results in a decreased ethane conversion. Furthermore, as reported by many researchers, the distributed oxygen dosing may lead to a high hydrocarbon to oxygen ratio close to the reactor inlet, which can result in catalyst deactivation, coking and further undesirable side reactions. The dosage of small amounts of oxygen in the primary gas flow or an increase in the secondary to primary gas flow ratio can counteract these effects. In this section the latter is investigated.

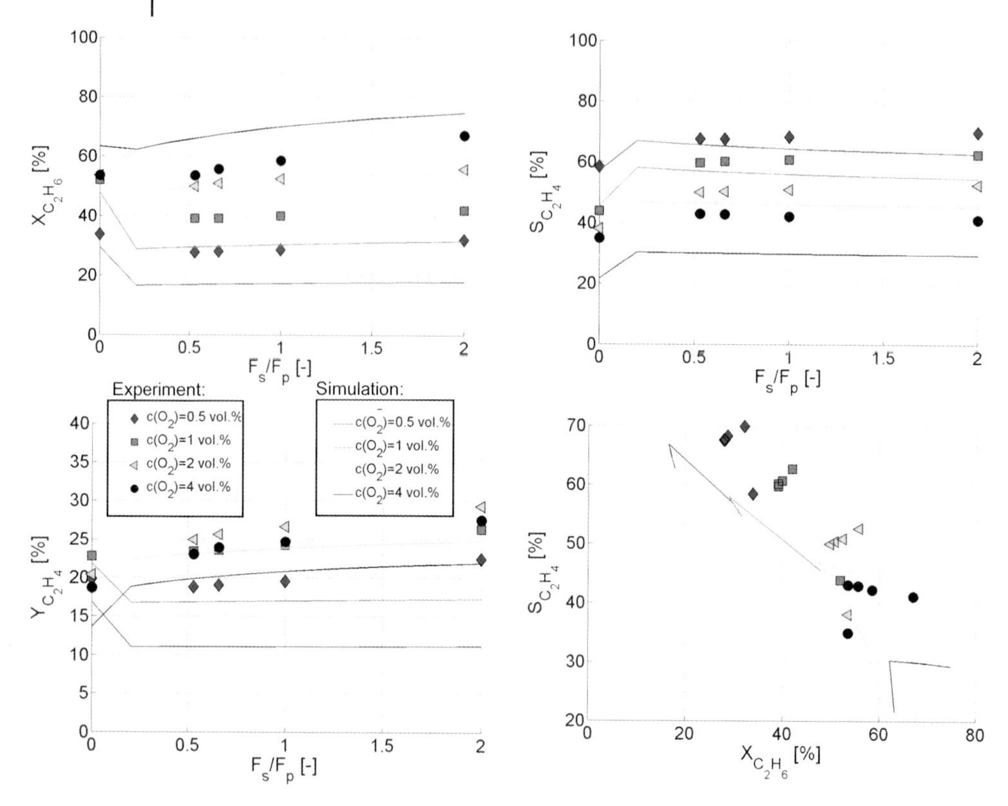

Figure 6.12 Variation of the secondary to primary gas flow ratio: $c(C_2H_6) = 1$ vol%, $T = 590\,°C$, $m_{Cat}/F = 1000\,kg\,s/m^3$.

The theoretical study shows that higher secondary gas fraction leads to an increase in the local oxygen concentration at the reactor inlet, which generates a higher average oxygen concentration throughout the bed. A fluidized bed is also characterized by back-mixing effects, which influences these profiles and which also depends on gas velocity, bed height, catalyst particles, etc.

The results in Figure 6.12 show the tendencies as function of the secondary gas fraction at different O_2 concentrations in a range from 0.5 to 4.0 vol%.

For comparison, the corresponding FLBR results are shown at $F_s/F_p = 0$ (co-feed). The conversion shows the typical increase for all feed ratios with increasing feed concentration. Depending on the oxygen concentration, one can distinguish between two ranges. Under oxygen excess conditions (2 and 4 vol%) the ethane conversion rises with increasing secondary gas fraction. The FLBR shows lower conversions (53%) than the FLBMR (56 and 67%, for 2 and 4 vol%, respectively) at a flow ratio of 2/1. By contrast, in the range of stoichiometric feed concentrations the conversion is higher in the case of co-feeding. The decrease in conversion in the FLBMR is significant, and it can be reduced by increasing the secondary gas flow, but it remains below the values of the FLBR even at high (2/1) flow ratios. This diagram shows once again the strong dependence of conversion on the local reactant concentration (see Section 6.3.4).

A strong selectivity increase can be reached by distributed oxygen dosing at stoichiometric feed concentrations (70% selectivity at 0.5 vol% O_2)., However, unlike the ethane conversion, the ethylene selectivity is hardly affected by the flow ratio. At 4 vol% oxygen concentration, where the change in the selectivity is most pronounced, increasing the flow ratio results in small decrease of only 2%.

A high secondary to primary gas flow ratio seems to be beneficial for the ODHE reaction. At the same total gas feed, 6–9% higher ethylene yields can be reached by utilizing the flow ratio as an additional degree of freedom in the FLBMR in comparison to the conventional reactor.

6.3.8
Influence of Distributed Reactant Dosing with Oxygen in the Primary Gas Flow

The previous parameter studies have shown a significant improvement of the ehtylene selectivity and ethylene yield in the FLBMR in comparison with the FLBR. Nevertheless, ethane conversions in the FLBMR were lower in some cases even under oxygen excess conditions. It has been shown that increasing the secondary gas fraction leads to higher conversions under otherwise identical overall feed compositions. Alternatively, a combined dosing strategy can be applied by feeding part of the total oxygen amount at the reactor inlet. Besides the two studied dosing variants, which in fact can be regarded as extremes, FLBMR and FLBR, an additional configuration was examined, which had an oxygen distribution between the primary and membrane feeds with a ratio of 1/1 (comb. FLBMR). For all dosing strategies the concentration of oxygen in the overall feed was the same. The secondary to primary gas flow ratio was set to 1/1. The results are given in Figure 6.13.

It can be seen that, in the absence of oxygen in the primary gas (FLBMR), the ethane conversion is lower, especially under oxygen excess conditions. The combined dosing strategy shows a significant conversion increase even compared to the FLBR. In this case only enough oxygen is fed through the membranes to compensate the dilution of the primary gas. In the absence of chemical reactions a constant oxygen profile over the bed height would result, similar to a FLBR. The ethane conversion is up to 8% higher, which is a result of the longer contact time and the lower back-mixing. As expected, the highest ethylene selectivity can be observed in the experiments in the FLBMR without primary oxygen feeding. By dosing oxygen at the inlet, the local oxygen concentration rises, which leads to a higher oxygen to hydrocarbon ratio and thus to a decreased ethylene selectivity. However, the measured selectivity is still up to 10% higher for the combined dosing than for the FLBR. From the selectivity versus conversion plot it seems that the combined dosing strategy has no significant advantage, compared to a FLBMR. It should be however taken into consideration that a less reduced catalyst shows a lower affinity for coking and hence can reach a longer operation time.

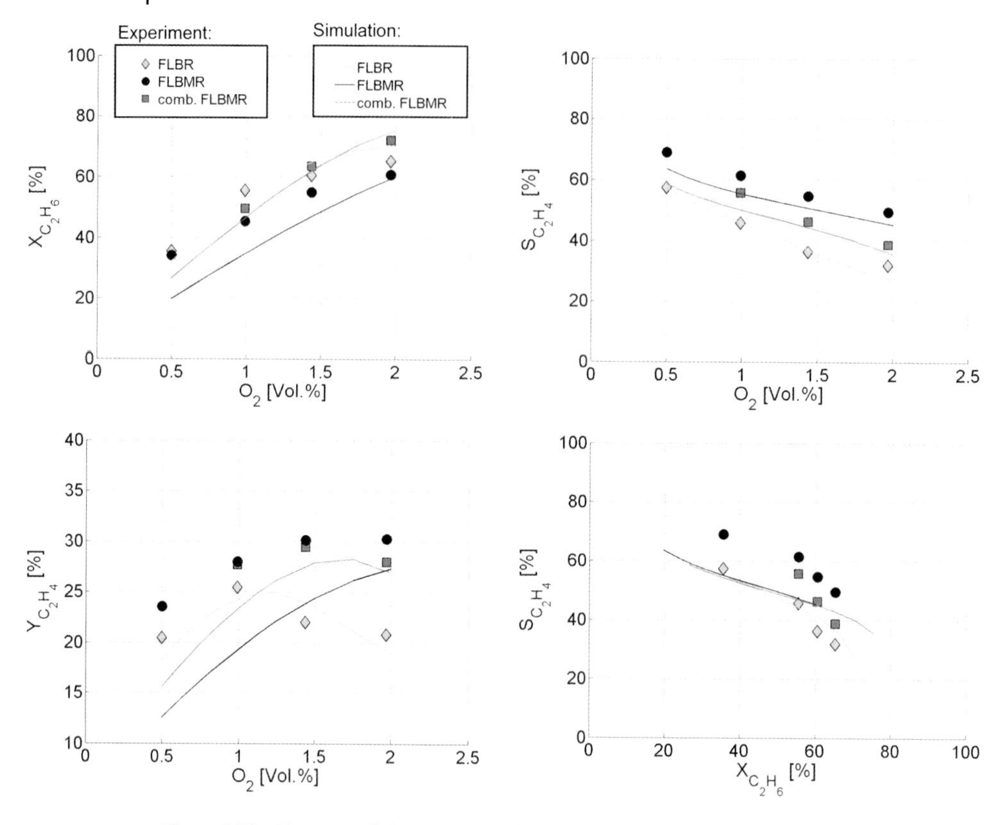

Figure 6.13 Variation of the oxygen conditions in the primary gas flow: $c(C_2H_6) = 1$ vol%, $T_{bed} = 590\,°C$, $m_{Cat}/F = 1400\,kg\,s/m^3$, $F_s/F_p = 1.0$.

6.4 Conclusions

The performance of distributed reactant dosing in a FLBMR was studied in comparison to conventional co-feeding in a fluidized bed. Both reactant dosing strategies were evaluated for the model reaction of oxidative dehydrogenation of ethane to ethylene under various experimental conditions.

The beneficial effect of oxidant dosing via membranes was most pronounced at high temperatures and moderate oxygen excess. Significantly higher values for selectivity and yield of ethylene were achieved, however, at a moderate lower level of ethane conversion in comparison to co-feeding. To reach even higher conversions by distributed dosing, a larger oxygen concentration or a higher O_2/C_2H_6 ratio than in the FLBR is needed. The higher amount of oxygen compensates the shorter residence time of the oxygen molecules in the reaction zone.

Under oxygen excess conditions the ethylene yield is up to 9% higher in the FLBMR than in the FLBR. This could be due to the higher average contact time of the ethane and less mass transfer limitations because of reduced bubble size and bubble fraction (Deshmukh, 2004; Kleijn van Willigen *et al.*, 2005; Christensen, Coppens, and Nijenhuis, 2005). The latter aspect also explains the fact that the co-feeding shows a more pronounced senstivity regarding the fluidization velocity, where increasing the gas flow leads to a stronger ethane conversion decrease in the FLBR than in the FLBMR.

Furthermore it was demonstrated that an alternative to improve the conversion in the FLBMR is the increase of the secondary to primary gas flow ratio or the dosage of small amounts of oxygen in the primary gas flow. The combined dosing strategy does not lead to an improvement in the ethylene yield but can be useful to avoid the catalyst coking at low oxygen concentrations.

A phenomenological two-phase dispersion model was applied to describe the effects of segregated reactant dosage on the reactor performance in fluidized beds and to identify attractive parameter ranges for reactor operation. The results show a good agreement between the simulations and the experiments for the co-feeding and reproduce the main tendencies for distributed reactant feeding. The discrepancies are the consequence of the model simplifications and the coarse empirical correlation for the fluid dynamic parameters of the conventional fluidized bed.

Special Notation not Mentioned in Chapter 2

Latin Notation

a	m^3/m^2	specific surface area
D	m	diameter
F	m^3/s	volumetric flow rate
H	m	height
j	kg/m^2s	mass flux
k	m/s	mass transfer coefficient
$\dot{m}$	kg/s	mass flow rate
m	kg	mass of solids
M	g/mol	molar mass
n		number
Pe		Péclet number
R	J/mol K	universal gas constant
S	%	selectivity
u	m/s	superficial velocity
u	m	circumference
U_m	m	perimeter
X	%	conversion
Y	%	yield
z	m	height coordinate

Greek Notation

$\mathcal{D}$	m^2/s	diffusion coefficient
ε_b		bubble fraction
ε_s		void fraction in the emulsion
ρ	kg/m^3	density
ω		mass fraction

Subscripts and Superscripts

0	at superficial velocity conditions
app	apparatus
b	bubbles, bypass
bed	bed
elu	at elutriation point
mf	at minimum fluidization
p	primary
Q	convective flow
rel	relative
s	suspension gas, secondary

References

Adris, A.M., Lim, C.J., and Grace, J.R. (1993) The effect and implications of gas volume increase due to reaction on bed expansion, bubbling and overall conversion in a fluidized bed reactor. A.I.Ch.E. Annual Meeting, St. Louis, November 7–12, 1993.

Adris, A.M., Lim, C.J., and Grace, J.R. (1997) The fluidized-bed membrane reactor for steam methane reforming: model verification and parametric study. *Chem. Eng. Sci.*, **52** (10), 1609–1622.

Ahchieva, D., Peglow, M., Heinrich, S., Mörl, L., Wolff, T., and Klose, F. (2005) Oxidative dehydrogenation of ethane in a fluidized bed membrane reactor. *Appl. Catal. A Gen.*, **296** (2), 176–185.

Alonso, M., Lorences, M., Pina, M., and Patience, G. (2001) Butane partial oxidation in an externally fluidized bed membrane reactor. *Catal. Today*, **67** (1–3), 151–157.

Alonso, M., Lorences, M.J., Patience, G.S., Vega, A.B., Diez, F.V., and Dahl, S. (2005) Membrane pilot reactor applied to selective oxidation reactions. *Catal. Today*, **104** (2–4), 177–184.

Al-Sherehy, F., Grace, J.R., and Adris, A.E.M. (2004) Gas mixing and modeling of secondary gas distribution in a bench-scale fluidized bed. *AIChE J.*, **50** (5), 922–936.

Al-Sherehy, F., Grace, J.R., and Adris, A.E.M. (2005) The influence of distributed reactant injection along the height of a fluidized bed reactor. *Chem. Eng. Sci.*, **60** (24), 7121–7130.

Bokkers, G.A., Laverman, J.A., van Sint Annaland, M., and Kuipers, J.A.M. (2006) Modelling of large-scale dense gas-solid bubbling fluidized beds using a novel discrete bubble model. *Chem. Eng. Sci.*, **61** (17), 5590–5602.

Chen, Z., Yan, Y., and Elnashaie, S.S.E.H. (2003) Novel circulating fast fluidized-bed membrane reformer for efficient production of hydrogen from steam reforming of methane. *Chem. Eng. Sci.*, **58**, 4335–4349.

Christensen, D.O., Coppens, M.O., and Nijenhuis, J. (2005) Residence times in fluidized beds with secondary gas injection. Proceedings of the AIChE Annual Meeting, Cincinnati, OH.

Deshmukh, S. (2004) Membrane assisted fluidized bed reactor. University of Twente, PhD thesis, The Netherlands.

Deshmukh, S.A.R.K., Laverman, J.A., Cents, A.H.G., van Sint Annaland, M., and Kuipers, J.A.M. (2005a) Development of a membrane assisted fluidized bed reactor. 1: gas phase back-mixing and bubble to emulsion mass transfer using tracer injection and ultrasound. *Ind. Eng. Chem. Res.*, **44** (16), 5955–5965.

Deshmukh, S.A.R.K., Laverman, J.A., van Sint Annaland, M., and Kuipers, J.A.M. (2005b) Development of a membrane-assisted fluidized bed reactor. 2. Experimental demonstration and modeling for the partial oxidation of methanol. *Ind. Eng. Chem. Res.*, **44** (16), 5966–5976.

Deshmukh, S.A.R.K., Heinrich, S., Mörl, L., van Sint Annaland, M., and Kuipers, J.A.M. (2007) Membrane assisted fluidized bed reactors: Potentials and hurdles. *Chem. Eng. Sci.*, **62** (1–2), 416–436.

Dixon, A. (2003) Recent research in catalytic membrane reactors. *Int. J. Chem. Reactor Eng.*, Review **R6**, 1.

Hilligardt, K. (1986) Zur Strömungsmechanik von Grobkornwirbelschichten. PhD thesis, TU Hamburg-Harburg.

Hilligardt, K., and Werther, J. (1986) Local bubble gas hold-up expansion of gas/solid fluidized beds. *Ger. Chem. Eng.*, **9**, 215–221.

Jin, Y., Wei, F., and Wang, Y. (2003) Effect of internal tubes and baffles, in *Handbook of Fluidisation and Fluid-Particle Systems* (ed. W.-C. Yang), Marcel Dekker, pp. 171–200.

Kleijn van Willigen, F., Christensen, D., Ommen, J.R.V., and Coppens, M.O. (2005) Imposing dynamic structures on fluidized beds. *Catal. Today*, **105** (3–4), 560–568.

Klose, F., Wolff, T., Thomas, S., and Seidel-Morgenstern, A. (2004a) *Appl. Catal. A Gen.*, **257**, 193–199.

Klose, F., Joshi, M., Hamel, C., and Seidel-Morgenstern, A. (2004b) Selective oxidation of ethane over a $VO_x/\text{-}Al_2O_3$ catalyst–investigation of the reaction network. *Appl. Catal. A Gen.*, **260** (1), 101–110.

Kunii, D., and Levenspiel, O. (1992) *Fluidization Engineering*, John Wiley & Sons, Inc., New York.

Lee, G.S., and Kim, S.D. (1989) Gas mixing in slugging and turbulent fluidized beds. *Chem. Eng. Commun.*, **86**, 91–111.

Levenspiel, O. (1999) *Chemical Reaction Engineering*, John Wiley & Sons, Inc., New York.

Marcano, J.G.S., and Tsotsis, T.T. (2002) *Catalytic Membranes and Membrane Reactors*, Wiley-VCH Verlag GmbH, Weinheim.

Miracca, I., and Capone, G. (2001) The staging in fluidized bed reactors: from CSTR to plug-flow. *Chem. Eng. J.*, **82**, 259–266.

Mleczko, L., Schweer, D., Durjanova, Z., Andorf, R., and Baerns, M. (1994) Reaction engineering approaches to the oxidative coupling of methane to C2 hydrocarbons. *Stud. Surf. Sci. Catal.*, **8**, 155–164.

Patil, C.S., van Sint Annaland, M., and Kuipers, J.A.M. (2005) Design of a novel autothermal membrane-assisted fluidized-bed reactor for the production of ultrapure hydrogen from methane. *Ind. Eng. Chem. Res.*, **44** (25), 9502–9512.

Pena, M., Carr, D., Yeung, K., and Varma, A. (1998) Ethylene epoxidation in a catalytic membrane reactor. *Chem. Eng. Sci.*, **53** (22), 3821–3834.

Poling, B., Prausnitz, J.M., and O'Connell, J.P. (2001) *Properties of Gases and Liquids*, 5th edn, McGraw-Hill Book Company, New York.

Ramos, R., Pina, M., Menendez, M., Santamaria, J., and Patience, G. (2001) Oxidative dehydrogenation of propane to propene, 2: Simulation of a commercial inert membrane reactor immersed in a fluidized bed. *Can. J. Chem. Eng.*, **79** (6), 902–912.

Saracco, G., Neomagus, H.W.J.P., Versteeg, G.F., and Swaaij, W.P.M. (1999) High-temperature membrane reactors. Potential and problems. *Chem. Eng. Sci.*, **54** (13–14), 1997–2018.

Tonkovich, A.L., Zilka, J., Jimenez, D., Roberts, G., and Cox, J. (1996) Experimental investigations of inorganic membrane reactors: a distributed feed approach for partial oxidation reactions. *Chem. Eng. Sci.*, **53** (5), 789–806.

Volk, W., Johnson, C.A., and Stotler, H.H. (1962) Effect of reactor internals on quality of fluidization. *Chem. Eng. Prog.*, **58** (3), 44–47.

Werther, J., and Schössler, M. (1999) Modeling catalytic reactions in bubbling fluidized beds of fine particles. Proceedings of the 6th International Conference on Circulating Fluidized Beds, 357–368, Würzburg, Germany.

Westermann, T., and Melin, T. (2005) Membrane reactors. *Chemie-Ingenieur-Technik*, **77** (11), 1655–1668.

7

Solid Electrolyte Membrane Reactors

Liisa Rihko-Struckmann, Barbara Munder, Ljubomir Chalakov, and Kai Sundmacher

7.1
Introduction

The present chapter introduces a specific type of membrane reactor only briefly mentioned in Chapter 1. The solid electrolyte membrane reactors (SEMR)–or electrochemical membrane reactors as they also are called–are equipped with ion-conducting membranes, which ideally are impermeable for non-charged reaction species. These reactors operate as electrochemical cells, in which the oxidation and reduction reactions are carried out separately on catalyst/electrodes layers located on the opposite sides of the electrolyte. The development of solid electrolyte membrane reactors has reached a semi-commercial stage in fuel cells, in which the maximal generation of electric energy by the total oxidation of hydrogen or hydrocarbon feeds is the primary goal of operation. Research on chemical reactor applications is strongly concentrated in the high-temperature range using either oxygen ion- or proton-conducting inorganic membranes. Some interesting examples were published recently, investigating proton-conducting polymeric membranes for the production of chemicals.

This chapter gives a brief overview on the current status and future trends in the development and application of solid electrolyte reactors equipped with solid electrolyte (SE) materials used as membranes in these reactors. Initially the working principle of a solid electrolyte membrane reactor and material aspects are discussed. The second part of the chapter gives a detailed description of the special aspects concerning the modeling of solid electrolyte membrane reactors, as well as the application examples of maleic anhydride (MA) synthesis and oxidative dehydrogenation of ethane. Finally the chapter reviews recent papers concerning solid electrolyte membrane reactors applying:

- high-temperature oxygen ion conductors,
- high-temperature proton conductors,
- low-temperature proton conductors.

There are some recent review papers, especially written for high-temperature applications, which we would like to recommend to readers interested in more

Membrane Reactors: Distributing Reactants to Improve Selectivity and Yield.
Edited by Andreas Seidel-Morgenstern

ISBN: 978-3-527-32039-4

detailed information on these materials (Marnellos and Stoukides, 2004). In 2000, Stoukides gave an outlook on the applications of high-temperature oxygen SE membrane reactors (Stoukides, 2000). Iwahara *et al.* (2004) and Kokkofitis *et al.* (2007) published reviews of hydrogen technology using proton-conducting ceramics. Bredesen, Jordal, and Bolland (2004) presented a survey of high-temperature membranes and applications of membrane reactors for integration in power generation cycles with CO_2 capture, which is an important aspect in the power plant engineering. The cited reviews elucidate SE membranes from the chemical engineers' point of view. The excellent reviews by Goodenough (2003) and Kreuer (2003) are recommended to readers who would like to be informed about material aspects of oxide ion-conducting or proton-conducting electrolytes, respectively.

7.2 Operational and Material Aspects in Solid Electrolyte Membrane Reactors

7.2.1 Classification of Membranes

Aside from the porous membranes, for which the characteristics and aspects of modeling are described in detail in previous chapters in this book, two categories of gas-dense membranes can be classified, namely mixed ion/electron conductors (MIEC) and ion conductors (Sundmacher, Rihko-Struckmann, and Galvita, 2005). MIEC are those in which the values of ionic and electronic conductivity are comparable, whereas the latter is referred to solid electrolytes (SE) in which the ionic transfer numbers, that is, the fraction of electrical current transferred by the ions, is two or more orders of magnitude higher than that for electrons. The transport mechanisms of the electron (MIEC) and ion conductors are illustrated in Figure 7.1 using oxygen transport as example. MIEC have both high ionic and high electronic conductivities, whereas SE exhibit only high ionic but very low electronic conductivities. MIEC membrane reactors are simpler in design and construction than the SE reactors because both ions and electrons are transported internally, that is, inside the membrane material. SE membrane reactors require the external circuit for electron recycle to the cathode, but they provide therefore some specific benefits, as discussed later in detail.

The transfer of oxygen occurs in the dense oxygen ion-conducting membranes in the form the ionic species, O^{2-}, moving from vacancy to vacancy in the lattice of the solid material, driven by the electrostatic potential difference. The vacancies are created in a solid solution of oxides of di- and trivalent cations with oxides of tetravalent metals. Due to this transport mechanism, the permeability of oxygen is usually remarkably lower than in porous materials, which often limits the technical applicability of oxygen ion conducting solid electrolyte membrane reactors. But, as outlined later in this chapter, the permeability is strongly dependent on the operating temperature, whereas a high operating temperature enhances the

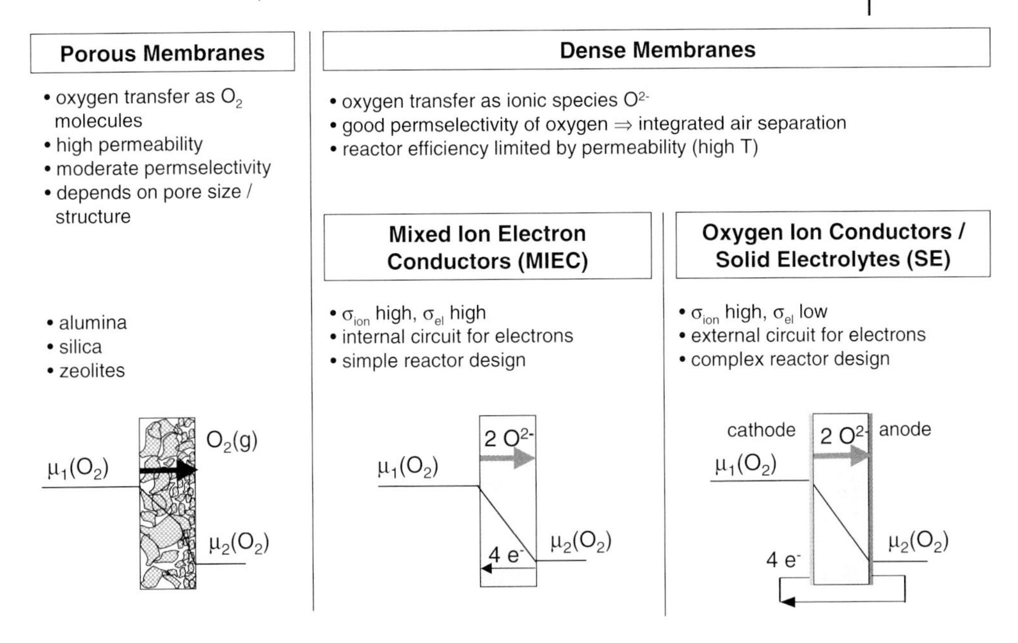

Figure 7.1 Classification of ceramic membranes for oxygen transport. Reproduced from (Sundmacher, Rihko-Struckmann, and Galvita, 2005), reprinted with permission from Elsevier.

permeability of the SE material. Simultaneously, the permselectivities of SE membranes are excellent. When using air, that is, a mixture of oxygen and nitrogen, oxygen is transferred exclusively from one side to the other. Consequently, SE membranes allow a direct integration of air separation in the process, and any additional air separation is unnecessary. This in turn reduces the operating and investment costs of the oxidation process.

Due to the electrochemical ion/electron transfer reactions in SE membranes, the electrode layers (anode, cathode) have to be constructed directly adjacent to the ion-conducting membrane. The design of SE reactors – especially the construction and fabrication of membrane–electrode interfaces – is complicated. But, the ionic conduction of the SE materials provides a unique tool for the control of SE membrane reactors. Due to the faradaic coupling of oxygen flux and cell current, the galvanostatic control of the external flux of electrons or, alternatively, the potentiostatic control of electrode potentials offers the opportunity to drive the reactions at the two electrodes in the desired direction.

7.2.2
Ion Conductivity of Selected Materials

Figure 7.2 shows a map of selected SE materials and their ionic conductivities as a function of temperature. In the group of proton conductors there are two main classes: (a) polymeric materials which can only be operated in the low-temperature

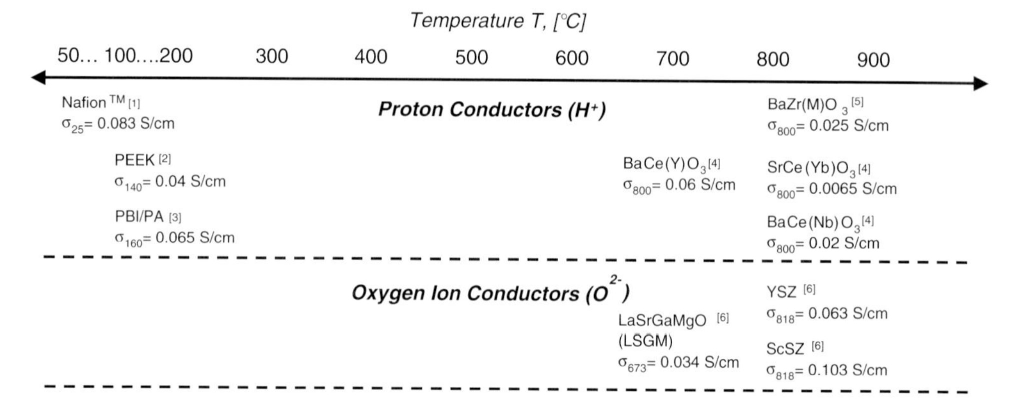

Figure 7.2 Ion conductivities of selected SE materials. Reproduced from (Sundmacher, Rihko-Struckmann, and Galvita, 2005), reprinted with permission from Elsevier. For references [1]–[6], see the original paper.

range up to 200 °C and (b) certain mixed oxides which show reasonable proton conductivities at about 500–900 °C. Typical representatives of low-temperature proton conductors are Nafion, polybenzimidazole (PBI) and polyetheretherketones (PEEK). Currently investigated high-temperature proton conductors are Ba–Zr, Sr–Ce and Ba–Ce mixed oxides. Oxygen ion conductors can be only operated in the high-temperature range. The most important materials here are yttria-stabilized zirconia (YSZ), scandia-stabilized zirconia (ScSZ), and – nowadays extensively investigated – perovskite materials such as Sr/Mg-doped lanthan gallat (LSGM). It is interesting to note that the mentioned materials in the low and high-temperature range all have ion conductivities of similar magnitude, that is, around $\sigma = 0.01–0.1\ S\ cm^{-1}$.

When applying SE materials in catalytic membrane reactors, it is important that the materials show suitable oxygen transport rates within the catalyst temperature operating window. As an example, for butane partial oxidation (POX) with the VPO catalyst, a SE membrane owing high oxygen permeability in the range between 400 and 600 °C would be optimal. Table 7.1 gives typical oxygen flux densities for YSZ, ScSZ (with 9% Sc) and LSGM (Sundmacher, Rihko-Struckmann, and Galvita, 2005). As discussed in the previous section, the flux densities of solid electrolytes are generally lower than those of typical porous membranes (supported microporous SiO_2, mesoporous Al_2O_3), at the same temperature.

7.2.3
Membrane–Electrode–Interface Design in Solid Electrolyte Membrane Reactor

The schematic illustration in Figure 7.1 is a highly simplified representation of SE membrane reactors applying ion conductors. In reality SE membrane reactors have complicated multi-layered construction. The electrodes, where the electron/ion transfer reactions occur, have to be fixed on both sides of the ion-conducting

Table 7.1 Total conductivity of solid electrolytes and calculated oxygen flux for a membrane thickness of 250 μm ($U = 1.0$ V).

Temperature	420 °C	624 °C
Material	**Total conductivity**	**$S\ m^{-1}$**
YSZ[a)]	0.018	0.80
ScSZ[b)]	0.011	1.34
LSGM[c)]	0.092	2.32
	Oxygen flux density	**$mmol\ m^{-2}s^{-1}$**
YSZ[a)]	0.188	8.3
ScSZ[b)]	0.114	13.9
LSGM[c)]	0.949	24.1

a) yttria (13%).
b) scandia (9%) stabilized zirconia.
c) $La_{0.9}Sr_{0.1}Ga_{0.85}Mg_{0.15}O_{3-\delta}$.

membrane to be able close the charge circuit. They necessitate a close contact not only to the membrane but also to the electronic current collectors, which are directly connected to the external electron circuit. As a summary, in order to work optimally the electrode/catalyst layer has to fulfill the following requirements:

- a catalytic activity for the electrochemical and catalytic reactions,
- an electronic conductivity – electrons which are released in the anodic reaction or consumed in the cathodic reaction at the reaction sites, have to be collected,
- an ionic conductivity – the ions transferred through the membrane have to be conducted towards the reaction sites,
- porous structure inside of the electrode – the non-charged reactants have to be transported to the reaction sites.

Theoretically, the following four operations (electron conduction, ion conduction, catalytic activity, free access for gaseous feed) have to be possible in one location for the electrochemical reaction to occur. In order to combine the described operations, electrodes in SE membrane reactors are typically designed as gas diffusion electrodes (GDE), or more generally speaking, fluid diffusion electrodes. Figure 7.3 shows a schematic illustration of a typical GDE. Optimal electrode design requires a perfectly executed balance of the different functions. This is often achieved by preparing mixtures of ion conducting particles (made of the membrane material), particles of electron conductor and particles with catalytic activity. In an optimal case, one metallic phase possesses both high electronic conductivity and high catalytic activity. Furthermore, by using well defined particle size fractions of the materials or by carrying out thermal treatment, one can adjust

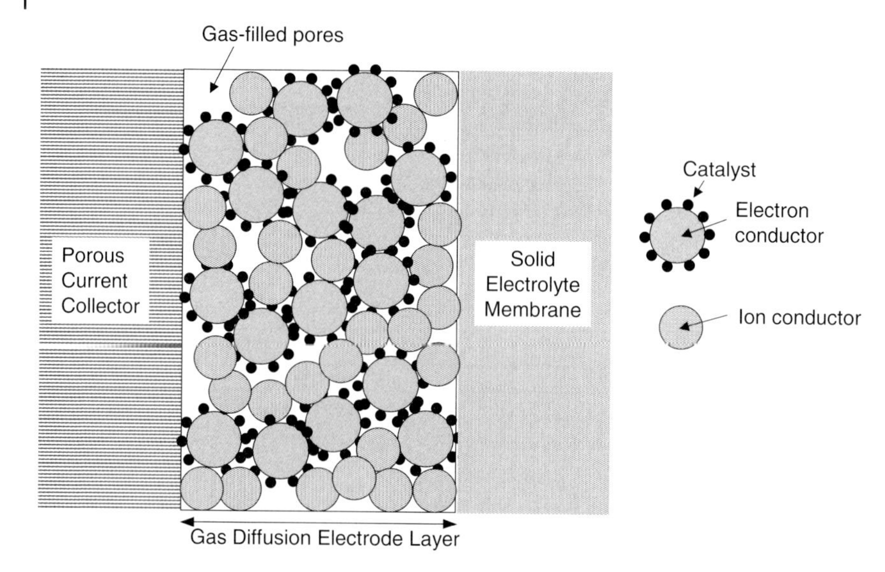

Figure 7.3 Schematic illustration of gas diffusion electrode (GDE). Reproduced from (Sundmacher, Rihko-Struckmann, and Galvita, 2005), reprinted with permission from Elsevier.

the electrode pore structure. This in turn offers the possibility to optimize the transport properties of the GDE with respect to the non-charged reactants.

7.2.4 Operating Modi of Solid Electrolyte Membrane Reactors

Solid electrolyte membrane reactors can be operated in a variety of modi which are illustrated in Figure 7.4 using the hydrogen–oxygen reaction combined with an oxygen ion-conducting membrane as example. In the open circuit mode ($i_{cell} = 0$), the membrane reactor is operated potentiometrically as a sensor without any net current through the electrolyte. In this mode, the activity difference, which in most cases describes us the concentration difference of reactants, on both sides of the solid electrolyte can be monitored via the open-circuit cell voltage E^0_{cell}, often abbreviated as OCV, based on Nernst's law.

In the fuel cell mode, the cell current is positive ($i_{cell} > 0$) and the cell voltage is below the OCV level due to internal cell resistances. A precise control of the oxygen permeation rate is possible due to the faradaic coupling of oxygen flux and the cell current. Cathodic reactants are reduced to ionic species at the cathodic electrode, and the ions are transferred through the membrane to the anodic electrode where they react with the anodic reactants. The DC power density, $p = E_{cell} \cdot i_{cell}$, is positive, that is, one obtains electric power output from the electrochemical process ($P > 0$). In this mode, the negative $\Delta_R G$ of the chemical reaction is converted directly into electrical energy, thereby reducing the amount of thermal energy being released during the reaction. The highly negative $\Delta_R G$ of hydrocarbon oxidation reactions

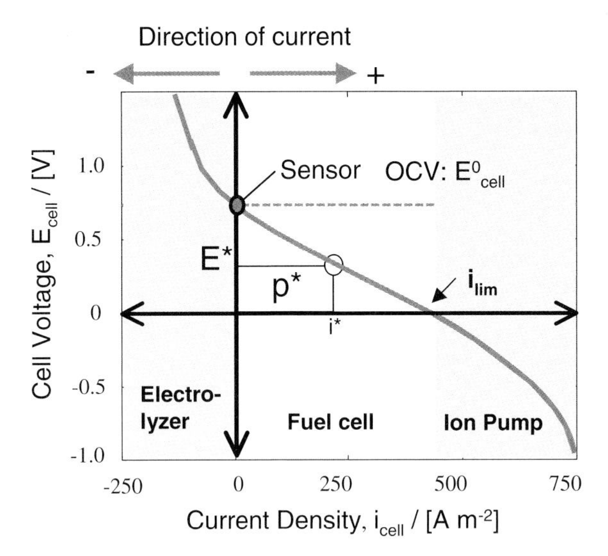

Figure 7.4 Operating modi of SE membrane reactors. Electrolyzer: $i_{cell} < 0$, $E_{cell} > E^0_{cell}$, $P < 0$. Sensor: $i_{cell} = 0$, $E_{cell} = E^0_{cell}$, $P = 0$. Fuel cell: $i_{cell} > 0$, $0 < E_{cell} < E^0_{cell}$, $P > 0$. Ion pump: $i_{cell} > 0$, $E_{cell} < 0$, $P < 0$. Reproduced from (Sundmacher, Rihko-Struckmann, and Galvita, 2005), reprinted with permission from Elsevier.

enables high driving force and a positive cell voltage, positive power output and therefore allows the process to be operated in the self-driven fuel cell mode. In the case of partial oxidation reactions of, for example, hydrocarbons, the simultaneous generation of valuable chemicals and energy is possible in the fuel cell mode.

If the current density is forced to exceed a certain limiting value $i_{lim} > 0$, it is possible only as the cell voltage is negative. In this special form of electrolysis, in the ion pump mode, the direction of current is equal to the fuel cell mode, but the transfer of charged species through the solid electrolyte is supported with external electric energy input to the SE reactor ($p < 0$).

Finally, the direction of the overall electrochemical reaction can be changed to the opposite one if negative cell currents are applied. In this classic electrolysis mode, the cell voltage exceeds the OCV level ($E > E^0_{cell}$) and the power is negative ($p < 0$). Then, the direction of the electronic current and the ionic flux are opposite to the fuel cell mode.

7.2.5
Cell Voltage Analysis

The overall cell voltage E_{cell} of a solid electrolyte membrane reactor can be easily measured in experiments (Figure 7.5). However, the E_{cell} consists of several physical and electrochemical contributions and a detailed deconvolution of the cell voltage into:

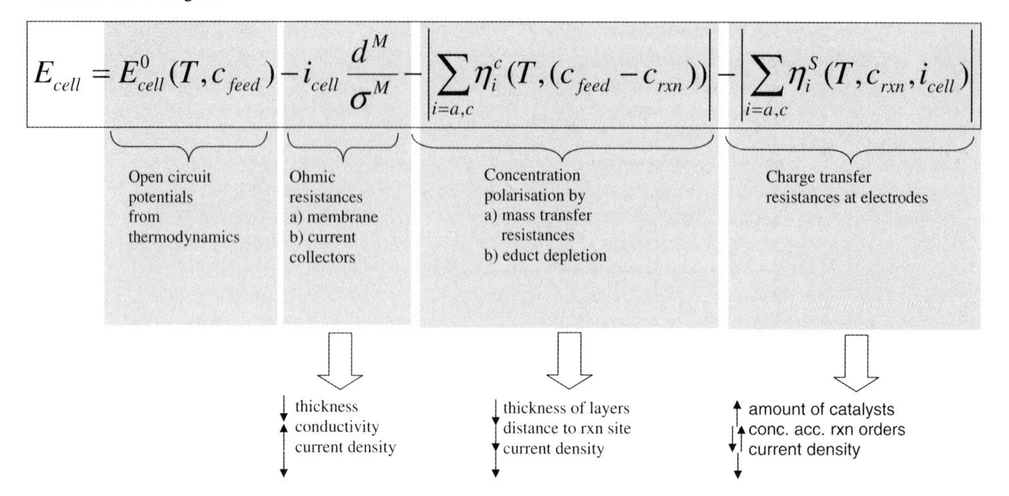

Figure 7.5 Decomposition of cell voltage into main resistances. Reproduced from (Sundmacher, Rihko-Struckmann, and Galvita, 2005), reprinted with permission from Elsevier.

- open circuit cell voltage E^0_{cell} (OCV),
- voltage drop due to the ohmic resistances (electrolyte membrane, the electrodes and the current collectors),
- concentration polarization due to mass transfer resistances within the electrode and due to the depletion of reactants from the reactor inlet up to the electrode,
- charge transfer resistances at the electrodes.

gives valuable information on how to improve the reactor performance. As listed in Figure 7.5, one system has various cell design parameters which should be changed in order to reduce the contribution of certain resistances. The most important factors are the thicknesses of the different cell layers, the intrinsic conductivities of the materials used, the amount of the catalyst applied and the construction of the electrode layer to quarantee a good contact between the ionic carrier, electron carrier as well as the catalyst.

7.2.6 Non-Faradaic Effects

An overview on SE membrane reactors cannot be written without–at least briefly–mentioning the non-electrochemical modification of catalyst activity effect (NEMCA). Normally the rates of electrochemical reactions taking place at the electrodes obey Faraday's law, that is, the current flowing through the membrane and the reaction rate at the considered electrode are proportional to each other. However, there are a couple of reactions where deviations from Faraday's law were observed when the process is operated in a special mode–a mixed feed–in which one reactant, for example, oxygen, is simultaneously fed both gaseous as well as electrochemical pumping. If gaseous ethane and oxygen are fed and they react on a catalyst which is placed on top of an YSZ membrane which provides additional electrochemical oxygen supply, the catalytic rate increase during polarization can be higher than the calculated one from Faraday's law. This is due to the electrochemical promotion of the catalytic reaction caused by the change of the electrostatic potential of the catalyst, as outlined in a series of papers by Vayenas and coworkers (see, e.g., Vayenas, Bebelis, and Ladas, 1990; Bebelis and Vayenas, 1989).

7.3
Modeling of Solid Electrolyte Membrane Reactors

The electrochemical reactions in solid electrolyte membrane reactors bring additional aspects to the system which have to be considered during modeling and which increase the complexity of the models of such reactors. Besides the heterogeneously catalyzed reactions and the possible mass transport resistances, the electrochemical charge transfer reactions on the elctrodes as well as the charge transfer in solid electrolytes have to be included in the model. In contrast to porous membranes where the driving force for the mass transfer through the membrane is related to the Δp over the membrane, in solid electrolyte membrane reactors the ion conduction through the membrane is driven by the electrical potential difference across the membrane.

Membrane reactors are especially favorable for reaction kinetics in which the partial pressure of one component appears in a lower order in the desired reaction rate than in the rate expressions for the undesired reactions. In such a case a selectivity advantage can be achieved by a controlled distributed feed of the limiting component across an inert membrane. For example in the partial oxidation reactions, the local oxygen partial pressure could be lowered over the whole length of the reactor and the total oxidation reactions suppressed. In a solid electrolyte membrane reactor the faradaic coupling provides the unique tool for the precise dosing of the critical reagent (e.g., oxygen or hydrogen).

Besides their concentration, the type and state of the oxygen species available at the active catalyst sites influence significantly the reaction selectivity. This is especially cruicial in solid electrolyte reactors, where aside from gaseous O_2 being released on the anode, other more active oxygen spieces might also exist and influence selectivity either positively or negatively. Eng and Stoukides (1991) and Wang and Lin (1995) reported that, in solid electrolyte membrane reactors, the oxygen anions (O^{2-}) transported across the membrane might directly form selective oxygen species on the membrane or catalyst surface while gas-phase oxygen, which would rather give rise to unselective reactions, could be excluded from the reaction zone. Ye *et al.* (2006) and Chalakov *et al.* (2007) observed high reactivities for the desired intermediate, which was likely due to the additional electrochemical reactions between oxygen ions on the anode and the intermediate, as discussed later in detail.

This chapter presents a general model for solid electrolyte membrane reactors which is capable of describing the prevailing physical phenomena taking place in such reactors (Munder *et al.*, 2005). The partial oxidation of *n*-butane to maleic anhydride and the oxidative dehydrogenation of ethane are selected as model reactions and the kinetic models for both reactions are integrated into the developed reactor model. The partial oxidation of *n*-butane to maleic anhydride, on the one hand, is still the only industrial application of the direct partial oxidation of alkanes (Centi, Cavani, and Trifiro, 2001) and, on the other hand, it exhibits the typical properties of selective oxidations. In our laboratory we (Ye *et al.*, 2004) studied experimentally the feasibility of electrocatalytic synthesis for this reaction. However, theoretical analysis and model simulations are necessary in order to evaluate the reactor performance and to provide an efficient design and optimal operation conditions.

7.3.1
Reactor Model for Systems Containing Solid Electrolyte Membranes

The following physical phenomena have to be considered in the modeling of an oxygen ion-conducting electrolyte membrane reactor for a heterogeneously catalyzed oxidation reaction. The charge in the form of oxygen anions, O^{2-}, is transported by migration through the solid electrolyte membrane (SE, see Figure 7.6). Due to geometrical reasons – the membrane thickness is typically some orders of magnitude smaller than the surface dimensions – it can usually be assumed that the current flows predominantly in the radial direction.

On the surface between the electrolyte and the anodic catalyst layer (AC), both the electrochemical charge transfer reactions and the heterogeneously catalyzed

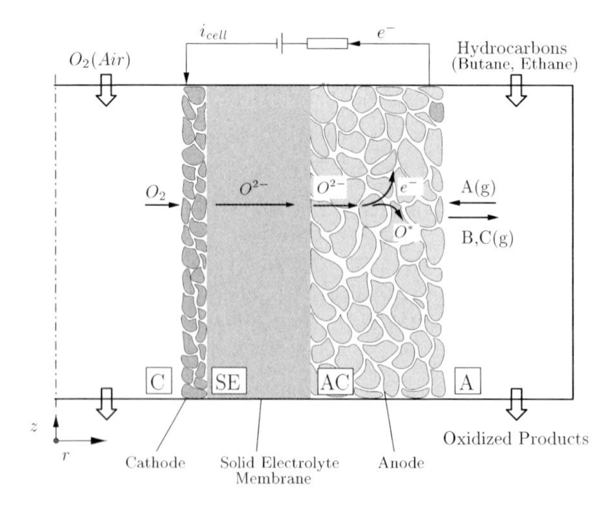

Figure 7.6 Schematic diagram of the SEMR. Reproduced from (Munder *et al.*, 2005), reprinted with permission from Elsevier.

gas phase reactions might take place. The reactions are coupled with diffusive mass transport through the catalyst pores for gaseous O_2 and with charge transport in form of oxygen anions, O^{2-}, and electrons, e^-, through the solid phase of the catalyst layer, respectively. Both gas phase diffusion and oxygen ion transport occur predominantly in radial direction, whereas released electrons move in an axial direction toward the external electric circle. Finally, in the anodic gas compartment mass transport can be convective and diffusive. Assuming plug flow along the anodic gas compartment without axial dispersion and neglecting possible radial concentration profiles in a first approach, the complete system could be modeled one-dimensionally with respect to the anodic gas compartment (A) and one-plus-one dimensionally with respect to the solid electrolyte membrane (SE) and the porous anodic catalyst layer (AC).

However, at this stage, in order to facilitate further model simplifications, it is reasonable to estimate the influence of the internal mass transfer resistances within the anodic catalyst layer (AC) in terms of dimensionless numbers. The Damköhler number of the second order, $\mathrm{Da_{II}}$, which is also known as the square of the Thiele modulus, gives an estimate of the ratio of the reaction rates to the diffusion rate under characteristic conditions and is defined as:

$$\mathrm{Da_{II}} = \Phi^2 = \frac{(d^{\mathrm{AC}})^2 \cdot r_{\mathrm{ref}}(T) \cdot M_{\mathrm{Cat}}}{D^{\mathrm{eff}} \cdot \tilde{c}_t \cdot V} \tag{7.1}$$

where r_{ref} stands for a (intrinsic) reference reaction rate. In the operating conditions typically applied for maleic anhydride synthesis (T = 350–550°C), using the kinetic parameters as given in the literature (Hess, 2002) [0.1–10.0 $\mathrm{mol\,kg_{cat}^{-1}\,s^{-1}}$], and a catalyst layer thickness in the range 0.1–1.0 mm, the value for $\mathrm{Da_{II}}$ is less than 10^{-2}. This means, that due to the slow catalytic reactions, the gas diffusion resistance inside the catalyst pores could be neglected and the process is likely kinetically controlled.

A second Damköhler number for electrochemical systems, $\mathrm{Da_{II}^{el}}$, can be written as the ratio of a characteristic electrochemical reaction rate to a characteristic O^{2-} migration rate through the solid phase of the anodic catalytic layer (Munder *et al.*, 2005). It can be calculated according to Equation 7.2:

$$\mathrm{Da_{II}^{el}} = \frac{d^{\mathrm{AC}} \cdot r_{\mathrm{el,ref}} \cdot F}{\sigma_{\mathrm{O2-}}^{\mathrm{eff}} \cdot \Delta\Phi_{\mathrm{max}}} \tag{7.2}$$

Even if the ion-conductivity were estimated to be as high as 0.01 $\mathrm{S\,m^{-1}}$, the values for $\mathrm{Da_{II}^{el}}$ higher than 10 would be reached due to the low operation temperature. Therefore, it is reasonable to assume here that, in the electrochemically supported MA synthesis, the charge transfer reactions take place within a very narrow region close to the membrane surface.

From the above considerations, one can then conclude that, under the given operation conditions (Ye *et al.*, 2004), the electrochemical MA synthesis can be practically divided into two locally separated reaction steps. The first step comprises all charge transport processes and the electrochemical charge transfer reactions within the SE membrane and a narrow surface region of the electrodes. The second covers all gas phase transport processes and heterogeneously catalyzed gas phase reactions. Due to the low diffusion resistances, the anode gas compartment (A) and the anodic catalyst layer (AC) are assumed to be one phase without concentration gradients in the radial direction and this can be modeled by a one-dimensional (axial direction) pseudo-homogeneous approach. The processes in this combined phase are coupled with the electrochemical processes via the radial oxygen flow.

Further assumptions can be made as follows. Due to the low reactant concentrations, the slow rate of reaction and the temperature control in the experimental setup, the reactor might be assumed to work under isothermal conditions. Furthermore, the pressure drop along the gas compartment is neglected, and the gas phase obeys the ideal gas law. The cathodic side is assumed to be a quasi-steady-state oxygen reservoir without concentration change, which can be experimentally established by a sufficiently high air flow rate through the cathode channel (C). Interfacial mass transfer resistances between the gas phases and solid surfaces of the membrane layers are assumed to be smaller than the diffusion resistances in the catalytic layers and are therefore neglected as well. Adsorption and desorption processes might have an influence but are not taken into consideration here.

On the basis of the above assumptions, the following model (Equations 7.3 and 7.4) are obtained in addition of the general model equations in Chapter 2. The component balances for the gas phase species i on the anode side yield:

$$\tilde{c}_t \frac{\partial \tilde{y}_i}{\partial t} = -\frac{1}{A_z}\frac{\partial (F^A \tilde{y}_i)}{\partial z} + \frac{A_s}{V} j_i + \frac{M_{\text{cat}}}{V}\sum_j \nu_{ij} r_j \tag{7.3}$$

The assumed isobaric conditions and the ideal gas law lead to a quasi-stationary total mass balance for the anodic gas phase:

$$0 = -\frac{1}{A_z}\frac{\partial F^A}{\partial z} + \frac{A_s}{V}\sum_i j_i + \frac{M_{\text{cat}}}{V}\sum_i \sum_j \nu_{ij} r_j \tag{7.4}$$

The flow density $j_i(z)$ obeys Faradays law for oxygen, and is zero for all other gas phase components.

The electrochemical processes can be represented by an equivalent circuit, as illustrated in Figure 7.7. The charge balance equations being specific for solid electrolyte membrane reactors, which were not discussed in Chapter 2, are presented in the following. Charge balances describe the anodic and the cathodic potential difference across the interfaces between the electrolyte and the respective electrode.

$$\begin{aligned} \frac{\partial \Delta\Phi^{\mathrm{A}}}{\partial t} &= \frac{\partial \eta^{\mathrm{A}}}{\partial t} = \frac{1}{C_{DL}}\left(i_{\text{cell}} - 2Fr_{\text{el}}^{\mathrm{A}}\right) \\ \frac{\partial \Delta\Phi^{\mathrm{C}}}{\partial t} &= \frac{\partial \eta^{\mathrm{C}}}{\partial t} = \frac{1}{C_{DL}}\left(-i_{\text{cell}} - 2Fr_{\text{el}}^{\mathrm{C}}\right) \end{aligned} \tag{7.5, 7.6}$$

Ohm's law describes the potential drop across the solid electrolyte as well as along the electrodes.

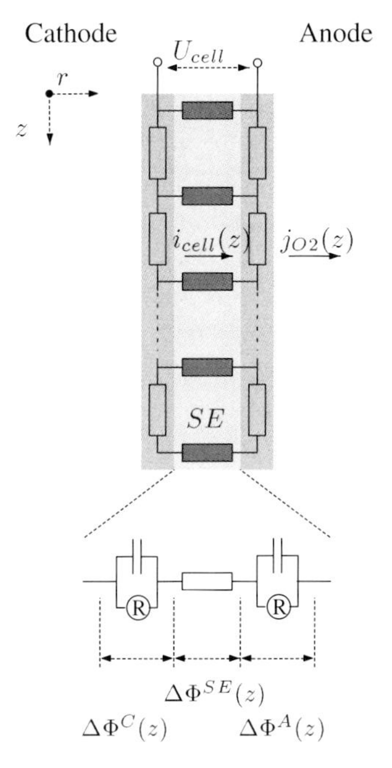

Figure 7.7 Equivalent circuit representing the electrochemical model of the SEMR. Reproduced from (Munder *et al.*, 2005), reprinted with permission from Elsevier.

$$\Delta\Phi^{\text{SE}} = \frac{d^{\text{SE}}}{\sigma^{\text{SE}}(T)}\, i_{\text{cell}} \tag{7.7}$$

$$\frac{\partial \Delta\Phi^{\text{CA}}}{\partial z} = \left(R^{\text{A}} + R^{\text{C}}\right)\frac{A_s}{L^2}\int_{\tilde{z}=z}^{\tilde{z}=L} i_{\text{cell}}\, d\tilde{z} \tag{7.8}$$

The local potential difference between cathode and anode at each point along the z-coordinate is calculated according to Kirchhoff's law (Figure 7.9)

$$\begin{aligned}\Delta\Phi^{\text{CA}} &= \Delta\Phi^{\text{C}} - \Delta\Phi^{\text{SE}} - \Delta\Phi^{\text{A}} \\ &= \Delta\Phi_{00}^{\text{CA}} + \eta^{\text{C}} - \Delta\Phi^{\text{SE}} - \eta^{\text{A}}\end{aligned} \tag{7.9}$$

The cathodic and anodic overpotentials, η^{A} and η^{C}, include both the activation and concentration effects.

Last, the integral of current densities over the reactor length is equal to the total cell current:

$$I_{\text{cell}} = \frac{A_s}{L}\int_{z=0}^{z=L} i_{\text{cell}}\, dz \tag{7.10}$$

and a modified Arrhenius type equation can be applied for the temperature dependence of the SE membrane conductivity:

$$\sigma^{\text{SE}}(T)\cdot T = C^{\text{SE}}\cdot \exp\left(-\frac{E_{\text{A}}^{\text{SE}}}{RT}\right) \tag{7.11}$$

7.3.2
Kinetic Equations for Charge Transfer Reactions

The reaction equations for the oxygen charge transfer reactions on the cathode and anode (Equations 7.12 and 7.13), are as follows:

$$\text{Cathode:}\quad 0.5O_2 + 2\,e^- \rightleftarrows O^{2-} \tag{7.12}$$

$$\text{Anode:}\quad O^{2-} \rightleftarrows 0.5O_2 + 2\,e^- \tag{7.13}$$

The rates of the electrochemical charge transfer reactions can be described by the Butler–Volmer equation (7.14):

$$r_{\mathrm{el}}^{\mathrm{A/C}} = k_0^{\mathrm{A/C}} \exp\left(-\frac{E_{\mathrm{A}}^{\mathrm{A/C}}}{RT}\right)\left[\exp\left(\frac{\alpha^{\mathrm{A/C}}\; nF}{RT}\;\eta^{\mathrm{A/C}}\right) - (\tilde{y}_{\mathrm{O2}})^{0.5} \exp\left(-\frac{(1-\alpha^{\mathrm{A/C}})nF}{RT}\eta^{\mathrm{A/C}}\right)\right] \tag{7.14}$$

7.3.3
Parameters for Charge Transfer and Solid Electrolyte Conductivity

In order to validate the model and to estimate the unknown reaction rate constants and SE conductivity parameters, a series of experiments was conducted. The details of the experimental setup are given by Ye *et al.* (2004) and Ye (2006).

The experimental data published by Ye *et al.* (2004) were used to determine the conductivity constant, C^{SE}, and the activation enthalpy, $E_{\mathrm{A}}^{\mathrm{SE}}$, by a least square fit of the modified Arrhenius-type equation (7.11). Table 7.2 lists the obtained parameter values; Figure 7.8 shows the experimental data and the simulated conductivity as a function of the inverse temperature. From these results, the value for the oxygen conductivity of the SE membrane between 0.04 and 0.24 S m^{-1} within the temperature range 450–550 °C is applied here in the modeling of butane oxidation. The maximal potential difference across the SE membrane is assumed to be 2.0 V, and therefore the highest molar oxygen fluxes that can be reached are 0.5 and 3.0 mmol $m^{-2}\,s^{-1}$, respectively.

Table 7.2 Model parameters for SE conductivity (7.11) and kinetic parameters for the electrochemical reactions (7.14).

Parameter	Value	Unit
C^{SE}	2.8×10^8	$\mathrm{S\ K\ m^{-1}}$
$E_{\mathrm{A}}^{\mathrm{SE}}$	96.9×10^3	$\mathrm{J\ mol^{-1}}$
k_0^{A}	1.46×10^3	$\mathrm{mol\ m^{-2}\ s^{-1}}$
k_0^{C}	1.46×10^4	$\mathrm{mol\ m^{-2}\ s^{-1}}$
$E_{\mathrm{A}}^{\mathrm{A/C}}$	200×10^3	$\mathrm{J\ mol^{-1}}$
$\alpha^{\mathrm{A/C}}$	0.3	

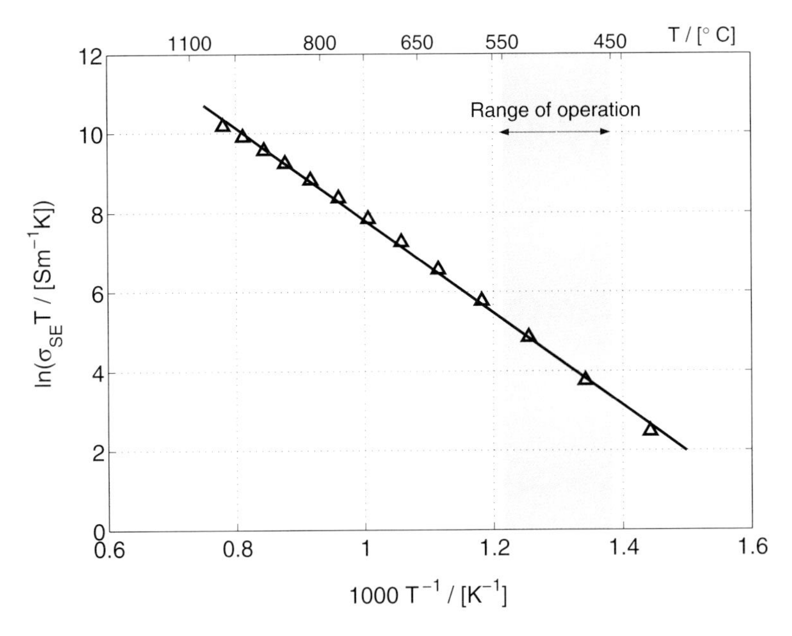

Figure 7.8 Modified Arrhenius plot of measured and simulated solid electrolyte conductivity. Reproduced from (Munder *et al.*, 2005), reprinted with permission from Elsevier.

The characteristic isothermal current–voltage behavior obtained from oxygen-pumping experiments (Ye *et al.*, 2004) were used in order to determine the unknown kinetic parameters for the electrochemical reactions (7.12, 7.13). The experimental setup did not include a reference electrode and therefore the cathodic and the anodic charge transfer processes could not be measured separately. A value of 10 for the ratio of the cathodic to the anodic exchange current density and equal activation energies and charge transfer coefficients were assumed.

Under these assumptions, the parameters of Butler–Volmer kinetics, Equation 7.14, $k_0^{A/C}$, $E_A^{A/C}$, $\alpha^{A/C}$, were determined by a least-square optimization using the SEMR model developed above. Figure 7.9 compares the calculated isothermal current–voltage behavior of the cell at 473 °C and 520 °C to the experimental data and Table 7.2 lists the obtained parameter values.

At this point, some qualitative conclusions can be drawn. High voltages were necessary in the MA synthesis to attain a moderate current density (10–20 A m^{-2}) along the reactor length. For comparison, in SOFCs operating at clearly higher temperatures, the current density might be two orders of magnitude higher. The main reason for the low current density is the low operating temperature, which has to be applied for the maleic anhydride synthesis in the SEMR but which leads to an exponential rise in both the SE membrane conduction resistance and the resistance of charge transfer reactions. Figure 7.9b shows the simulation of partial cell overvoltages at the operating temperature of 473 °C. It was found that not only SE ohmic resistance but also anodic and cathodic charge transfer reactions contributed significantly to the voltage drop in the investigated system.

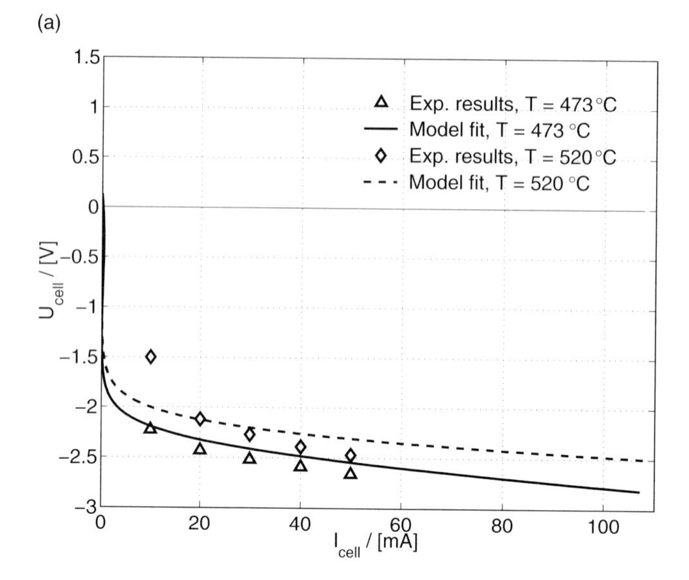

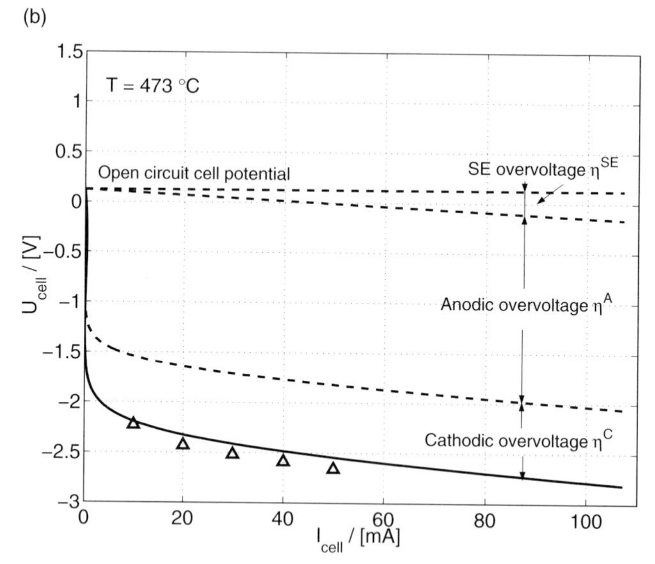

Figure 7.9 Characteristic isothermal current-voltage behavior of the SEMR obtained from oxygen pumping experiments: a) at two different temperatures, and b) analysis of partial cell overvoltages (geometrical surface area $A_s = 23\,cm^2$). Reproduced from Munder *et al.* (2005), reprinted with permission from Elsevier.

7.3.4
Analysis of Maleic Anhydride Synthesis in Solid Electrolyte Membrane Reactor

The partial oxidation of *n*-butane to maleic anhydride (MA) is a complicated synthesis reaction, with the overall reaction given in Equation 7.15.

$$C_4H_{10} + 3.5O_2 \xrightarrow{\text{VPO}} C_4H_2O_3 + 4H_2O \tag{7.15}$$

The VPO catalyst that promotes the above reaction (7.15) is vanadylpyrophosphate $(VO)_2P_2O_7$, which to our knowledge does not exhibit electrocatalytic properties and which has very low electric conductivity (Rihko-Struckmann *et al.*, 2006). Thus, as a first approach it was assumed that the maleic anhydride synthesis in a SEMR takes place on the same route as in conventional catalytic reactors. Preceding molecular oxygen is released to the gas phase by the electrochemical charge transfer reaction as declared in the previous section.

A simple Mars–van Krevelen type kinetic was assumed for the synthesis of MA. The reaction pathway is of the parallel-series type as shown in Figure 7.10.

In detail, the selective formation of MA, Equation 7.16, as well as the butane and MA decomposition to CO_x, Equations 7.17 and 7.18, take place by consuming catalyst lattice oxygen. Only one active site, namely Cat_{ox} is assumed to be active in the oxidation reactions (7.16–7.18)

$$C_4H_{10} + 7\,Cat_{ox} \rightarrow C_4H_2O_3 + 4H_2O + 7\,Cat_{red} \tag{7.16}$$

$$C_4H_{10} + q_2\,Cat_{ox} \rightarrow 4\,CO_x + 5H_2O + q_2\,Cat_{red} \tag{7.17}$$

$$C_4H_2O_3 + q_3\,Cat_{ox} \rightarrow 4\,CO_x + H_2O + q_3\,Cat_{red} \tag{7.18}$$

The reduced catalyst is afterwards re-oxidized by gaseous oxygen, Equation 7.19:

$$Cat_{red} + O_2 \rightarrow Cat_{ox} \tag{7.19}$$

For reactions (7.16–7.19), rate expressions are derived according to the Mars–van Krevelen mechanism, Equation 7.20, assuming initially a quasi-steady-state for the intermediate lattice oxygen species, Cat_{ox}, as well as a first-kinetic order with respect to its concentration, x_{ox}.

$$r_j = k_j^0 \exp\left(\frac{E_A^j}{RT}\right)\tilde{y}_{HC}x_{ox} = k_j y_{HC}\frac{1}{1+\sum_j K_j \dfrac{\tilde{y}_{HC}}{\tilde{y}_{O2}}} \tag{7.20}$$

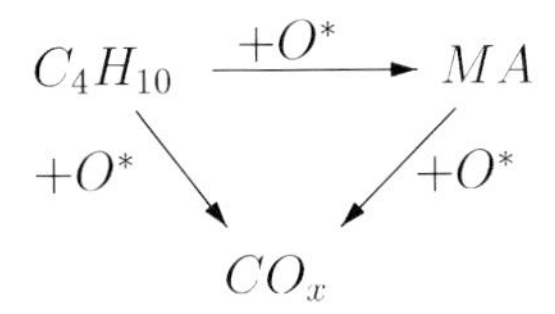

Figure 7.10 Reaction scheme for the synthesis of maleic anhydride (MA).

However, such simple kinetics was not able to predict the experimental results and therefore it was necessary to further modify the model equations for MA synthesis. Retaining the assumed reaction mechanism, the Mars–van Krevelen type rate expressions given in Equation 7.20 for the reactions in Equations 7.16–7.19 were modified by introducing reaction orders, β_j, deviating from 1.0 with respect to the concentration of the oxygen species, x_{ox}, as shown in Equation 7.21. The assumption of a first-order reaction with respect to the hydrocarbon species was kept as was suggested, for example, by Lorences *et al.* (2003) for low butane and MA concentrations.

$$r_j = k_j \tilde{y}_{HC} x_{ox}^{\beta j} \tag{7.21}$$

Furthermore, the quasi-steady-state assumption for the intermediate lattice oxygen species was adopted from the original Mars–van Krevelen approach, and the following implicit algebraic equation was added to the model equations:

$$\underbrace{k_4 \tilde{y}_{O2}^{0.5} (1 - x_{ox})^{\beta 4}}_{\text{Catalyst reoxidation}} = \underbrace{7 \cdot k_1 \tilde{y}_{Bu} x_{ox}^{\beta 1}}_{\text{MA formation}} + \underbrace{q_2 \cdot k_2 \tilde{y}_{Bu} x_{ox}^{\beta 2}}_{\text{Butane total oxidation}} + \underbrace{q_3 \cdot k_3 \tilde{y}_{MA} x_{ox}^{\beta 3}}_{\text{MA total oxidation}} \tag{7.22}$$

With this modified kinetic approach, the differential selectivity for MA is given by:

$$S_{MA,diff}(z) = \frac{r_1 - r_3}{r_1 + r_2} = \frac{1 - \dfrac{k_3\ \tilde{y}_{MA}(z)}{k_1\ \tilde{y}_{Bu}(z)} x_{ox}^{(\beta 3 - \beta 1)}(z)}{1 + \dfrac{k_2}{k_1} x_{ox}^{(\beta 2 - \beta 1)}(z)} \tag{7.23}$$

The results of steady-state butane oxidation experiments (Ye *et al.*, 2004) were used to estimate the kinetic parameters. In order to reduce the number of adjustable parameters, the activation enthalpies, $E_{A,j}$, for all reactions were initially adopted from Hess (2002). By fitting the experimental reactor outlet concentrations of butane and maleic anhydride simultaneously for all experiments, the values for the remaining constants were determined. Table 7.3 summarizes the obtained parameters. Figure 7.11 shows the model predictions for butane conversion, MA selectivity and MA yield, together with experimental results as a function of the oxygen-to-butane feed ratio. Additionally, the selectivity curve for

Table 7.3 Kinetic parameters for the gas phase reactions in MA synthesis (7.16–7.19).

Reaction i (r_i)	k_i^0 (kmol g^{-1} s^{-1})	$E_{A,i}$ (kJ mol^{-1})	β_i
$i = 1$ (7.16)	0.512	80.0[a)]	0.71
$i = 2$ (7.17)	8.605	92.0[a)]	1.03
$i = 3$ (7.18)	227.5	113.0[a)]	0.48
$i = 4$ (7.19)	10.0	102.0[a)]	1.0

a) Adapted from (Hess, 2002).

(a)

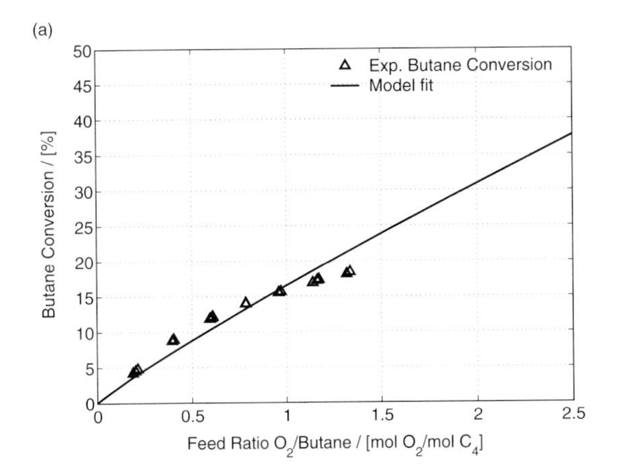

(b)

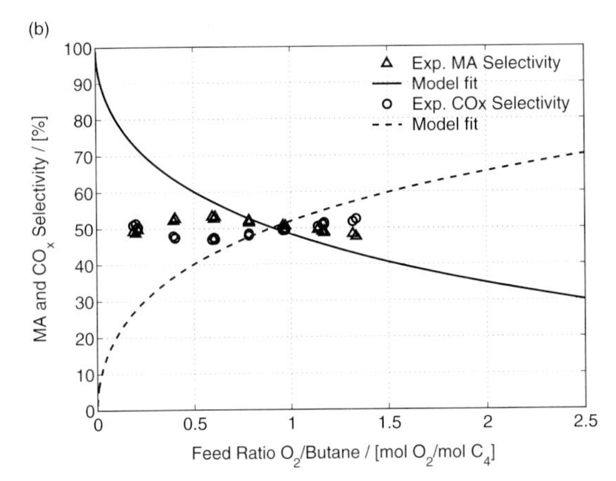

(c)

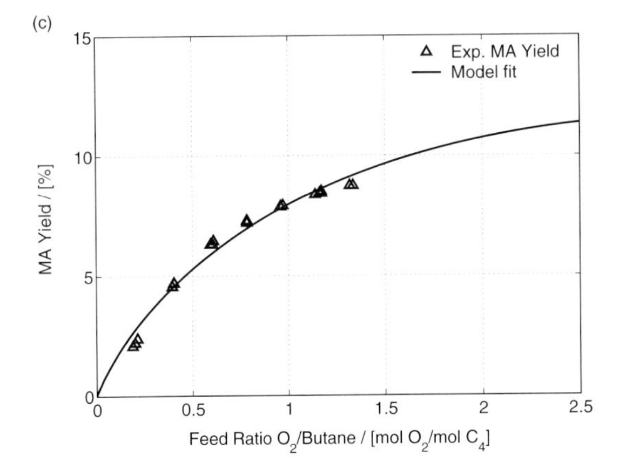

Figure 7.11 Experimental and simulated results as function of the oxygen-to-butane feed ratio ($T = 480\,°C$, $\dot{V}^{A,0} = 32\,ml\,min^{-1}$, $\tilde{y}^0_{Bu} = 0.0056$, $M_{Cat} = 165\,mg$). Reproduced from Munder *et al.* (2005), reprinted with permission from Elsevier.

CO_x, which was calculated assuming the carbon balance to be completely fulfilled, is added to Figure 7.11b and compared with measured results.

Generally, the simulations agreed well with the experimentally determined butane conversion and MA yield. The increasing butane conversion and, as a result, the increasing MA yield for rising oxygen-to-butane feed ratios up to 1.4 are well reproduced. However, there is a notable difference between measured and simulated selectivity to MA and CO_x for oxygen-to-butane feed ratios below 0.5. The decline of the integral MA selectivity for low oxygen-to-butane feed ratios could not be reproduced by the model, as seen in Figure 7.11.

Both a more detailed description of the mass transfer in the catalyst layer (1-D + 1-D) and the implementation of the transient kinetics presented by Huang *et al.* (2002) with re-estimated parameters provided an excellent basis for further analysis of the system (Munder, Rihko-Struckmann, and Sundmacher, 2007). By extending the Huang model with two reactions: (a) the adsorption of carbon species at reduced VPO lattice and (b) the subseqent oxidation of the species:

$$C_4H_{10} + (L) \xrightarrow{r_9} C_4(L) \tag{7.24}$$

$$C_4(L) + qO(S) \xrightarrow{r_{10,11}} 4CO_x + 5H_2O + (L) + q(S) \tag{7.25}$$

the model predictions could be improved, especially for the co-feed conditions.

7.3.5
Analysis of Oxidative Dehydrogenation of Ethane in a Solid Electrolyte Packed-Bed Membrane Reactor

The oxidative dehydrogenation of ethane has been discussed extensive in the previous Chapters. In the present Chapter 7 we discuss the characteristics of carrying out the reaction in a solid electrolyte packed-bed membrane reactor. In the analysis, the basic reactor model including the electron transfer reactions and the ion transfer in the electrolyte as presented above for the electrochemical synthesis of MA was applied also here for the oxidative dehydrogenation of ethane (ODHE) with a $VO_x/\gamma\text{-}Al_2O_3$ catalyst (Chalakov *et al.*, 2007). However, in the present ODHE studies the catalyst particles were placed in a packed-bed solid electrolyte membrane reactor and not fixed with the electrode, as in the example of MA synthesis. The charge transfer reactions–oxygen reduction to ions and the release of gaseous dioxygen on the anode, as well as ion conduction–are comparable to those presented in the previous chapter for MA synthesis. The pseudo-homogeneous, one-dimensional approach for the anodic catalyst layer is fully applicable here. In the reaction network (see Chapter 3, Figure 3.1) ethane reacts in two parallel pathways to ethylene and CO_2 by Equations 3.1 and 3.2. Furthermore, the formed ethylene can be oxidized by two parallel pathways to CO or CO_2 according to Equations 3.3 and 3.4. and the consecutive CO to CO_2 according Equation 3.5. The kinetic model for the oxidative dehydrogenation of ethane originally proposed by Klose *et al.* (2004) and presented in Chapter 3 is applied here also for heterogeneously catalyzed reactions, because it is based on measurements with the same $VO_x/\gamma\text{-}Al_2O_3$ catalyst at the same temperature range as applied in our experiments.

The Mars–van Krevelen mechanism is assumed as well here for the dehydrogenation step, in which ethane reacts with lattice oxygen to form ethylene. The total oxidation of ethane and ethylene (3.2–3.4) is assumed to occur only in the presence of gaseous oxygen by the Langmuir–Hinshelwood mechanism, assuming non-competitive adsorption of the reactants. The last reaction (3.5) in the network is described by a Langmuir–Hinshelwood mechanism with competitive adsorption of the reactants. According to the model presented by Klose *et al.* (2004) it was assumed that oxygen participates in all reactions in a dissociated form giving in the kinetic equations the reaction order of 0.5 for oxygen whereas for all other compounds the reaction order is one. As discussed above for the MA synthesis, the equations of the electrochemical operation (the charge transfer reactions of oxygen on the cathode and the anode by Butler–Volmer kinetics, Equations 7.12–7.14) are included here as well in the OHDE model.

In the experimental investigation of ODHE was found that the ethylene selectivity was not positively influenced in the solid electrolyte membrane reactor operation compared to the packed-bed membrane reactor operation (Chalakov *et al.*, 2007). The selectivity ratio S_{CO_2}/S_{CO} was found to depend on the type of oxygen supply, so that with a gaseous oxygen supply the ratio was always lower than that with an electrochemically supplied oxygen.

The model presented in Chapter 3 (Table 3.3) with the parameter values from Table 3.4 was not directly able to predict the obtained results in co-feed mode. A re-estimation for the parameters in the non-electrochemical reactions was necessary in order to decribe the experimental results for OHD of ethane in the membrane reactor operated in the co-feed mode. After this re-estimation, the conversion, yield and CO_x selectivities were well predicted in co-feed condition without electrochemically supplied oxygen. However, the model applicability was still limited for conditions in oxygen pumping mode (electrochemical supply). The model, including the equations for charge transfer reactions of oxygen on the cathode and the anode (7.12–7.14), was not able to predict the experimental results of the EMR operation conditions with electrochemically supplied oxygen. One reason for the discrepancies between the experimental data and the simulated was concuded to be the additional electrochemically induced side reactions, as suggested already by Chalakov *et al.* (2007), Ye *et al.* (2004) and Ye *et al.* (2005). In contrary to the co-feed mode, where the oxygen on the anode exists only in form of gaseous O_2, additional oxygen species, e.g., O^{2-}, O_2^-, O_2^{2-}, O^-, O (adsorbed) might be present in EMR operation during the electrochemical pumping. Therefore the model was completed finally by implementing two electrochemical side reactions into the reaction network. The charged oxygen species are likely highly active in oxidation reactions, as concluded by Chalakov *et al.* (2007) and therefore two further reactions were added to the model describing the ethylene reaction with O^{2-} either to CO (7.26) or CO_2 (7.27) according to equations:

Table 7.4 Kinetic equations for the electrochemical reactions in the electrochemically supported oxidative dehydrogenation of ethane.

Reaction	Equation
3el	$r_{el}^{CO} = k_{el}^{CO}\left[\exp\left(\frac{\alpha_a^A F}{RT}\Delta\Phi^A\right)\right] y_{C_2H_4}^A$
4el	$r_{el}^{CO_2} = k_{el}^{CO_2}\left[\exp\left(\frac{\alpha_a^A F}{RT}\Delta\Phi^A\right)\right] y_{C_2H_4}^A$
6	$r_{el}^{C} = k_0^C \exp\left(-\frac{E_A^C}{RT}\right)\left[\exp\left(\frac{\alpha_a^C F}{RT}\Delta\Phi^C\right) - (y_{O_2})^{0.5}\exp\left(-\frac{\alpha_c^C F}{RT}\Delta\Phi^C\right)\right]$
7	$r_{el}^{A} = k_0^A \exp\left(-\frac{E_A^A}{RT}\right)\left[\exp\left(\frac{\alpha_a^A F}{RT}\Delta\Phi^A\right) - (y_{O_2})^{0.5}\exp\left(-\frac{\alpha_c^A F}{RT}\Delta\Phi^A\right)\right]$

$$C_2H_4 + 4O^{2-} \rightarrow 2CO + 8e^- + 2H_2O \tag{7.26}$$

$$C_2H_4 + 6O^{2-} \rightarrow 2CO_2 + 12e^- + 2H_2O \tag{7.27}$$

The backward reactions were assumed to be unlikely and therefore the kinetics of the reactions in Equations 7.26 and 7.27 were described by the simple Tafel equation.

The kinetic equations for the electrochemical charge transfer and side reactions (7.14, 7.26, 7.27) are summarised in Table 7.4 and all parameter values in Table 7.5. The kinetic equations for the heterogeneously catalyzed gas phase reactions were taken directly from Table 3.3. The model predictions calculated with this extended model, including the electrochemical side reactions, are presented on Figure 7.12. After the implementation of the additional electrochemical reactions (7.26 and 7.27), the model simulations agreed well with the experimental results. The selectivities towards the three products and the ethylene yield were well predicted with the extended model. Due to the oxygen-consuming, competing electrochemical side reactions, the reaction with ethane is depressed to some extent.

As seen in the examples of the maleic anhydride synthesis and the oxidative dehydrogenation of ethane in solid electrolyte membrane reactors, conventional reaction models and the equations which are presented in literature are usually not directly applicable to conditions where reagents are supplied electrochemically. The models describing solid electrolyte membrane systems become more complex as the electrochemical aspects have to be included. The kinetic equations for charge transfer reactions of the reagents on the cathode and the anode described by Buttler–Volmer equations have to be included in the model. The charge balances characterizing the anodic and the cathodic potential difference across the interfaces between the electrolyte and the electrodes have to be taken into consideration. The ohmic potential loss in the electrolyte and in the electrodes is described by the corresponding equations, and the temperature

Table 7.5 Kinetic parameters for the gas phase reactions in oxidative dehydrogenation of ethane (Equations in Table 7.4).

Reaction	k_{0i}	$E_{A,i}$ (kJ mol^{-1})
Ox	1.9×10^3 (mol kg^{-1} s^{-1})	94
Red	1.7×10^3 (mol$^{0.5}$ m$^{1.5}$ kg^{-1} s^{-1})	

Reaction i (r_i)	k_{0i} (mol kg^{-1} s^{-1})	$E_{A,i}$ (kJ mol^{-1})
$i = 2$	$4.4 \cdot \times 10^3$	114
$i = 3$	19.6	51
$i = 4$	0.3	51
$i = 5$	45.1×10^3	118

Electrochemical reactions	k_{0i} (mol m^{-2} s^{-1})	$E_{A,i}$ (kJ mol^{-1})	$\alpha_{a/c}^{A/C}$
$i = 3$el	1.1×10^{-2}	220	0.3
$i = 4$el	0.9×10^{-2}	180	0.3
C (reaction 6)	1.46×10^3	130	0.3
A (reaction 7)	1.46×10^4	130	0.3

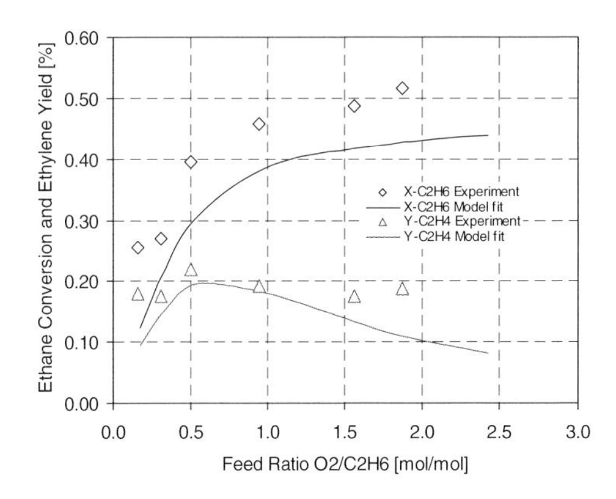

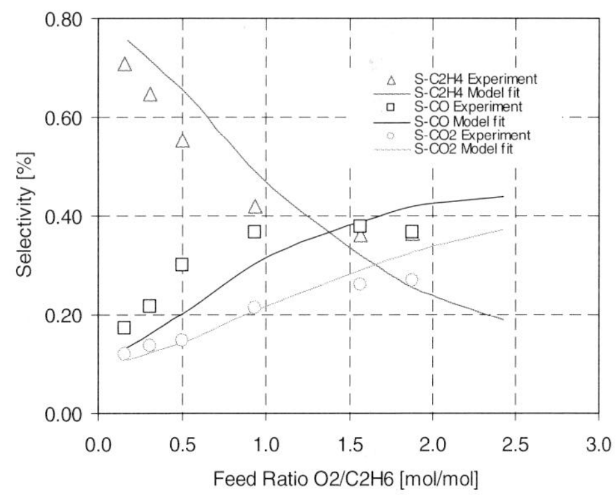

Figure 7.12 Experimentally obtained and model predicted ethane conversion, ethylene yield and selectivities as a function of the oxygen-to-ethane feed ratio during EMR operation ($T = 580\,°C$; $M_{Cat}/\dot{V} = 0.013$ g min^{-1} cm^{-3}). The predictions with the extended model include the additional electrochemical side reactions. Reproduced from (Chalakov *et al.*, 2009), reprinted with permission from Elsevier.

dependence of the electrolyte conductivity is described by modified Arrhenius type equations. The overall cell potential might be calculated according to Kirchoff's law. As clearly seen in the both examples, additonal electrochemical reactions are likely occuring in the reaction system where one reagent is supplied electrochemically. The existence of more active reagent species on the electrode is difficult to confirm experimentally, but they are likely and the reactions due to these species are to be considered in the model. However, the engineering approaches as presented here in the above examples and in the cited publications (Munder *et al.*, 2005; Munder, Rihko-Struckmann, and Sundmacher, 2007; Chalakov *et al.*, 2007) including both the electrochemical and heterogeneously catalyzed reactions as well as the reaction engineering aspects are to date rare in the literature, but are necessary for determining the optimal operation conditions in such complex systems.

7.4
Membrane Reactors Applying Ion-Conducting Materials

In the last part of this chapter some examples of various systems are given where, according to the literature, either proton or oxygen ion-conducting materials have been used as a membrane electrolyte. When one considers financial funding, most resources have been focused on the development of fuel cells. Due to the abundant amount of publications – review articles as well as books describing the recent development of different kind of fuel cells (Larminie and Dicks, 2003; Srinivasan, 2006; Basu, 2007; Spiegel, 2007; Barbir, 2005) – we discuss here the aspects of fuel cells only briefly.

7.4.1
High-Temperature Oxygen Ion Conductors

7.4.1.1 Solid Oxide Fuel Cell for Electrical Energy Production

A technology which uses oxygen ion-conducting membranes semi-commercially is the solid oxide fuel cell (SOFC) for electrical energy production. Electrochemical energy conversion can be carried out highly efficiently in solid oxide fuel cells. The most valuable benefits of SOFC technology compared to the other fuel cells are that a wide variety of fuels can be processed and the catalyst tolarates sulfur due to the high operating temperatures (>800 °C). A significant benefit is also the high achievable electrical net efficiency, which for small 1 kW units is about 50%; for large pressurized SOFC/gas turbine systems, electrical efficiencies greater than 65% are expected (Weber and Ivers-Tiffée, 2004; Williams, 2007).

The developers of SOFC technology among others are Siemens–Westinghaus with the 250 kW multi-tubular units, Rolls–Royce with a segmented series arrangement of individual cells (Gardner *et al.*, 2000), Allied–Signal Aerospace Company with the monolith concept (Larminie and Dicks, 2003) and a joint venture company Staxera (2009) and Hexis (2009) with planar technology providing sophisticated heat management with integrated indirect reforming. So far the development of SOFC has concentrated mostly on stationary units, but a new, attractive field for this technology is SOFC-based auxiliary power units (APU) for mobile applications (Pfafferodt *et al.*, 2005; Grube, Hohlein, and Menzer, 2007).

Regardless of the present success of pilot tests with pre-commercial SOFC units, there still exist severe challenges of the SOFC technologies. The present high operating temperature results in many inherent problems, such as low long-term stability of the materials, cell degradation due to mechanical stress, electrode sintering and slow start-up. The thermal expansion coefficients of the fuel cell components – electrolyte, electrode layers and interconnections – have to match well with each other, otherwise the thermal stress causes delamination at the unit interfaces or cracking of the electrolyte. At the prevailing temperature further demands are high chemical compatibility and stability, as discussed in detail in the review by Weber and Ivers-Tiffée (2004). Increased long-term stability and a decrease in system cost would be possible, if the development of low or intermediate temperature SOFC technologies (600–800 °C) is successful in the near future.

One idea in the field of SOFC technology based on the oxidation of hydrocarbons deviates from the strict separation of fuel and oxidant (Hibino *et al.*, 2000a, 2002b; Yano *et al.*, 2007; Buergler, Grundy, and Gauckler, 2006). The working principle of the single chamber cell is based on the highly selective catalytic activity of the electrodes. The criteria for the electrodes are: (a) one electrode has to be electrochemically active for the oxidation of the fuel but should be inert to oxygen reduction, (b) the other electrode (cathode) should be highly active for the reduction of oxygen but inert in the oxidation reaction. High concentrations of the partial oxidation products, hydrogen and CO, cover the immediate vicinity of the anodic electrode and lead the electrochemical oxidation of these gases. At present, ideally selective anode and cathode materials for fuel oxidation and oxygen reduction are yet not available for single cell fuel cells, and further efforts are needed to realize these. The energy conversion efficiency of such systems is still low, due to the large amount of unreacted fuel.

7.4.1.2 Oxidative Coupling of Methane to C_2 and Syngas from Methane

The conversion of methane into higher hydrocarbons, such as the C_2 coupled products, ethane and ethylene, or syngas production by partial oxidation, has significant commercial importance. Methane is an abundant natural resource, but its direct transport is not highly economical and its conversion to upgraded products is highly challenging. Recent reviews cover the research on methane coupling and hydrogen production in the past 20 years with the use of solid electrolyte membane reaction appyling either O^{2-}, H^+ or MIEC (O^{2-}, e^-) electrolytes (Athanassiou *et al.*, 2007; Stoukides, 2000). The option of using solid oxide membranes for supplying the necessary oxygen for methane activation has several economical and environmental advantages over the direct use of air as oxidant. The membrane is impermeable to nitrogen providing only oxygen for the reactions, thus avoiding NO_x formation. In addition, in a solid electrolyte process, there are the combined possibilities of electrochemical enhancement of products selectivity, simultaneous generation of electrical power and, in general, more control over the reaction pathway.

In the electrocatalytic partial oxidation of methane to synthesis gas in a SE membrane reactor, the methane feed stream does not contain oxygen. It is transferred directly into the reaction zone through the membrane by passing an anodic current through the cell. The electrocatalytic oxidation of methane has been intensively investigated both in co-generation mode as well as in oxygen pumping conditions with various catalysts (Athanassiou *et al.*, 2007). The main advantage in such systems is that the SE membrane prevents the reaction mixture from explosion, since CH_4 and O_2 (air) are separated by the YSZ electrolyte.

Research has been focused also towards the production of partially oxidized higher hydrocarbons, e.g., ethane, propane and propene (Takehira *et al.*, 2002, 2004; Hellgardt, Cumming, and Al-Musa, 2005), as well as butane as detailed above (Ye *et al.*, 2004, 2006). Under oxygen-pumping conditions with MoO_3 as catalyst, the evolution of gaseous oxygen was observed in the operation with alkanes (Takehira *et al.*, 2004). Alkenes were more active and they were oxidized with high selectivity without oxygen evolution. Ethane and propane were found to be inert in a cell configuration MoO_3/Au|YSZ|Ag, while isobutane was partially oxidized to methacrolein. The highest selectivity (73%) was obtained for methacrolein from isobutene. In a comparison of experiments using V_2O_5 as catalyst in the same cell under oxygen-pumping conditions, both alkanes and alkenes were oxidized. Isobutane was oxidized to methacrolein with low selectivity, and propane formed propene by oxidative hydrogenation. Ethane was slowly oxidized to CO_2. With V_2O_5 as catalyst the evolution of gaseous oxygen was observed in all reactions.

7.4.1.3 **Dry Reforming of Methane**

The dry reforming of methane with CO_2 provides the possibility to enhance natural gas utilization and to convert the carbon resources inherently contained in CO_2 and CH_4 into synthesis gas. Besides of the most conventional steam reforming of methane for hydrogen production, there are hundereds of publications concerning the catalytic aspects of dry reforming. The reforming of CH_4 by CO_2 in a solid electrolyte membrane reactor has some advantages over catalytic reforming. Catalyst deactivation by coking is the most severe limitation of the process, and it can be suppressed to some extent by the oxygen ions being directly supplied to the catalyst being fixed on the electrode layer of the SE membrane.

The effect of electrochemically supplied oxygen in the reforming of CH_4 with CO_2 have been investigated, among others, by Belyaev *et al.* (1998) and by Moon and Ryu (2003) in a Pt|YSZ|Pt or a NiO–MgO|YSZ|(La,Sr)MnO_3 cell, respectively. During oxygen pumping conditions, the system stability was improved as the catalyst stability was increased by electrochemically supplied oxygen to the anodic catalyst layer. The observed current was generated with the reactions of the electrochemically supplied oxygen with CH_4, CO, H_2 or surface carbon formed during the internal reforming of CO_2.

7.4.2
High-Temperature Proton Conductors

A typical proton-conducting ceramic material such as $SrCe_{0.95}Yb_{0.05}O_{3-\alpha}$ is a solid solution based on the perovskite-type oxide $SrCeO_3$, in which Ce is partly replaced by Yb. Other perovskite-type oxides – based on $SrCeO_3$ or $BaCeO_3$ – in which some trivalent cations are partially substituting cerium, show protonic conductivity, too. The general formulas are $SrCe_{1-x}MxO_{3-\alpha}$ or $BaCe_{1-x}MxO_{3-\alpha}$ where M stands for a certain rare earth element, x is less than its upper limit of solid solution formation range (usually less than 0.2) and α represents the oxygen deficiency per unit formula. The ceramics of these perovskite-type oxide solid solutions exhibit p-type electronic (hole) conduction under oxidizing atmosphere free of hydrogen or water vapor at high temperatures (Iwahara *et al.*, 2004).

Proton conductivity of these oxides can be obtained if protonic defects exist. There are two mechanisms for proton conduction. The first is the Grotthus mechanism, in which the proton jumps between adjacent oxygen ions. The second mechanism is hydroxyl ion migration, also called the "vehicle" mechanism.

Protonic defects can be formed by reaction between water molecules and oxygen vacancies according to the equation:

$$O_O^x + V_O^{\bullet\bullet} + \cdot H_2O(g) \rightarrow 2OH_O^{\bullet} \quad (7.28)$$

where two effectively positive hydroxyl groups on regular oxygen positions are formed. Another important mechanism forming protonic defects is the reaction of hydrogen with electron holes according to:

$$2h^{\bullet} + 2O_O^x + H_2 \rightarrow 2OH_O^{\bullet} \quad (7.29)$$

for which the presence of excess holes is obviously necessary.

The conductivities in hydrogen atmosphere are in the order of 10^{-3}–10^{-2} S cm^{-1} at 600–1000 °C. Proton conduction has been measured by electrochemical hydrogen transport experiments in hydrogen- or water vapor-containing atmosphere. The role of water vapor can be seen in Figure 7.13 (Matsumoto *et al.*, 2001). It was found that the hydrogen evolution rate obeyed Faraday's law up to very high current densities, using humidified cathode carrier gases. Within the examined range of water vapor pressures, 6.6×10^2 to 2.3×10^3 Pa, the current efficiency was almost unity until current densities of 450–600 mA cm^{-2} were reached, which was about ten times larger than in the operating case where dry carrier gas was used.

A high-temperature proton-conducting membrane can be used in various applications, as a sensor of hydrogen or hydrocarbons, for the separation of hydrogen, energy conversion and the synthesis of chemicals, as discussed briefly in the following sections. Recent studies also clarify the opportunitities of using proton conductors for nuclear fusion process, where hydrogen isotopes (deuterium and tritium) must be extracted from reactor core exhaust gas containing He and other elements (Iwahara *et al.*, 2004).

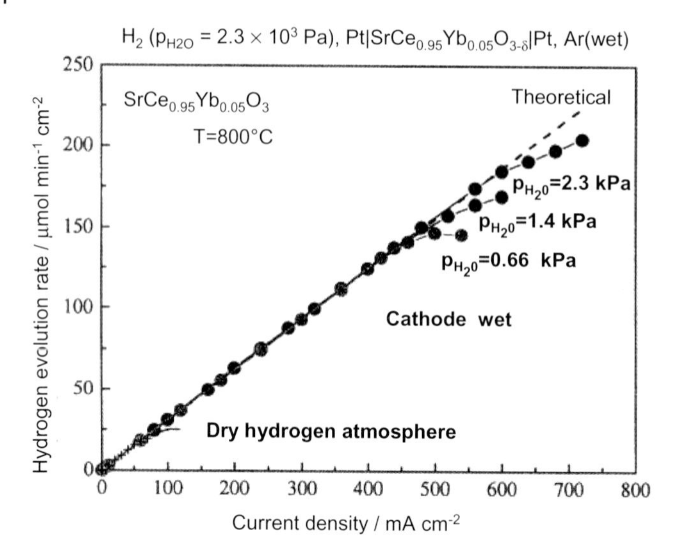

Figure 7.13 Hydrogen pumping using proton-conducting ceramic: H_2 evolution rate at cathode versus cell current density. Reproduced from (Matsumoto *et al.*, 2001), reprinted with permission from the Electrochemical Society.

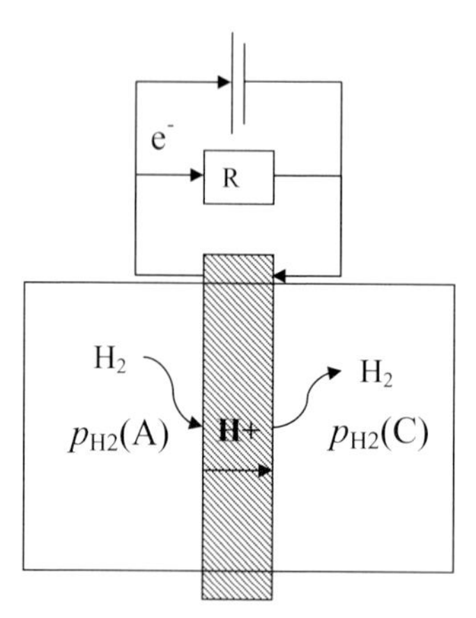

Figure 7.14 Working principle of hydrogen concentration cell using a proton-conducting membrane.

7.4.2.1 Hydrogen Sensors and Pumps

In the open circuit mode (OCM), proton-conducting electrolyte membranes can be used to detect hydrogen, steam, alcohol or hydrocarbons (e.g., a leak detector for chemical plants or coal mines). The working principle of a sensor with proton-conducting membrane is based on the principle of the electrochemical hydrogen concentration cell (see Figure 7.14) where the theoretical open circuit cell voltage (OCV) is $E^0_{cell} \sim \ln[P_{H2}(A)/P_{H2}(C)]$, where $P_{H2}(A)$ and $P_{H2}(C)$ are the partial pressures of hydrogen in each electrode compartment. Therefore, the OCV can be used as a signal for hydrogen activity if P_{H2}(A) or P_{H2}(C) is known. For hydrocarbon or alcohol sensors both single- and two-chamber constructions are possible (Iwahara *et al.*, 2004). In the single-chamber sensor, the electrodes possess different activity. Only one electrode is active for hydrocarbon oxidation in air. Therefore, the separation of electrode compartments is not necessary and no standard gas is needed, which are the major advantages of this sensor variant.

As a further application of proton-conducting membranes, hydrogen can be selectively separated from gas mixtures, for example, containing compounds such as water and hydrogen disulfide, by proton pumping at close-circuit conditions.

7.4.2.2 Fuel Cells

During the past two decades, researchers in the field of solid oxide fuel cells paid much attention to the preparation and characterization of solid oxide materials with protonic conductivity (Grover Coors, 2003; Hassan, Janes, and Clasen, 2003; Fehringer *et al.*, 2004; Taherparvar *et al.*, 2003). Historically, one of first works on high-temperature proton-conducting fuel cells was published by Iwahara *et al.* (1981).

The interest in protonic conductors and their utilization in fuel cells is high because complete hydrogen utilization could easily be attained in a SOFC based on a protonic electrolyte. Protonic ceramic fuel cells are targeted for operation at 55–65% electrical efficiency with pipeline natural gas as feed. This can only be achieved with greater than 90% direct methane fuel utilization. Such high fuel utilization is made possible by two major factors. First, high thermochemical efficiency of reforming and water shift reactions at the anode is possible at the high operating temperatures of 700–800 °C. Second, water vapor is produced at the cathode where it is subsequently swept away by the air flow, rather than at the anode where it would dilute the fuel (carbon dioxide is the only anode exhaust gas).

7.4.2.3 Electrocatalytic Membrane Reactors

Moreover, proton-conducting membranes are applicable in electrocatalytic reactors for hydrogenations and dehydrogenations of organic compounds, methane activation, steam electrolysis, water gas shift, ammonia reactions as listed in the review by Kokkofitis *et al.* (2007). Several reactor concepts are illustrated in Figure 7.15. Unique features of proton-conducting membrane reactor concepts compared to traditional catalytic reactors are:

- hydrogen and the compounds to be hydrogenated or dehydrogenated are kept separated by the membrane,
- the chemical potential of hydrogen at the reaction sites and the reaction rate can be controlled via the electrode potential or via the electric current,
- hydrogenation and dehydrogenation of organic compounds on either side of the membrane can be carried out simultaneously in a single unit.

In the methane coupling, the formation of ethane and ethylene was enhanced by applying an electric potential difference to the reactor (Hamakawa, Hibino, and

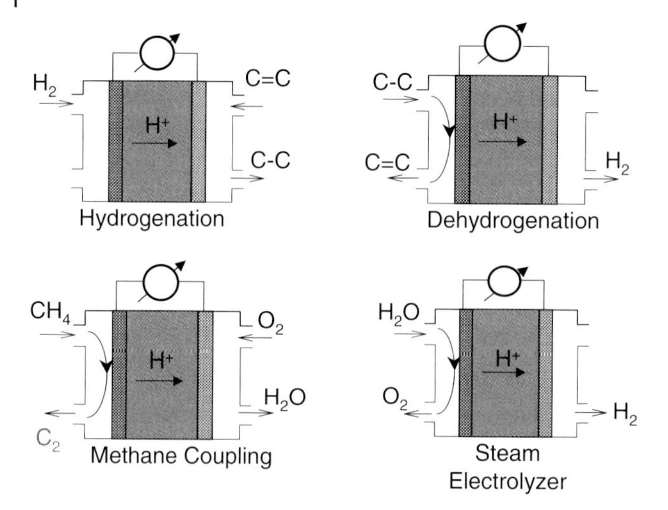

Figure 7.15 Examples for the use of proton-conducting membranes in electrocatalytic membranes reactors. Reproduced from (Sundmacher, Rihko-Struckmann, and Galvita, 2005), reprinted with permission from Elsevier.

Iwahara, 1993). Another application was the reduction of NO occurring in automobile exhaust lines (Kobayashi *et al.*, 2002). The reduction of NO by hydrogen, which was produced by a steam electrolysis cell, was tested with different catalysts on the cathode side. A mixture of Pt-sponge and Sr/Al_2O_3 was found to be the most active catalyst for the preferred reduction of NO in excess of O_2. Recently, the hydrogenation of CO_2 was demonstrated successfully with proton-conducting membranes strontia–zirconia–yttria perovskite and the working (cathodic) electrode was a polycrystalline copper film (Karagiannakis, Zisekas, and Stoukides, 2003). The observed reaction rates are about one order of magnitude higher than under normal catalytic conditions, when hydrogen is supplied electrochemically. Recently proposed application of a high-temperature proton-conducting membrane is the dehydrogenation of propane over Pt and Pd to produce propylene and hydrogen (Karagiannakis, Kokkofitis, and Zisekas, 2005).

7.4.3 Low-Temperature Proton Conductors

The most established low-temperature proton conductor is the polymer electrolyte membrane (PEM) material Nafion, which is commercially available from DuPont. This membrane is a fully fluorinated polymer-ether backbone having sulfonic acid groups. After solvatization of the acid groups with water, the PEM exhibits protonic conductivity. Due to the necessity of humidification and due to limitations regarding maximum operating temperatures (<120 °C) of Nafion-type materials, intense research activities aim to develop new proton conductors. The operation of PEM-based reactors, especially fuel cells in transportation, would be clearly simplified if the membranes could work without any humidification up to temperatures 200 °C. A promising material being suitable for higher temperatures is polybenzimidazole (PBI) (He *et al.*, 2007; Li *et al.*, 2004). The humidification-free operation of PBI membranes at higher temperatures allows a higher CO concentration in hydrogen fuel cells (PEMFC) fed with reformate gas because at higher temperatures CO adsorption at the anode catalyst (Pt) is of less importance. Other possible membranes for PEM fuel cells are polymer–ceramic composite protonic conductors (Savadogo, 2004; Alberti and Casciola, 2003) and polyaromatic polyheterocyclic materials such as polysulfones (PSU), polyethersulfone (PES), polyetherketone (PEK), polyetheretherketone (PEEK) and polyphenyl quinoxaline (PPQ) (Roziere and Jones, 2003). The latter materials have to be doped with appropriate acids to achieve the desired proton conductivity.

7.4.3.1 PEM Fuel Cells

Today, PEM materials are used as low-temperature proton-conducting membranes in energy conversion where the total oxidation of the fuel with maximal electrical energy production is the primary goal. The applications of low-temperature fuel cells have reached semi-commercial stage and several test units have been launched to the market. The operation of low-temperature proton-conducting membrane fuel cell using hydrogen as feed is effective, and high current densities can be reached. However, the use of gaseous hydrogen as a feed brings many logistic and safety problems to solve and therefore research efforts have been focused over several decades to develop safe storage and transportation systems for various forms of hydrogen.

Due to the mentioned difficulties in the operation of hydrogen fuel cell, a competitive fuel cell technology operating at low temperature is the direct methanol fuel cell (DMFC), where a liquid methanol–water solution is directly used as the anode feed. Handling liquid methanol is less complicated than handling gaseous hydrogen, and therefore the DMFC is a very promising low-temperature fuel cell technology, especially for transportation applications. The working principle of this fuel cell is presented in Figure 7.16. A review by Schultz, Zhou, and Sundmacher (2001) discussed the status and trends of DMFC technology in detail. A detailed analysis of the kinetic aspects of the electrochemical oxidation of methanol were carried out by Vidakovic, Christov, and Sundmacher (2005); Vidakovic *et al.* (2007); and Krewer *et al.* (2006). A severe limitation of DMFC operation is caused by deactivation of the anodic electrode catalyst (Pt/Ru) due to the irreversible adsorption of the reaction intermediate CO on catalyst active sites. Another limitation comes from the undesired transport of methanol from the anodic compartment through the membrane to the cathodic side (methanol cross-over). Cross-over and the direct oxidation of methanol on the cathode lead to a reduced cathode potential and thereby to a reduced overall cell voltage. A detailed mathematical analysis of methanol cross-over and the transport mechanisms in PEM was reported by Sundmacher *et al.* (2001) and Schultz *et al.* (2007).

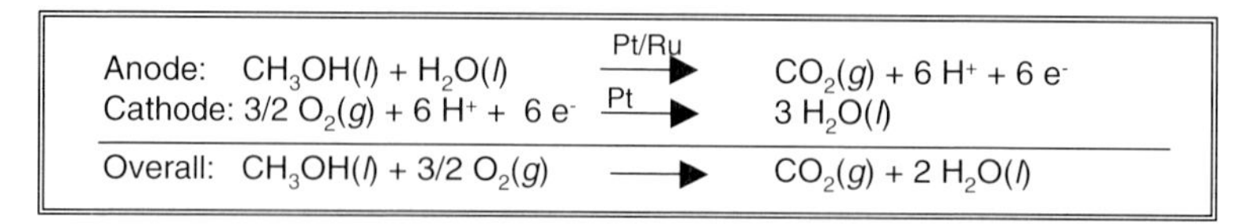

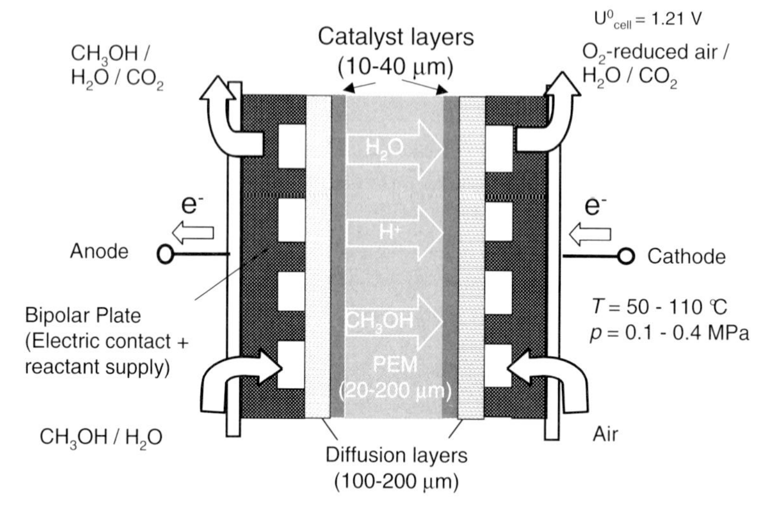

Figure 7.16 Working principle of direct methanol fuel cell (DMFC). Reproduced from (Sundmacher, Rihko-Struckmann, and Galvita, 2005), reprinted with permission from Elsevier.

7.4.3.2 **Proton Exchange Membrane Reactors**

The application of polymeric proton exchange membranes (PEM) in chemical reactors at low temperatures (<120 °C) is not common. Only a few examples of such PEM reactors–operated in electrolysis mode or in fuel cell mode–can be found in the open literature. Theoretically these reactors can be used for specific oxidation, dehydrogenation and hydrogenation reactions, as some interesting examples in the literature show. The operating modes of these reactors depend on the reactions applied. In an optimal case, the co-generation of electrical energy and valuable chemical products has been successful. Figure 7.17a,b present schematic illustrations of PEM reactor configurations which are discussed in more detail in the following.

Electrolysis Operating Mode The oxidation of various aliphatic alcohols with oxygen produced by *in situ* water electrolysis can be taken as an example of electrolysis carried out in a polymer electrolyte membrane reactor (Simond and Comninellis, 1997). Water was fed together with the various aliphatic alcohols to the anodic side of Nafion 117 membrane reactor, where IrO_2 deposited on the porous titanium layer worked as anode. The hydroxyl radicals formed in the electrolysis were interacting with the anode forming the higher oxide IrO_3, which was either reactive in the alcohol oxidation or evolved gaseous oxygen in the catalyst redox reaction (IrO_3/IrO_2). The tested alcohols showed remarkable differences in

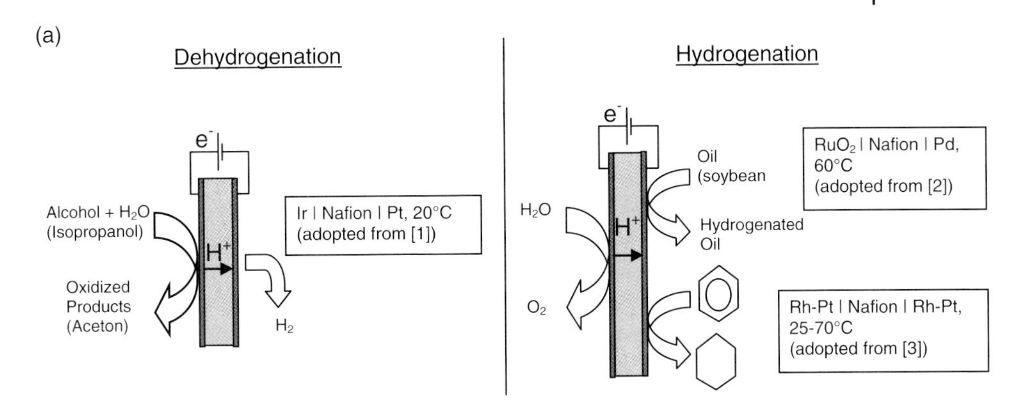

[1][Simond and Comninellis, 1997], [2][An et al., 1999], [3][Itoh et al., 2000]

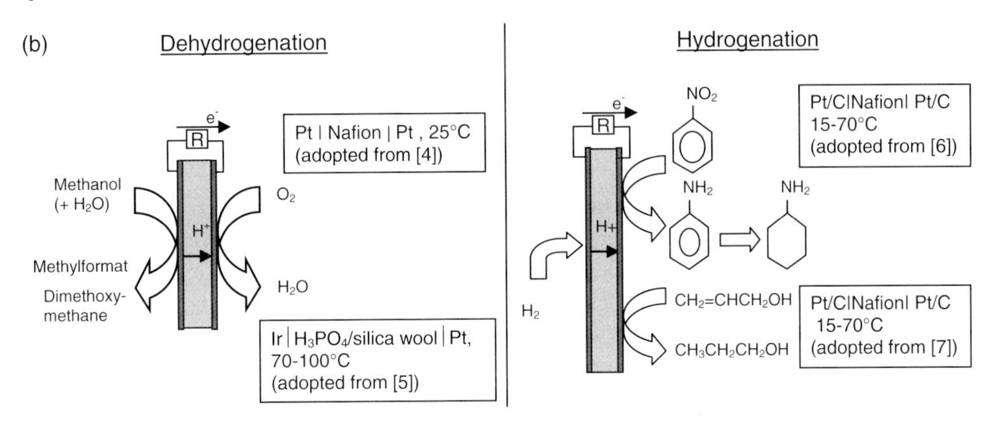

[4][Yamanaka and Otsuka, 1988], [5][Otsuka et al.. 2003], [6][Yuan et al., 2001], [7][Yuan et al., 2003]

Figure 7.17 a) Proton exchange membrane (PEM) reactors: examples for electrolysis operating modus. Reproduced from (Sundmacher, Rihko-Struckmann, and Galvita, 2005), reprinted with permission from Elsevier. b) Proton exchange membrane (PEM) reactors: examples for fuel cell operating modus. Reproduced from (Sundmacher, Rihko-Struckmann, and Galvita, 2005), reprinted with permission from Elsevier.

reactivities, secondary alcohol, isopropanol being the most reactive, followed by ethanol and methanol, with *n*-propanol having clearly the lowest reactivity.

The electrolysis of water can be applied also in hydrogenation reactions, as two examples show: the hydrogenation of benzene (Itoh *et al.*, 2000) or soybean oil (An *et al.*, 1999; An, Hong, and Pintauro, 1998). In both systems, the electrolysis of water was carried out on the anode, where O_2 and H^+ were formed electrochemically. Protons migrated through the membrane, and on the cathode either atomic or molecular hydrogen was consumed for hydrogenations. Initially electrochemical and non-electrochemical hydrogenation of benzene (co-feed) were compared

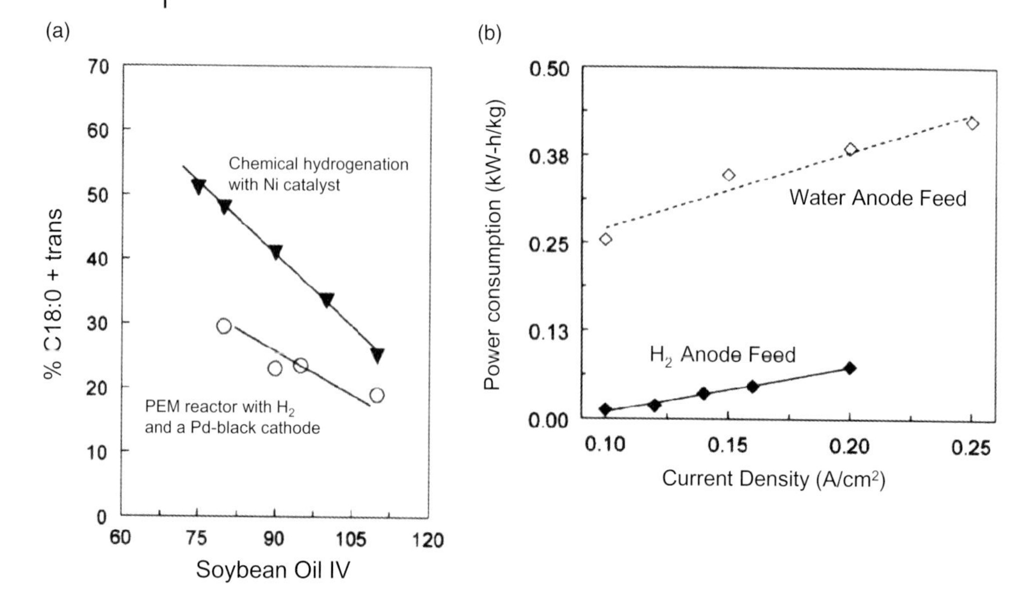

Figure 7.18 a) Comparison of reactor type (PEM reactor with Pd black cathode versus chemical hydrogenation with Ni) in soybean oil hydrogenation. Reprinted with permission from (Pintauro *et al.*, 2005). Copyright American Chemical Society. b) Dependence of the PEM reactor power consumption from the anode feed (H_2O vs H_2) for partially hydrogenated soybean oil. Reprinted with permission from (Pintauro *et al.*, 2005). Copyright American Chemical Society.

when hydrogen was pumped as protons through the membrane, and it was evident that the production rate of cyclohexane was much higher during solid electrochemical operation.

The direct electrochemical pumping of hydrogen in a PEM was applied for the selective hydrogenation of soybean oil (Pintauro *et al.*, 2005). In the partially hydrogenated oil product, the nutrition quality of the product was improved by PEM reactor processing, as a lower percentage of *trans*-fatty acid isomers and saturated stearic acids was detected in the product oil after hydrogenation in a PEM reactor. These undesired components existed in higher concentrations after a comparable conventional high-temperature chemical hydrogenation with a Ni catalyst (see Figure 7.18a). Furthermore, it was found that the utilization of gaseous hydrogen clearly decreased the power consumption compared to the operation including water electrolysis, as seen in Figure 7.18b.

Fuel Cell Operating Mode The contributions of the research group of Otsuka has been considerable in the field of proton-conducting membrane reactors, see for example, (Otsuka, Yamanaka, and Hosokawa, 1990; Otsuka and Yamanaka, 1990, 2000; Yamanaka and Otsuka, 1991; Otsuka, Hashimoto, and Yamanaka, 2002). In their first study in 1988, Otsuka *et al.* used a Pt-bounded Nafion 117 membrane as electrolyte at room temperature in a PEM-type membrane reactor. The studied reaction was the partial oxidation of methanol in the gas phase (Yamanaka and

Otsuka, 1988). Under open circuit conditions (E^0_{cell}) the main product was CO_2 (>95% selectivity). In the electrochemical operation the valuable intermediates methyl formate and methylal were formed as main products, and only traces of CO_2 were observed. However, due to high internal cell resistances, the attained electric current and accordingly the rate of oxygenate formation remained low.

The low productivity due to the low conductivity of solid polymer electrolyte membranes in the preliminary study gave the researchers the impulse for developing a new ion-conducting membrane from phosphoric acid-impregnated silica wool. They reported the successful oxidation of both alkanes (Yamanaka, Hasegawa, and Otsuka, 2002) and methanol in this system (Otsuka, Ina, and Yamanaka, 2003). The configuration allowed them to increase the temperature, which resulted in increased conductivity. The partial oxidation of methanol was studied at temperatures of 70–100 °C with a cell configuration of CH_3OH|noble metal|H_3PO_4 on silica wool|Pt|O_2.

The operation was carried out under short circuit conditions without a resistance in the external circuit. The cell performance was strongly deteriorated by the cross-over of methanol from the anode to the cathode.

One successful co-generation of energy and valuable products was published by Yuan *et al.* (2001). The selective hydrogenation of nitrobenzene to cyclohexylamine was carried out in fuel cell mode, simultaneously producing electric power. The anode and cathode were prepared by hot pressing a carbon-supported Pt-catalyst on a Nafion 117 membrane. The measurements were carried out in batch recycle mode for nitrobenzene, obtaining an open circuit voltage of 0.32 V at 70 °C. Negligible amounts of undesired over-oxidation products, cyclohexylamine and aniline, were observed. In fuel cell operation the maximum power density was 1.5 mW cm^{-2} obtained at a current density of 15 mA cm^{-2}. A reaction time of 2 h gave 8.2% conversion of nitrobenzene, the selectivities being 57.3% and 28.2% to cyclohexylamine and aniline.

Another example of fuel cell operation in a chemical reactor from the same group is the hydrogenation of allyl alcohol to 1-propanol accompanied by the co-generation of electrical energy (Yuan *et al.*, 2003). The selected hydrogenation reaction might not be reasonable in the economic sense, but it can be seen as an interesting model reaction for co-generation in a PEM reactor. The open circuit potential was experimentally determined to be between $E^0_{cell} = 0.23$ to 0.27 V although the standard open circuit potential was calculated to be 0.477 V. The maximum power density was 6.2 mW cm^{-2} at a current density of 66 mA cm^{-2}. Rather low conversions of allyl alcohol were reported (2.22% in 6 h), but the selectivity to 1-propanol was very high.

One special application of membrane reactors working in a fuel cell mode was reported by Sundmacher and Hoffmann (1999). The electrochemical chlorine separation from a nitrogen stream was carried out in a PEM reactor. The reactor operated as a H_2/Cl_2 fuel cell having an open circuit voltage of 1.36 V. A thin polymer electrolyte membrane layer was applied as a barrier layer between the anode and the liquid electrolyte (HCl) to prevent the break-through of H_2 bubbles into the liquid electrolyte layer. High mass transfer rates and current efficiencies were obtained when withdrawing the product HCl continuously from the electrolyte layer.

7.5 Conclusions

Solid ion conductors can be used as gas-dense membranes in various technical applications. The most important fields are electrochemical gas sensors, fuel cells, electrolyzers and solid electrolyte reactors in which ions are transferred through the membrane by an electric field applied between the two electrodes. Solid electrolytes (SE) are distinguished by a very high selectivity with respect to the mass transport of the ionic species. Concerning permeability, these materials have to compete with porous membranes and with mixed ion electron conductors (MIEC).

With respect to high-temperature oxygen ion conductors, solid oxide fuel cells (SOFC) are still the most important field of research and application. However, nowadays new applications are investigated intensively, such as the partial oxidation of light hydrocarbons to oxygenates. High-temperature proton-conducting membranes offer new possibilities for designing solid electrolyte reactors. Therefore, there is need for research on stable protonic electrolytes having high conductivity and active electrodes working in atmospheres with a low level of humidity, both in hydrogen (anodes) and in air (cathodes).

Low-temperature proton conductors (PEM) are well established membrane materials in hydrogen fuel cells (PEMFC) and in direct methanol fuel cells (DMFC). In this area, material optimization has to focus on the reduction of the undesired membrane cross-over fluxes of water and methanol. Composite membranes will play an important role for future generations of fuel cells. Moreover, new interesting applications have been reported where PEM reactors are used to carry out selective hydrogenation or selective oxidation reactions in mild conditions. In the future, PEMs might be also good candidates for the realization of microbial fuel cells (see, e.g., Kim *et al.*, 2002) based on mediator-less direct electron transfer. This will require intense collaboration between biochemists and electrochemical engineers.

Acknowledgement

The authors gratefully acknowledge the significant contribution of Prof. Helmut Rau during the research project.

Special Notation not Mentioned in Chapter 2

Latin Notation

A_s	m^2	geometrical surface area of Anode
A_z	m^2	cross-sectional area of anode gas channel (reactor shell side)
C_{DL}	$As\ V^{-1}\ m^{-2}$	anodic and cathodic double layer capacity
C^{SE}	$A\ V^{-1}\ m^{-1}$	SE conductivity constant

d^{AC}	m	thickness of anodic catalyst layer
d^{SE}	m	SE membrane thickness
$E_A^{A/C}$	J mol^{-1}	activation energy for anodic/cathodic charge transfer reaction
E_A^j	J mol^{-1}	activation energy for reaction j
E_A^{SE}	J mol^{-1}	activation energy for O^{2-} conduction within the SE membrane
F		Faraday constant, (96 485 As mol^{-1})
F^A	mol s^{-1}	total molar flow rate within the anode gas channel
i_{cell}	A m^{-2}	cell current density
I_{cell}	A	total cell current
$I_{cell,ref}$	A	reference cell current
j_i	mol m^{-2} s^{-1}	molar flow density of species i across the SE membrane
k_j	mol kg^{-1} s^{-1}	reaction rate constant
k_j^0	mol kg^{-1} s^{-1}	pre-exponential reaction rate constant for reaction j
$k_0^{A/C}$	mol m^{-2} s^{-1}	pre-exponential reaction rate constant for anodic/cathodic charge transfer reaction
$q_{2/3}$		moles of oxygen required to oxidize 1 mol of n-butane/MA to CO_x
$r_{el}^{A/C}$	mol m^{-2} s^{-1}	electrochemical charge transfer reaction rate at anode/cathode
$r_{el,ref}$	mol m^{-2} s^{-1}	reference charge transfer reaction rate, $(I_{cell,ref}/A_S F)$
r_j	mol kg^{-1} s^{-1}	reaction rate of reaction j
$R^{A/C}$	Ω	ohmic resistance of anodic/cathodic current collector
x_{ox}, x_{red}		mole fraction of oxidized/reduced catalyst, $(x_{ox} + x_{red} = 1)$

Greek Notation

$\alpha^{A/C}$		anodic/cathodic charge transfer coefficient
β_j		kinetic order of the oxygen species in reaction j
$\Delta\Phi^{A/C}$	V	anodic/cathodic potential difference
$\Delta\Phi^{CA}$	V	local cell voltage, $(\Phi^C - \Phi^A)$
$\Delta\Phi^{SE}$	V	potential difference across the SE membrane
$\eta^{A/C}$	V	anodic/cathodic overvoltage
ν_{ij}		stoichiometric coefficients of species i in reaction j
σ^{SE}	A V^{-1} m^{-1}	conductivity of SE membrane for O^{2-} ions

Characteristic Dimensionless Numbers

Da		Damköhler number, $(m_{cat} \cdot r_{ref}/F^{A,0})$
Da_{II}		second kind of Damköhler number
Φ		Thiele modulus

Superscripts

A/C	anodic/cathodic
SE	solid electrolyte

References

Alberti, G., and Casciola, M. (2003) Composite membranes for medium-temperature PEM fuel cells. *Annu. Rev. Mater. Res.*, **33**, 129–154.

An, W., Hong, J., and Pintauro, P. (1998) Current efficiency for soybean oil hydrogenation in a solid polymer electrolyte reactor. *J. Appl. Electrochem.*, **28**, 947–954.

An, W., Hong, J., Pintauro, P., Warner, K., and Neff, W. (1999) The electrochemical hydrogenation of edible oils in a solid polymer electrolyte reactor. II. Hydrogenation selectivity studies. *J. Am. Oil Chem. Soc.*, **76**, 215–222.

Athanassiou, C., Pekridis, G., Kaklidis, N., Kalimeri, K., Vartzoka, S., and Marnellos, G. (2007) Hydrogen production in solid electrolyte membrane reactors (SEMRs). *Int. J. Hydrogen Energy*, **32**, 38–54.

Barbir, F. (2005) *PEM Fuel Cells. Theory and Practice*, vol. XX, Academic Press.

Basu, S. (2007) *Recent Trends in Fuel Cell Science and Technology*, Springer, Berlin.

Bebelis, S., and Vayenas, C.G. (1989) Non-faradaic electrochemical modification of catalytic activity: 1. The case of ethylene oxidation on Pt. *J. Catal.*, **118**, 125–146.

Belyaev, V., Galvita, V., and Sobyanin, V. (1998) Effect of anodic current on carbon dioxide reforming of methane on Pt electrode in a cell with solid oxide electrolyte. *React. Kinet. Catal. Lett.*, **63**, 341–348.

Bredesen, R., Jordal, K., and Bolland, O. (2004) High-temperature membranes in power generation with CO_2 capture. *Chem. Eng. Process*, **43**, 1129–1158.

Buergler, B.E., Grundy, A.N., and Gauckler, L.J. (2006) Thermodynamic equilibrium of single-chamber SOFC relevant methane-air mixtures. *J. Electrochem. Soc.*, **153**, A1378.

Centi, G., Cavani, F., and Trifiro, F. (2001) *Selective Oxidation by Heterogeneous Catalysis*, Kluwer Academic/Plenum Publishers, New York.

Chalakov, L., Rihko-Struckmann, L., Munder, B., and Sundmacher, K. (2007) Feasibility study of the oxidative dehydrogenation of ethane in an electrochemical packed-bed membrane reactor. *Ind. Eng. Chem. Res.*, **46**, 8665–8673.

Chalakov, L., Rihko-Struckmann, L., Munder, B., and Sundmacher, K. (2009) Oxidative dehydrogenation of ethane in an electrochemical packed-bed membrane reactor: model and experimental validation. *J. Chem. Eng.*, **145**, 385–392.

Eng, D., and Stoukides, M. (1991) Catalytic and electrocatalytic methane oxidation with solid oxide membranes. *Catal. Rev. Sci. Eng.*, **33**, 375–412.

Fehringer, G., Janes, S., Wildersohn, M., and Clasen, R. (2004) Proton conducting ceramics as electrode/electrolyte materials for SOFCs: preparation, mechanical and thermal-mechanical properties of thermal sprayed coatings, material combination and stacks. *J. Eur. Ceram. Soc.*, **24**, 705–715.

Gardner, F.J., Day, M.J., Brandon, N.P., Pashley, M.N., and Cassidy, M. (2000) SOFC technology development at Rolls-Royce. *J. Power Sources*, **86**, 122–129.

Goodenough, J.B. (2003) Oxide-ion electrolytes. *Annu. Rev. Mater. Res.*, **33**, 91–128.

Grover Coors, W. (2003) Protonic ceramic fuel cells for high-efficiency operation with methane. *J. Power Sources*, **118**, 150–156.

Grube, T., Hohlein, B., and Menzer, R. (2007) Assessment of the application of fuel cell APUs and starter-generators to reduce automobile fuel consumption. *Fuel Cells*, **7**, 128–134.

Hamakawa, S., Hibino, T., and Iwahara, H. (1993) Electrochemical methane coupling using protonic conductors. *J. Electrochem. Soc.*, **140**, 459–462.

Hassan, D., Janes, S., and Clasen, R. (2003) Proton-conducting ceramics as electrode/electrolyte materials for SOFC's part I: preparation, mechanical and thermal properties of sintered bodies. *J. Eur. Ceram. Soc.*, **23**, 221–228.

He, R., Li, Q., Jensen, J., and Bjerrum, N. (2007) Doping phosphoric acid in polybenzimidazole membranes for high temperature proton exchange membrane fuel cells. *J. Polym. Sci. Part A*, **45**, 2989–2997.

Hellgardt, K., Cumming, I., and Al-Musa, A. (2005) The effect of electrochemical oxygen on the selectivity of the partial oxidation of propene over silver catalysts. *Solid State Ionics*, **176**, 831–835.

Hess, S. (2002) Kinetische Unterschungen zur getrennten Reaktionsführung der selektiven Butanoxidation, Thesis, Univ. Erlangen.

Hexis (2009) Dezentral Strom und Wärme gewinnen, http://www.hexis.com (accessed 20 April 2009).

Hibino, T., Hashimoto, A., Inoue, T., Tokuno, J., Yoshida, S., and Sano, M. (2000a) A low-operating-temperature solid oxide fuel cell in hydrocarbon-air mixtures. *Science*, **288**, 2031–2033.

Hibino, T., Tsunekawa, H., Tanimoto, S., and Sano, M. (2000b) Improvement of a single-chamber solid-oxide fuel cell and evaluation of new cell designs. *J. Electrochem. Soc.*, **147**, 1338–1343.

Huang, X.-F., Li, C.-Y., Chen, B.-H., and Silveston, P.L. (2002) Transient kinetics of *n*-Butane oxidation to maleic anhydride over a VPO catalyst. *AIChE J.*, **48**, 846–855.

Itoh, N., Xu, W.C., Hara, S., and Sakaki, K. (2000) Electrochemical coupling of benzene hydrogenation and water electrolysis. *Catal. Today*, **56**, 307–314.

Iwahara, H., Esaka, T., Uchida, H., and Maeda, N. (1981) Proton conduction in sintered oxides and its application to steam electrolysis for hydrogen production. *Solid State Ionics*, **3–4**, 359–363.

Iwahara, H., Asakura, Y., Katahira, K., and Tanaka, M. (2004) Prospect of hydrogen technology using proton-conducting ceramics. *Solid State Ionics*, **168**, 299–310.

Karagiannakis, G., Zisekas, S., and Stoukides, M. (2003) Hydrogenation of carbon dioxide on copper in a H^+ conducting membrane-reactor. *Solid State Ionics*, **162–163**, 313–318.

Karagiannakis, G., Kokkofitis, C., and Zisekas, S. (2005) Catalytic and electrocatalytic production of H-2 from propane, decomposition over Pt and Pd in a proton-conducting membrane-reactor. *Catal. Today*, **104**, 219–224.

Kim, H., Park, H., Hyun, M., Chang, I., Kim, M., and Kim, B. (2002) A mediator-less microbial fuel cell using a metal reducing bacterium, Shewanella putrefaciens. *Enzyme Microb. Technol.*, **30**, 145–152.

Klose, F., Joshi, M., Hamel, C., and Seidel-Morgenstern, A. (2004) Selective oxidation of ethane over a $VO_x/\gamma\text{-}Al_2O_3$ catalyst – investigation of the reaction network. *Appl. Cat. A Gen.*, **260**, 101–110.

Kobayashi, T., Abe, K., Ukyo, Y., and Iwahara, H. (2002) Performance of electrolysis cells with proton and oxide-ion conducting electrolyte for reducing nitrogen oxide. *Solid State Ionics*, **154–155**, 699–705.

Kokkofitis, C., Ouzounidou, M., Skodra, A., and Stoukides, M. (2007) High temperature proton conductors: applications in catalytic processes. *Solid State Ionics*, **178**, 507–513.

Kreuer, K.D. (2003) Proton-conducting oxides. *Annu. Rev. Mater. Res.*, **33**, 333–359.

Krewer, U., Christov, M., Vidakovic, T., and Sundmacher, K. (2006) Impedance spectroscopic analysis of the electrochemical methanol oxidation kinetics. *J. Electroanal. Chem.*, **589**, 148–159.

Larminie, R., and Dicks, A. (2003) *Fuel Cell Systems Explained*, John Wiley & Sons, Ltd, Chichester, p. 141.

Li, Q., He, R., Jensen, J., and Bjerrum, N. (2004) PBI-based polymer membranes for high temperature fuel cells – preparation, characterisation, and fuel cell demonstration. *Fuel Cells*, **4**, 147–159.

Lorences, M.J., Patience, G.S., Diez, F.V., and Coca, J. (2003) Butane oxidation to maleic anhydride: kinetic modeling and byproducts. *Ind. Eng. Chem. Res.*, **42**, 6730–6742.

Marnellos, G., and Stoukides, M. (2004) Catalytic studies in electrochemical membrane reactors. *Solid State Ionics*. **175**, 597–603.

Matsumoto, H., Hamajima, S., Yajima, T., and Iwahara, H. (2001) Electrochemical hydrogen pump using $SrCeO_3$-based proton conductor: effect of water vapor at the cathode on the pumping capacity. *J. Electrochem. Soc.*, **148**, D121–D124.

Moon, D., and Ryu, J. (2003) Electrocatalytic reforming of carbon dioxide by methane in SOFC system. *Catal. Today*, **87**, 255–264.

Munder, B., Ye, Y., Rihko-Struckmann, L., and Sundmacher, K. (2005) Solid electrolyte membrane reactor for controlled partial oxidation of hydrocarbons: model and experimental validation. *Catal. Today*, **104**, 138–148.

Munder, B., Rihko-Struckmann, L., and Sundmacher, K. (2007) Steady-state and forced-periodic operation of solid electrolyte membrane reactors for selective oxidation of *n*-butane to maleic anhydride. *Chem. Eng. Sci.*, **62**, 5663–5668.

Otsuka, K., and Yamanaka, I. (1990) One step synthesis of hydrogen peroxide through fuel cell reaction. *Electrochim. Acta*, **35**, 319–322.

Otsuka, K., and Yamanaka, I. (2000) Oxygenation of alkanes and aromatics by reductively activated oxygen during H_2–O_2 cell reactions. *Catal. Today*, **57**, 71–86.

Otsuka, K., Yamanaka, I., and Hosokawa, K. (1990) A fuel- cell for the partial oxidation of cyclohexane and aromatics at ambient temperatures. *Nature*, **345**, 697–698.

Otsuka, K., Hashimoto, T., and Yamanaka, I. (2002) Direct synthesis of hydrogen peroxide (>1 wt%) over the cathode prepared from active carbon and vapor-grown-carbon-fiber by a new H_2–O_2 fuel cell system. *Chem. Lett.*, **8**, 852–853.

Otsuka, K., Ina, T., and Yamanaka, I. (2003) The partial oxidation of methanol using a fuel cell reactor. *Appl. Catal.*, **247**, 219–229.

Pfafferodt, M., Heidebrecht, P., Stelter, M., and Sundmacher, K. (2005) Model-based prediction of suitable operating range of a SOFC for an Auxiliary Power Unit. *J. Power Sources*, **149**, 53–62.

Pintauro, P., Gil, M., Warner, K., List, G., and Neff, W. (2005) Electrochemical hydrogenation of soybean oil with hydrogen gas. *Ind. Eng. Chem. Res.*, **44**, 6188–6194.

Rihko-Struckmann, L., Ye, Y., Chalakov, L., Suchorski, Y., Weiss, H., and Sundmacher, K. (2006) Bulk and surface properties of a VPO catalyst used in an electrochemical membrane reactor: conductivity-, XRD-, TPO- and XPS-study. *Catal. Lett.*, **109**, 89–96.

Roziere, J., and Jones, D.J. (2003) Non-fluorinated polymer materials for proton exchange membrane fuel cells. *Annu. Rev. Mater. Res.*, **33**, 503–555.

Savadogo, O. (2004) Emerging membranes for electrochemical systems: part II. High temperature composite membranes for polymer electrolyte fuel cell (PEFC) applications. *J. Power Sources*, **127**, 135–161.

Schultz, T., Zhou, S., and Sundmacher, K. (2001) Current status of and recent developments in the direct methanol fuel cell. *Chem. Eng. Technol.*, **24**, 1223–1233.

Schultz, T., Krewer, U., Vidakovic, T., Pfafferodt, M., Christov, M., and Sundmacher, K. (2007) Systematic analysis of the direct methanol fuel cell. *J. Appl. Electrochem.*, **37**, 111–119.

Simond, O., and Comninellis, C. (1997) Anodic oxidation of organics on Ti/IrO_2 anodes using Nafion® as electrolyte. *Electrochem. Acta*, **42**, 2013–2018.

Spiegel, C. (2007) *Designing and Building Fuel Cells*, MacGraw-Hill Professional.

Srinivasan, S. (2006) *Fuel Cells. From Fundamentals to Applications*, Springer, Berlin.

Staxera (2009) Willkommen bei Staxera, www.staxera.de (accessed 20 April 2009).

Stoukides, M. (2000) Solid- electrolyte membrane reactors: current experience and future outlook. *Catal. Rev. Sci. Eng.*, **42**, 1–70.

Sundmacher, K., and Hoffmann, U. (1999) Design and operation of a membrane reactor for electrochemical gas purification. *Chem. Eng. Sci.*, **54**, 2937–2945.

Sundmacher, K., Schultz, T., Zhou, S., Scott, K., Ginkel, M., and Gilles, E. (2001) Dynamics of the direct methanol fuel cell

(DMFC): experiments and model-based analysis. *Chem. Eng. Sci.*, **56**, 333–341.

Sundmacher, K., Rihko-Struckmann, L.K., and Galvita, V. (2005) Solid electrolyte membrane reactors: status and trends. *Catal. Today*, **104**, 185–199.

Taherparvar, H., Kilner, J., Baker, R., and Sahibzada, M. (2003) Effect of humidification at anode and cathode in proton-conducting SOFCs. *Solid State Ionics*, **162–163**, 297–303.

Takehira, K., Shishido, T., Komatsu, T., Hamakawa, S., and Kajioka, H. (2002) YSZ aided oxidation of C_2–C_4 hydrocarbons into oxygenates over MoO_3 or V_2O_5. *Solid State Ionics*, **152–153**, 641–646.

Takehira, K., Sakai, N., Schimomura, J., Kajioka, H., Hamakawa, S., and Shishido, T. (2004) Oxidation of C_2–C_4 alkanes over MoO_3–V_2O_5 supported on a YSZ-aided membrane reactor. *Appl. Catal. A*, **277**, 209–217.

Vayenas, C.G., Bebelis, S., and Ladas, S. (1990) Dependence of catalytic rates on catalyst work function. *Nature*, **343**, 625–627.

Vidakovic, T., Christov, M., and Sundmacher, K. (2005) Rate expression for electrochemical oxidation of methanol on a direct methanol fuel cell anode. *J. Electroanal. Chem.*, **580**, 105–121.

Vidakovic, T., Christov, M., Sundmacher, K., Nagabhushana, K., Fei, W., Kinge, S., and Bonnemann, H. (2007) PtRu colloidal catalysts: characterisation and determination of kinetics for methanol oxidation. *Electrochim. Acta*, **52**, 2277–2284.

Wang, W., and Lin, Y.S. (1995) Analysis of oxidative coupling of methane in dense oxide membrane reactors. *J. Membr. Sci.*, **103**, 219–233.

Weber, A., and Ivers-Tiffée, E. (2004) Materials and concepts for solid oxide fuel cells (SOFCs) in stationary and mobile applications. *J. Power Sources*, **127**, 273–283.

Williams, M. (2007) Solid oxide fuel cells: fundamentals to systems. *Fuel Cells*, **7**, 78–85.

Yamanaka, I., and Otsuka, K. (1991) The partial oxidations of cyclohexane and benzene on the fecl3-embedded cathode during the O2-H2 fuel- cell reaction. *J. Electrochem. Soc.*, **97**, 1033–1038.

Yamanaka, I., and Otsuka, K. (1988) Partial oxidation of methanol using a fuel-cell system at room temperature. *Chem. Lett.*, **17**, 753–756.

Yamanaka, I., Hasegawa, S., and Otsuka, K. (2002) Partial oxidation of light alkanes by reductive activated oxygen over the (Pd-black + $VO(acac)_2$/VGCF) cathode of H_2–O_2 cell system at 298 K. *Appl. Catal. A: Gen.*, **226**, 305–315.

Yano, M., Tomita, A., Sano, M., and Hibino, T. (2007) Recent advances in single-chamber solid oxide fuel cells: a review. *Solid State Ionics*, **177**, 3351–3359.

Ye, Y. (2006) Experimental study of *n*-butane partial oxidation to maleic anhydride in a solid electrolyte membrane reactor, Dissertation, Otto von Guericke University Magdeburg.

Ye, Y., Rihko-Struckmann, L., Munder, B., Rau, H., and Sundmacher, K. (2004) Feasibility of an electrochemical membrane reactor for the partial oxidation of n-butane to maleic anhydride. *Ind. Eng. Chem. Res.*, **43**, 4551–4558.

Ye, Y., Rihko-Struckmann, L., Munder, B., and Sundmacher, K. (2005) Partial oxidation of *n*-butane in a solid electrolyte membrane reactor: Periodic and steady-state operations. *Appl. Catal. A*, **285**, 86–95.

Ye, Y., Rihko-Struckmann, L., Munder, B., and Sundmacher, K. (2006) Partial oxidation of *n*-butane in a solid electrolyte membrane reactor. *J. Electrochem. Soc.*, **153**, D21–D26.

Yuan, X., Ma, Z., Jiang, Q., and Wu, W. (2001) Cogeneration of cyclohexylamine and electrical power using PEM fuel cell reactor. *Electrochem. Commun.*, **3**, 599–602.

Yuan, X., Ma, Z., He, Q., Hagen, J., Drillet, J., and Schmidt, V.M. (2003) Electro-generative hydrogenation of allyl alcohol applying PEM fuel cell reactor. *Electrochem. Commun.*, **5**, 189–193.

8
Nonlinear Dynamics of Membrane Reactors

Michael Mangold, Fan Zhang, Malte Kaspereit, and Achim Kienle

8.1
Introduction

This chapter studies the nonlinear dynamic behavior of membrane reactors. Chapter 1 discusses typical types of membrane reactors. The following focuses on tubular membrane fixed-bed reactors with a nonreactive membrane and chemical reactions on the membrane's tubular side, as shown in Figure 8.1. The membrane is either used for the separation of products (see Figure 8.1a) or for the distributed injection of reactants (see Figure 8.1b). The objective is to obtain a qualitative understanding of the nonlinear effects in membrane reactors rather than detailed quantitative results. Therefore the analysis is mainly based on simple reactor models. The chapter is divided in two major parts. The first part considers the limiting case of infinitely fast reaction rates. This permits comparisons with other reactive and nonreactive separation processes and thus puts membrane reactors in a more general framework. Different wave phenomena are studied. One of the results of this section is the formation of discontinuous spatial temperature and concentration patterns. The second part of the chapter studies this pattern formation in more detail and abandons the assumption of reaction equilibrium.

8.2
Limit of Chemical Equilibrium

8.2.1
Reference Model

Chapter 2 contains a detailed discussion on how to derive membrane reactor models from first principles. As this chapter aims at a mainly qualitative understanding of membrane reactor dynamics, simplified models are considered as most appropriate for the following analysis. In order to obtain such a model for a membrane reactor as illustrated in Figure 8.1, it is assumed that the sweep gas flow is very high, so that all material transported across the membrane is readily

Membrane Reactors: Distributing Reactants to Improve Selectivity and Yield.
Edited by Andreas Seidel-Morgenstern

ISBN: 978-3-527-32039-4

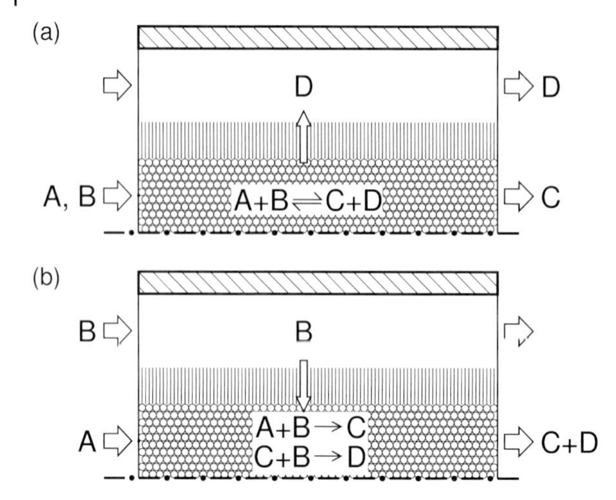

Figure 8.1 Operating principle of membrane reactors: a) selective withdrawal of products, b) distributed injection of reactants.

removed and thus the composition in the sweep gas channel does not change. Furthermore, the following assumptions are made:

- The axial dispersion of heat and mass is negligible.
- The total gas concentration is independent of the composition (ideal mixture).
- The total pressure is constant.
- All reactions are equimolar.
- The fluid and catalyst phases have the same temperature and hence can be modeled by a pseudo-homogeneous energy balance.
- All components in the mixture have the same heat capacity.

The assumption of equimolar reactions greatly simplifies the mathematics and therefore helps to focus more on the physical interpretation of the results.

Under the met assumptions the material balances read in dimensionless form:

$$\frac{\partial \vec{y}}{\partial t} + v\frac{\partial \vec{y}}{\partial z} = \overline{\overline{v}}\ \vec{r} - \vec{j} + \vec{Y} j_{\text{tot}} \tag{8.1}$$

with:

$$\vec{Y}, \vec{j} \in R^{N_C - 1}, \vec{r} \in R^{N_R}, \overline{\overline{v}} \in R^{(N_C - 1) \times N_R}.$$

Here, t and z are dimensionless time and space coordinates as defined by (Grüner, Mangold, and Kienle, 2006). N_C is the number of components in the mixture and $N_R$is the number of independent reaction rates. $\vec{y}$ is the vector of mole fractions, $\vec{r}$ is a vector of reaction rates and $\vec{j}$ denotes the fluxes across the membrane. It is worth noting that there are only $N_C - 1$ species but N_C independent fluxes, that is, one can either specify N_C component fluxes j_i or $N_C - 1$ component fluxes and the total flow rate j_{tot}. In the remainder, the first approach is used, that is, an extended flux vector is defined as:

$$\vec{j}_{ext} = \begin{pmatrix} \vec{j} \\ j_{tot} \end{pmatrix} \tag{8.2}$$

From an energy balance, the following equation for a dimensionless reactor temperature $\vartheta := (T - T_0)/T_0$ is obtained:

$$\frac{\partial \vartheta}{\partial t} + \nu \varepsilon \frac{\partial \vartheta}{\partial z} = \vec{B}^T \vec{r} - j_\vartheta \tag{8.3}$$

The parameter ε stands for the ratio between thermal capacity of the fluid and the total thermal capacity of the system; $\varepsilon - 1$ holds for an empty tubular reactor, $\varepsilon < 1$ for a fixed-bed reactor. $\vec{B}$ is a vector containing the scaled heat of reaction for each component. The flux j_ϑ describes the internal energy transported across the membrane, which is a combination of the enthalpy transported by the mass flux and of a heat flux j_Q caused by convection. j_ϑ can be written as:

$$j_\vartheta = \varepsilon \sum j_i (\vartheta - \vartheta_i^*) + j_Q \tag{8.4}$$

where ϑ_i^* is the temperature of the sweep gas side for $j_i < 0$; $\vartheta_i^* = \vartheta$ for $j_i > 0$.

From the total material balance and the thermal equation of state, the following differential equation results for the flow velocity:

$$\frac{\partial v}{\partial z} = -C_\vartheta \left(\frac{\partial \vartheta}{\partial t} + v \frac{\partial \vartheta}{\partial z} \right) - j_{tot} \tag{8.5}$$

C_ϑ is the scaled partial derivative of the total concentration with respect to the temperature; for an ideal gas $C_\vartheta = -1/(1 + \vartheta)$ holds.

The following focuses on infinitely fast chemical reactions. The reaction equilibria are given by N_R relations:

$$\vec{K}(\vec{y}, \vartheta) = \vec{0} \tag{8.6}$$

8.2.2
Isothermal Operation

In a first step, the case of isothermal operation is considered. This assumption simplifies the model to:

$$\frac{\partial \vec{y}}{\partial t} + v \frac{\partial \vec{y}}{\partial z} = \overline{\overline{\nu}}\, \vec{r}(\vec{y}) - \vec{j}(\vec{y}) + \vec{y} j_{tot} \tag{8.7}$$

$$\frac{\partial v}{\partial z} = -j_{tot} \tag{8.8}$$

8.2.2.1 Nonreactive Membrane Separation

In the nonreactive case the reaction rate r is equal to zero and the model equations read:

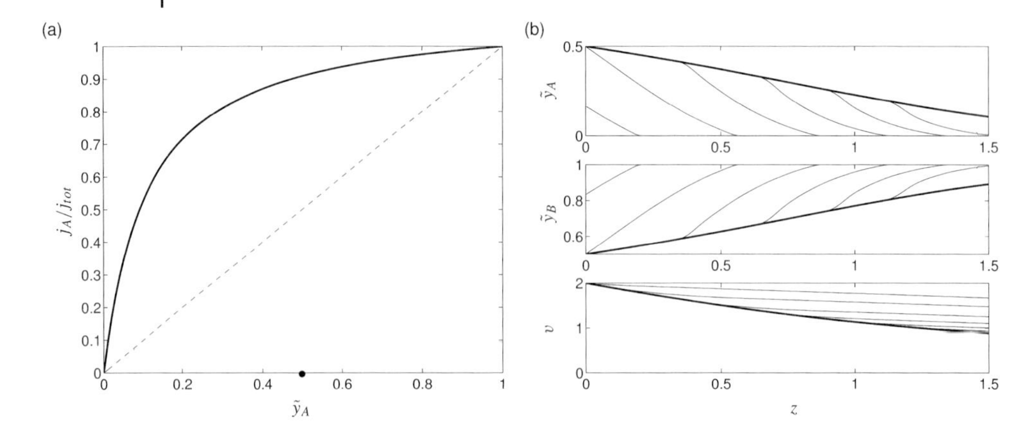

Figure 8.2 Nonreactive membrane separation. Parameters: $\beta_A = 2.0$, $\beta_B = 0.2$. a) Flux rate ratio of A through membrane versus $\tilde{y}_A$. b) Response to an input ramp of $\tilde{y}_{A,in}$, as shown in Figure 8.3. Thin lines = transient solutions at $t = 0.2, 0.4, \ldots, 1.2$; bold lines = steady state solution.

$$\frac{\partial \vec{\tilde{y}}}{\partial t} + v\frac{\partial \vec{\tilde{y}}}{\partial z} = -\vec{j}(\vec{\tilde{y}}) + \vec{\tilde{y}}\, j_{tot} \tag{8.9}$$

$$\frac{\partial v}{\partial z} = -j_{tot} \tag{8.10}$$

Due to the finite mass transport kinetics across the membrane this is, even in the nonreactive case, a system of nonhomogenous quasilinear partial differential equations whose properties crucially depend on the flux function $\vec{j}(\vec{\tilde{y}})$. For simplicity, we focus on a simple diagonal flux function according to:

$$j_i = \beta_i \tilde{y}_i,\ i = 1, \ldots, N_C. \tag{8.11}$$

A discussion of more complicated cases was given for example by Krishna and Wesselingh (1997).

Let us illustrate the behavior of the nonreactive membrane separation system with a simple example. Consider a binary mixture with components A, B. Let us assume that the membrane has highest permeability for component A and let us first focus on the steady state behavior of such a process. Let Equation 8.9 be the material balance of component A. According to Equation 8.9 the slope of the steady state profile depends on the relative flux ratio j_A/j_{tot} and its dependence on the mole fraction $\tilde{y}_A$. When j_A/j_{tot} is larger than $\tilde{y}_A$ then the concentration is decreasing and vice versa. Hence, the qualitative behavior is easily extracted from a diagram of the flux ratio j_A/j_{tot} versus concentration $\tilde{y}_A$, as illustrated on the left side of Figure 8.2.

This diagram is the analogon to the well known McCabe–Thiele diagram for distillation processes and nicely illustrates the influence of mass transfer resistance (Fullarton and Schlünder, 1986). For the simple relation in Equation 8.11 we find:

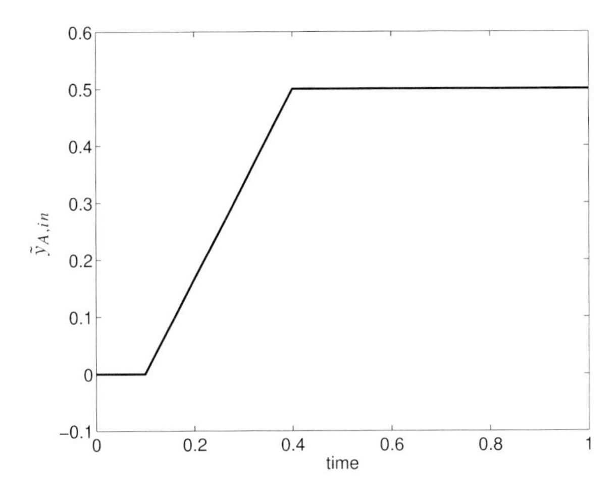

Figure 8.3 Amp-shaped input of concentration $\tilde{y}_{A,in}$ used for the startup scenarios in Figures 8.2, 8.4, 8.6.

$$\frac{j_i}{j_{tot}} = \frac{\beta_i \tilde{y}_i}{\sum_{k=1}^{N_C} \beta_k \tilde{y}_k}, i = 1, \ldots, N_C - 1, \tag{8.12}$$

which is equivalent to the vapor–liquid equilibrium for mixtures with constant relative volatilities. Since the membrane has highest permeability for component A the flux ratio j_A/j_{tot} is always larger than $\tilde{y}_A$. Therefore, the concentration of A is monotonically decreasing in the inner membrane tube and tends to zero for an infinitely long tube. In a similar way the flow rate in the inner tube decreases monotonically and tends to zero. For a tube of finite length a finite flow rate and a finite concentration is obtained.

The dynamic transient behavior during start-up is illustrated on the right side of Figure 8.2. Figure 8.3 shows the ramp-shaped inlet concentration belonging to this start-up scenario. Like in reactive distillation and reactive chromatography (Grüner, Mangold, and Kienle, 2006), the transient behavior is governed by traveling fronts. In the present case, these waves can only propagate downstream. Further, since the velocity in the inner tube is decreasing continuously from the inlet to the outlet due to material transport across the membrane, concentration values closer to the inlet move with higher velocity than concentration values closer to the outlet. Consequently, all waves are self-sharpening in Figure 8.2.

Alternatively, also inlet concentration disturbances of the steady-state profile may be considered. Due to the inhomogeneity in Equation 8.9 the concentration waves travel downstream on the nonconstant initial profile as the system undergoes a transient from the old to a new monotonically decreasing steady state.

The same type of dynamics can be observed in the multicomponent case. Any step disturbance is resolved into a single front traveling jointly for all components, due to the fact that all balance equations share the same transport velocity v. This simple behavior is in contrast to the more complex wave phenomena found in distillation and chromatographic processes (Grüner, Mangold, and Kienle, 2006; Kienle, 2000).

8.2.2.2 Membrane Reactor

In the reactive case, the full blown Equation 8.7 together with the total material balance (8.8) has to be considered. For fast chemical reactions, reaction equilibrium can be assumed. The chemical equilibrium conditions represent N_R algebraic constraints, which reduce the dynamic degrees of freedom of the system to $N_C - 1 - N_R$. In the limit of reaction equilibrium, the kinetic rate expressions for the reaction rates become indeterminate and have to be eliminated from the balance equations (8.7). Following the ideas of (Ung and Doherty, 1995), this is achieved by choosing N_R reference components and splitting the concentration vector accordingly into two parts:

$$\vec{y} = \left[\vec{y}^{\mathrm{I}}, \vec{y}^{\mathrm{II}}\right], \vec{y}^{\mathrm{I}} \in R^{N_R}, \vec{y}^{\mathrm{II}} \in R^{N_C-1-N_R} \tag{8.13}$$

An analogous splitting is introduced for the matrix of stoichiometric coefficients:

$$\overset{=}{\nu} = \left[\overset{=\mathrm{I}}{\nu}, \overset{=\mathrm{II}}{\nu}\right], \tag{8.14}$$

where $\overset{=\mathrm{I}}{\nu}$ has dimensions $N_R \times N_R$ and $\overset{=\mathrm{II}}{\nu}$ has dimensions $(N_C - 1 - N_R) \times N_R$. By solving the first N_R equations of (8.7) for the unknown reaction rates and substituting them into the remaining $N_C - 1 - N_R$ equations, the following reduced set of equations is obtained:

$$\frac{\partial \vec{Y}}{\partial t} + v\frac{\partial \vec{Y}}{\partial z} = -\vec{J}(\vec{y}) + \vec{Y} J_{\mathrm{tot}}. \tag{8.15}$$

$$\frac{\partial v}{\partial z} = -J_{\mathrm{tot}} \tag{8.16}$$

with transformed concentration and flux variables according to:

$$\vec{Y} = \vec{y}^{\mathrm{II}} - \overset{=\mathrm{II}}{\nu}\left(\overset{=\mathrm{I}}{\nu}\right)^{-1}\vec{y}^{\mathrm{I}}, \vec{J} = \vec{j}^{\mathrm{II}} - \overset{=\mathrm{II}}{\nu}\left(\overset{=\mathrm{I}}{\nu}\right)^{-1}\vec{j}^{\mathrm{I}}, \vec{J}_{\mathrm{ext}} = \vec{j}^{\mathrm{II}}_{\mathrm{ext}} - \overset{=\mathrm{II}}{\nu}_{\mathrm{ext}}\left(\overset{=\mathrm{I}}{\nu}\right)^{-1}\vec{j}^{\mathrm{I}}$$
$$\vec{Y} \in R^{N_C-1-N_R}, \vec{J} \in R^{N_C-1-N_R}, \vec{J}_{\mathrm{ext}} \in R^{N_C-N_R} \tag{8.17}$$

It is worth noting that, like in the nonreactive case, the dimension of the flux vector $\vec{J}_{\mathrm{ext}}$ exceeds the dimension of the concentration vector by one. Consequently, the dimension of $\vec{j}^{\mathrm{II}}_{\mathrm{ext}}$ exceeds the dimension of $\vec{y}^{\mathrm{II}}$ by one and the matrix $\overset{=\mathrm{II}}{\nu}_{\mathrm{ext}}$ consists of matrix $\overset{=\mathrm{II}}{\nu}$ with an additional row for component N_C.

In the above equations the relation $J_{\mathrm{tot}} = j_{\mathrm{tot}}$ is used. This follows from the definition of J_{tot}:

$$J_{\mathrm{tot}} = \sum_{i=1}^{N_C-N_R} J_i = \sum_{i=1}^{N_C-N_R}\left(j_i - \sum_{j=1}^{N_R} \overset{=\mathrm{II}}{\nu}_{\mathrm{ext},i}\left(\overset{=\mathrm{I}}{\nu}\right)^{-1}_j j_j\right) \tag{8.18}$$

and the constant total mole number for each reaction according to:

$$\sum_{i=1}^{N_C-N_R} \stackrel{=\text{II}}{\nu}_{\text{ext},i} = -\sum_{i=1}^{N_R} \stackrel{=\text{I}}{\nu}_i, \tag{8.19}$$

respectively. In the above equations $\stackrel{=\text{II}}{\nu}_{\text{ext},i}$, $\stackrel{=\text{I}}{\nu}_i$ represent the i-th row vector of the corresponding matrix and $\left(\stackrel{=\text{I}}{\nu}\right)^{-1}_j$ represents the j-th column vector of the corresponding inverse matrix. Hence:

$$\sum_{i=1}^{N_C-N_R} j_i - \sum_{j=1}^{N_R} \stackrel{=\text{II}}{\nu}_{\text{ext},i} \left(\stackrel{=\text{I}}{\nu}\right)^{-1}_j j_j = \sum_{i=1}^{N_C-N_R} j_i + \sum_{i=1}^{N_R} j_i = \sum_{i=1}^{N_C} j_i = j_{\text{tot}} \tag{8.20}$$

The reactive problem in transformed concentration variables (8.15), (8.16) is completely analogous to the corresponding nonreactive problem (8.9), (8.10) and we can proceed in a similar way like in the previous section to investigate the qualitative behavior. The same analogies can be drawn for reactive distillation and reactive chromatography, as discussed by (Grüner, Mangold, and Kienle, 2006).

In a first step we focus on a single reversible reaction of type $2A \leftrightarrow B + C$. The treatment for the corresponding problems in reactive distillation and reactive chromatography was given by (Grüner and Kienle, 2004). In particular, it was shown that total conversion in an infinitely long column is only possible if reactant A has intermediate affinity to the vapor or the solid phase, respectively. In the other cases, achievable product compositions are limited by reactive azeotropy. Conditions for total conversion in a membrane reactor, however, are more restrictive. In the present case the total conversion of reactant A is only possible if the membrane is not permeable at all for reactant A and at least one of the products is continuously removed to shift the equilibrium to the product side. If, in addition, the membrane has zero permeability for the other product, simultaneous separation of the products is achieved. This situation is illustrated in Figure 8.4 with the transformed concentration and flux variables as introduced in Equation 8.17. Therein, component A is taken as the reference, which leads to the following definitions:

$$Y_B = \tilde{y}_B + \frac{\tilde{y}_A}{2}, \; Y_C = 1 - Y_B, \tag{8.21}$$

$$J_B = j_B + \frac{j_A}{2}, \; J_C = j_C + \frac{j_A}{2},$$
$$J_{\text{tot}} = J_B + J_C = j_A + j_B + j_C = j_{\text{tot}}. \tag{8.22}$$

Further, it is assumed in Figure 8.4 that the membrane is only permeable for product B but not for product C.

For the simple system considered here, the flux rate ratio shown in Figure 8.4 can be calculated explicitly. It is equivalent to a reactive vapor–liquid equilibrium with zero volatility for reactant A due to zero permeability of the membrane for this component. In the transformed concentration variables a Y_B value of zero corresponds to pure product C, a value of 0.5 to pure reactant A and a value of 1.0 to pure product B. It should be noted that the diagram on the left of Figure 8.4 has the same structural properties as the corresponding diagram of Figure 8.2 for

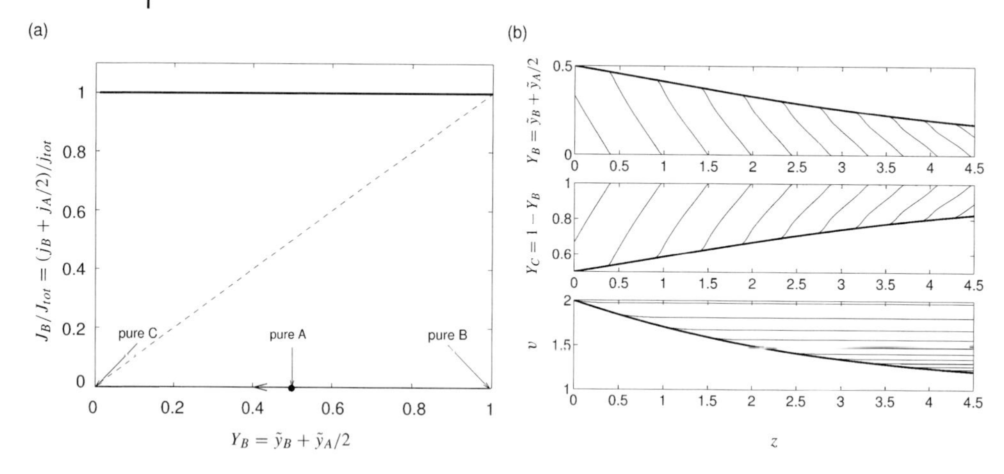

Figure 8.4 Membrane reactor with a reaction 2A ↔ B + C. Parameters: $\beta_A = \beta_C = 0$, $\beta_B = 1$, $K = 1$. a) Transformed flux ratio versus Y_B. b) Response to input ramp of $Y_{B,in}$ or $\tilde{y}_{A,in}$, respectively, as shown in Figure 8.3. Thin lines = transient solutions at $t = 0.3, 0.6, \ldots, 3.3$; bold lines = steady state solution.

the nonreactive binary case. In particular, the steady-state profile of Y_B in the inner tube will monotonically decrease and tend to $Y_B = 0$, which corresponds to pure product C in the inner tube. The dynamic transient behavior during start-up is also analogous to the nonreactive case considered in Figure 8.2.

In practice, the membrane however often has finite permeability for all components, which is considered next. In this case the enhancement of the reaction through the membrane strongly depends on the order of permeability of the membrane for the different components. In particular, three different cases are considered, where the membrane has highest, intermediate or lowest permeability for reactant A. The permeability for B is assumed to be higher than for C in all three cases. Like in the previous case an explicit calculation of the transformed flux rate ratio is possible. The results are shown in Figure 8.5. These diagrams are exactly the same as the McCabe–Thiele diagrams for the corresponding ternary reactive distillation processes (Grüner and Kienle, 2004), which are equivalent to McCabe–Thiele diagrams for binary nonreactive mixtures. The case with intermediate permeability corresponds to the nonreactive vapor liquid equilibrium of an ideal binary mixture, whereas the other two cases correspond to the nonreactive vapor liquid equilibrium of azeotropic binary mixtures. According to the terminology introduced by (Huang *et al.*, 2004), these 'azeotropic points' are called reactive arheotropes in the remainder. At these points the change of concentration through reaction and through mass transport across the membrane compensate each other and lead to constant composition. The corresponding profiles of the transformed concentration variables and the velocity v are shown in Figure 8.5. In the first case, the reactive arheotrope represents the maximum achievable composition which can be obtained in the inner tube of an infinitely

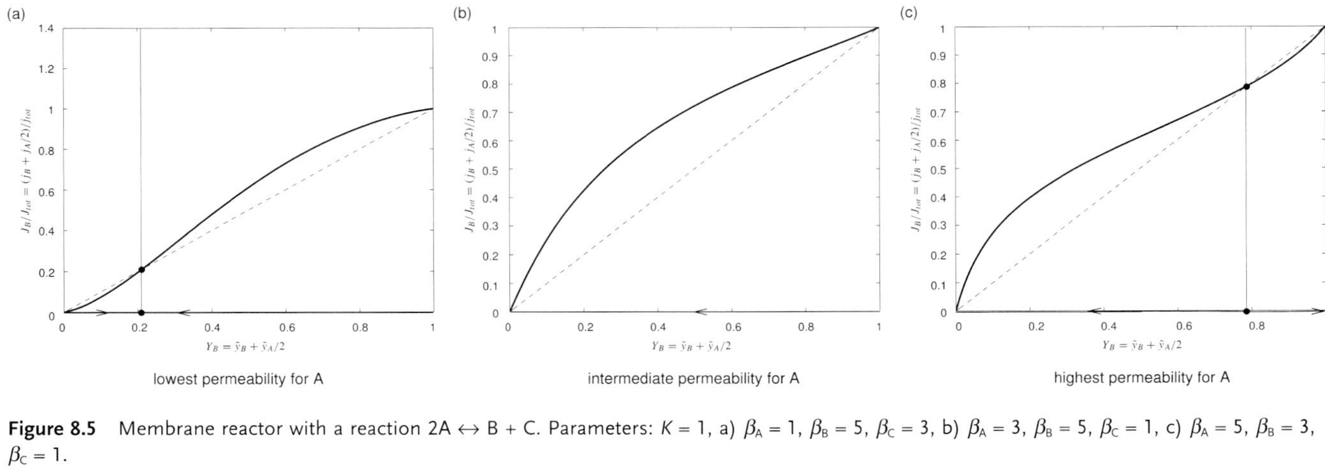

Figure 8.5 Membrane reactor with a reaction 2A ↔ B + C. Parameters: $K = 1$, a) $\beta_A = 1$, $\beta_B = 5$, $\beta_C = 3$, b) $\beta_A = 3$, $\beta_B = 5$, $\beta_C = 1$, c) $\beta_A = 5$, $\beta_B = 3$, $\beta_C = 1$.

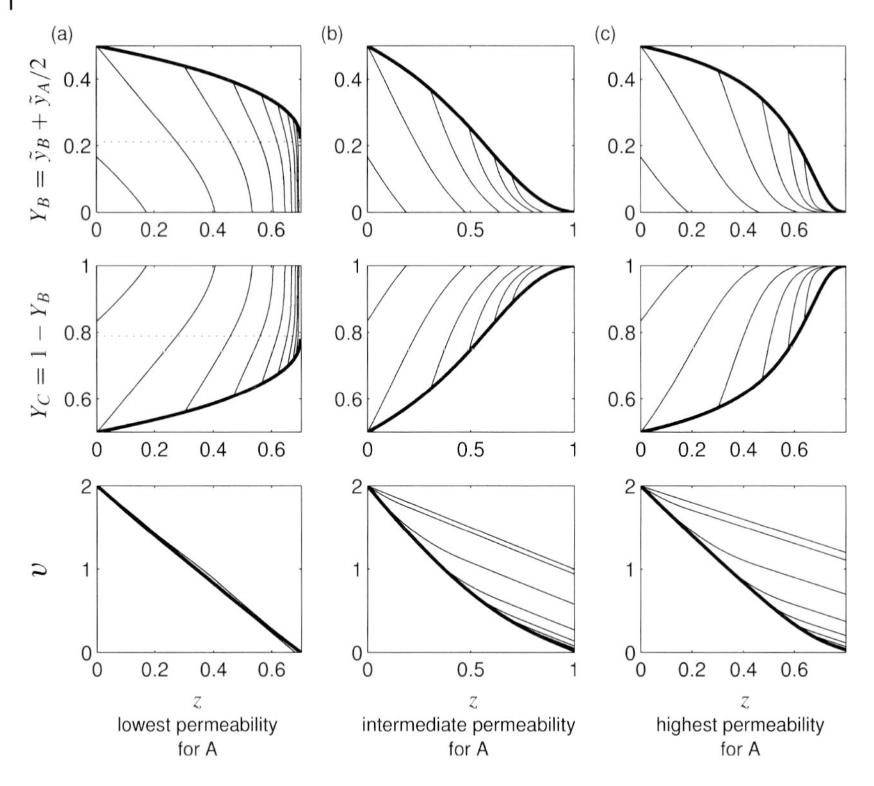

Figure 8.6 Membrane reactor with a reaction 2A ↔ B + C, response to input ramp of $Y_{B,in}$ or $\tilde{y}_{A,in}$, respectively, as shown in Figure 8.3.Thin lines = transient solutions at $t = 0.2, 0.4, ..., 1.2$, bold lines = steady state solution. a) $\beta_A = 1$, $\beta_B = 5$, $\beta_C = 3$, b) $\beta_A = 3$, $\beta_B = 5$, $\beta_C = 1$, c) $\beta_A = 5$, $\beta_B = 3$, $\beta_C = 1$. Dotted lines in a) mark composition at the arheotropic point.

long membrane reactor when the feed is pure reactant A. In the other two cases pure product C can be obtained in the inner tube. However, total conversion is of course not possible due to the loss of reactant across the membrane, due to the finite permeability for A. Again the qualitative dynamic behavior, which is illustrated in Figure 8.6 for a start-up scenario, is completely analogous to the nonreactive case. In particular, in the first case the self-sharpening characteristic of the traveling concentration fronts is very pronounced. Again, this comes from the decrease of the transport velocity from the entrance to the outlet.

An extension to multireaction and multicomponent systems with more than one independent variable is straight forward. Feasible products follow from the corresponding reactive residue curve maps for membrane processes, as introduced by (Huang *et al.*, 2004). Propagation dynamics follow from the variable transport velocity v, which is the same for all components. Step disturbances therefore also in the reactive case be resolved into a single wave traveling jointly for all (transformed) concentrations. The variable velocity v results in self-sharpening waves.

8.2.3 Nonisothermal Operation

8.2.3.1 Formation of Traveling Waves

First the case of an empty tubular reactor with $\varepsilon = 1$ is considered. In analogy to the isothermal case, the following reduced set of equations is obtained from (8.1) and (8.3) by a state transformation:

$$\frac{\partial}{\partial t}\begin{pmatrix}\vec{Y}\\ \Theta\end{pmatrix} + v\frac{\partial}{\partial z}\begin{pmatrix}\vec{Y}\\ \Theta\end{pmatrix} = -\begin{pmatrix}\vec{J}\\ J_\Theta\end{pmatrix} + \begin{pmatrix}\vec{Y}\\ 0\end{pmatrix} J_{\text{tot}} \tag{8.23}$$

The definition of $\vec{Y}, \vec{J}, J_{\text{tot}}$ is identical to the isothermal case (8.17). The temperature and heat fluxes are transformed according to:

$$\Theta = \vartheta - \vec{B}^T \left(\overset{=\text{I}}{v}\right)^{-1} \vec{\tilde{y}}^{\text{I}},\; J_\Theta = j_\vartheta - \vec{B}^T \left(\overset{=\text{I}}{v}\right)^{-1} \left(\vec{j}^{\text{I}} - \vec{y}^{\text{I}} j_{\text{tot}}\right) \tag{8.24}$$

To obtain the total material balance (8.5) in transformed coordinates, the expression $\frac{\partial \vartheta}{\partial t} + v\frac{\partial \vartheta}{\partial z}$ has to be computed from the transformation relation:

$$\begin{pmatrix} -\overset{=\text{II}}{v}\left(\overset{=\text{I}}{v}\right)^{-1} & \overset{=}{I} & \overset{=}{0} \\ \vec{B}\left(\overset{=\text{I}}{v}\right)^{-1} & 0 & 1 \\ \frac{\partial \vec{K}}{\partial \vec{\tilde{y}}^{\text{I}}} & \frac{\partial \vec{K}}{\partial \vec{\tilde{y}}^{\text{II}}} & \frac{\partial \vec{K}}{\partial \vartheta} \end{pmatrix} \begin{pmatrix} \frac{\partial \vec{\tilde{y}}^{\text{I}}}{\partial t} + v\frac{\partial \vec{\tilde{y}}^{\text{I}}}{\partial z} \\ \frac{\partial \vec{\tilde{y}}^{\text{II}}}{\partial t} + v\frac{\partial \vec{\tilde{y}}^{\text{II}}}{\partial z} \\ \frac{\partial \vartheta}{\partial t} + v\frac{\partial \vartheta}{\partial z} \end{pmatrix} = \begin{pmatrix} \frac{\partial \vec{Y}}{\partial t} + v\frac{\partial \vec{Y}}{\partial z} \\ \frac{\partial \Theta}{\partial t} + v\frac{\partial \Theta}{\partial z} \\ \vec{0} \end{pmatrix} \tag{8.25}$$

It is obvious from Equation 8.23 that–as in the isothermal case–in the nonisothermal case concentration and temperature waves also travel with an identical velocity, which is the flow velocity of the fluid. Sharpening or expansion of the waves may occur, if the flow velocity changes along the space coordinate.

The situation is different for fixed-bed membrane reactors, where $\varepsilon < 1$. In this case, thermal fronts may spread with a velocity different from the flow velocity of the fluid. This is discussed in the following for the special case of a single reaction, where it is easily possible to calculate the velocity of a thermal front explicitly. Using the reaction equilibrium (8.6) and the material balance (8.1) for $\tilde{y}^{\text{I}}$, the following relation for the reaction rate is obtained:

$$r = \left(v^{\text{I}}\right)^{-1}\left\{-\frac{\partial K/\partial \vec{\tilde{y}}^{\text{II}}}{\partial K/\partial \tilde{y}^{\text{I}}}\left(\frac{\partial \vec{\tilde{y}}^{\text{II}}}{\partial t} + v\frac{\partial \vec{\tilde{y}}^{\text{II}}}{\partial z}\right) - \frac{\partial K/\partial \vartheta}{\partial K/\partial \tilde{y}^{I}}\left(\frac{\partial \vartheta}{\partial t} + v\frac{\partial \vartheta}{\partial z}\right) + j^{\text{I}} - x^{\text{I}} j_{\text{tot}}\right\} \tag{8.26}$$

Inserting this into the material balances for $\vec{\tilde{y}}^{\text{II}}$ as well as into the energy balance (8.3) yields:

$$\overset{=}{M}\begin{pmatrix} \frac{\partial \vec{\tilde{y}}^{\text{II}}}{\partial t} + v\frac{\partial \vec{\tilde{y}}^{\text{II}}}{\partial z} \\ \frac{\partial \vartheta}{\partial t} + v\frac{\partial \vartheta}{\partial z} \end{pmatrix} = \begin{pmatrix} \vec{0} \\ -(\varepsilon - 1)v\frac{\partial \vartheta}{\partial z} \end{pmatrix} + \begin{pmatrix} \overset{=\text{II}}{v}/v^{\text{I}}\left(j^{\text{I}} - \tilde{y}^{\text{I}} j_{\text{tot}}\right) - \vec{j}^{\text{II}} + \vec{\tilde{y}}^{\text{II}} j_{\text{tot}} \\ B/v^{\text{I}}\left(j^{\text{I}} - x^{\text{I}} j_{\text{tot}}\right) - j_\vartheta \end{pmatrix} \tag{8.27}$$

The left-hand side matrix $\overline{\overline{M}}$ reads:

$$\overline{\overline{M}} = \overline{\overline{I}} - \underbrace{\begin{pmatrix} \overline{\overline{v}}^{II}/v^{I} \\ B/v^{I} \end{pmatrix}}_{=:\vec{l}} \underbrace{\begin{pmatrix} \dfrac{\partial K/\partial \vec{\tilde{y}}^{II}}{\partial K/\partial \tilde{y}^{I}}, \dfrac{\partial K/\partial \vartheta}{\partial K/\partial \tilde{y}^{I}} \end{pmatrix}}_{=:\vec{q}^{T}} \tag{8.28}$$

If $\overline{\overline{M}}$ is regular, the above equations can be decoupled by inversion of $\overline{\overline{M}}$. When doing this, one finds for the propagation velocity of a temperature front v_{ϑ}:

$$v_{\vartheta} = \varepsilon v - (1-\varepsilon) v \frac{B/v^{I} \cdot (\partial K/\partial \vartheta)/(\partial K/\partial \tilde{y}^{I})}{1 - \vec{l}^{T}\vec{q}} \tag{8.29}$$

Depending on the properties of the reaction equilibrium, expanding temperature waves, sharpening waves and discontinuous shocks can evolve. Shock solutions of this type were analyzed by (Glöckler, Kolios, and Eigenberger, 2003) for a reverse-flow fixed-bed reactor.

8.2.3.2 Formation of Discontinuous Patterns

Nonisothermal operation can lead to another interesting phenomenon when the membrane is used not for the separation but for the injection of reactants (Figure 8.1b). In this case, concentration and temperature patterns may form. This effect is studied in detail in the next section. Here, some preliminary considerations are made, again assuming that the reaction rate is infinitely fast. It is shown that spatiotemporal patterns can also occur in the limiting case of reaction equilibrium and that those patterns may contain discontinuities. As an example, the reversible reaction A $\leftrightarrow$ B is considered.

Additional model assumptions are:

- The reaction equilibrium reads:

$$\frac{\tilde{y}_{B}^{eq}}{\tilde{y}_{A}^{eq}} = k^{eq} \exp\left(\frac{\gamma\vartheta}{1+\gamma\vartheta}\right), \tag{8.30}$$

where γ is the dimensionless free enthalpy of reaction.

- The reaction rate is:

$$r = k_0 r'(\tilde{y}_A, \tilde{y}_B, \vartheta), \tag{8.31}$$

where k_0 is a rate constant that tends to infinity in the limiting case, and r' is some function that vanishes at the reaction equilibrium.

- The mixture in the reactor consists of the two reactants A and B and an inert I.
- The sweep gas side of the reactor contains pure A, that is, $\tilde{y}_A^S = 1$.
- The membrane is permeable for all three components, the flow rates given as: $j_i = \beta_i(\tilde{y}_i - \tilde{y}_i^S)$. The convective heat flux across the membrane is: $j_Q = \alpha(\vartheta - \vartheta^S)$.
- The reactor is a tubular reactor with $\varepsilon = 1$.

In the following, we consider the behavior of the system along a characteristic direction ξ, where:

$$\frac{dz}{d\xi} = v, \frac{dt}{d\xi} = 1. \tag{8.32}$$

From the material balances (8.1) and the energy balance (8.3), one obtains the differential equations:

$$\frac{d\tilde{y}_A}{d\xi} = -k_0 r' - j_A + \tilde{y}_A j_{tot} \tag{8.33}$$

$$\frac{d\tilde{y}_B}{d\xi} = k_0 r' - j_B + \tilde{y}_B j_{tot} \tag{8.34}$$

$$\frac{d\vartheta}{d\xi} = Bk_0 r' + j_\vartheta \tag{8.35}$$

By introducing the transformed variables:

$$Y_A := \tilde{y}_A, Y_B := \tilde{y}_A + \tilde{y}_B, \Theta := B\tilde{y}_A + \vartheta, \tag{8.36}$$

the above system of equations is converted to:

$$\frac{1}{k_0}\frac{dY_A}{d\xi} = -r' + \frac{1}{k_0}(-j_A + \tilde{y}_A j_{tot}) \tag{8.37}$$

$$\frac{dY_B}{d\xi} = -J_B + Y_B J_{tot} \tag{8.38}$$

$$\frac{d\Theta}{d\xi} = -J_\Theta \tag{8.39}$$

The reaction equilibrium, given by $r' = 0$ or (8.30), defines the quasi-stationary manifold of the system (8.37–8.39) for $k_0 \to \infty$. In the case of an infinitely fast reaction rate, the solutions of (8.37–8.39) move along that manifold. Discontinuities and relaxation oscillations may occur, if for given values of Y_B and Θ the value of Y_A on the quasi-stationary manifold is not unique. An example is shown in Figure 8.7. The trajectory remains on the quasi-stationary manifold until it reaches a minimum point on the manifold. At that point, the solution jumps to a different value of Y_A. The values of Y_B and Θ do not change at the discontinuity, that is:

$$Y_B^- = Y_B^+, \tag{8.40}$$

$$\Theta^- = \Theta^+, \tag{8.41}$$

where "–" indicates the state immediately before the discontinuity, and "+" the state immediately after the discontinuity. In original variables, (8.40) and (8.41) read:

$$\tilde{y}_A^- - \tilde{y}_A^+ = \tilde{y}_B^+ - \tilde{y}_B^-, \tag{8.42}$$

$$\vartheta^+ - \vartheta^- = B(\tilde{y}_A^- - \tilde{y}_A^+). \tag{8.43}$$

That is, (8.40) and (8.41) ensure the conservation of mass and the conservation of energy at the discontinuity. Figure 8.8 shows the relaxation oscillation in original variables.

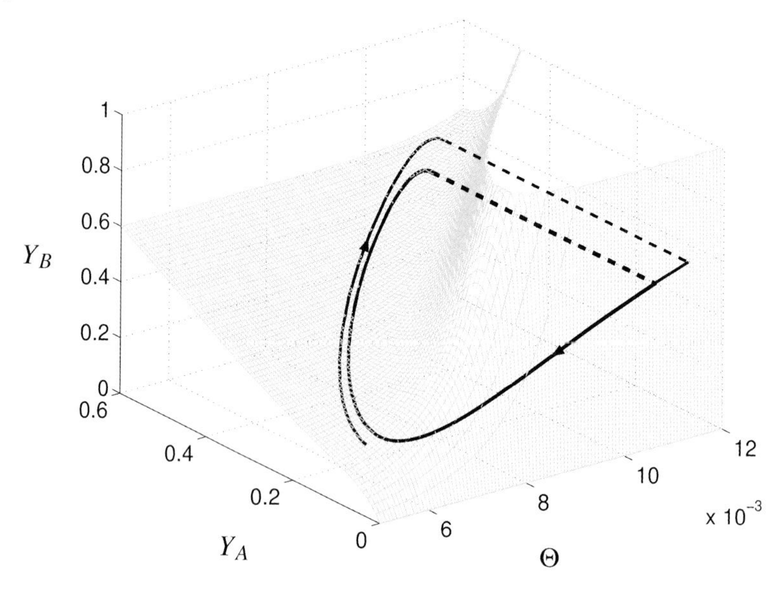

Figure 8.7 Transient of the nonisothermal membrane reactor model (8.37–8.39) to a relaxation oscillation in transformed variables. The gray mesh indicates the shape of the quasi-stationary manifold; black solid lines = trajectories on the manifold; black dashed lines = discontinuous parts of the solution. Parameter values: $\tilde{y}_A^S = 1$, $\vartheta^S = 0$, $\beta_A = \beta_I = 1$, $\beta_B = 10$, $\alpha = 3$, $\gamma = 1000$, $B = 0.0162$, $k^{eq} = 0.0143$.

8.3 Pattern Formation

The formation of spatial patterns in membrane reactors with side injection is studied in more detail in the following. A number of publications show that the side injection of reactants may cause quite complicated nonlinear behavior (Nekhamkina *et al.*, 2000a, 2000b; Sheintuch and Nekhamkina, 2003; Travnickova *et al.*, 2004). Sheintuch and co-workers pointed out an analogy between the dynamic behavior of a CSTR and the steady-state behavior of an ideal plug flow membrane reactor. They showed that for a small axial dispersion of heat spatially periodic patterns can emerge for an exothermic first-order reaction (Nekhamkina *et al.*, 2000a) and that complex aperiodic patterns may exist in the case of a simple consecutive reaction with two reaction steps (Sheintuch and Nekhamkina, 2003). (Travnickova *et al.*, 2004) found complex aperiodic spatiotemporal patterns in a cross-flow reactor with a first-order reaction and high axial dispersion of heat. In this section, we explore the possibility of obtaining and measuring stationary spatially periodic patterns in a laboratory membrane reactor. Different models of a membrane reactor with a porous membrane and a catalytic fixed bed are considered. All models use an experimentally validated reaction scheme for the partial oxidation of ethane. It should be noted that the partial oxidation reaction serves

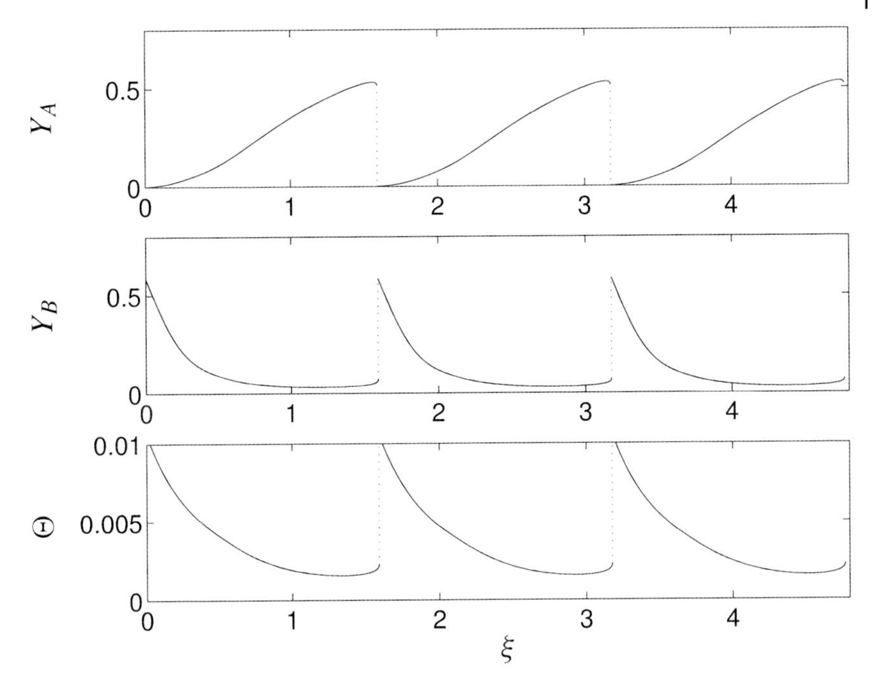

Figure 8.8 Discontinuous solutions of the nonisothermal membrane reactor model (8.37–8.39) along a characteristic ξ in original variables; parameter values as in Figure 8.7.

only as an example and similar qualitative behavior may be expected for other exothermic reactions in membrane reactors. The simplest model variant assumes ideal plug flow behavior in the fixed bed. It is studied in the first step by a bifurcation analysis and by dynamic simulations. In the second step, dispersive effects are included in the analysis. In the last step, the results are compared with a more detailed spatially distributed model of the membrane reactor.

8.3.1 Analysis of a Simple Membrane Reactor Model

(Klose *et al.*, 2004) investigated the oxidative dehydrogenation of ethane over a $VO_x/\gamma\text{-}Al_2O_3$ catalyst in a laboratory fixed-bed reactor. As discussed in Chapter 3, kinetic expressions were derived for this reaction. The reaction network proposed by (Klose *et al.*, 2004) consists of five reaction steps (see Figure 8.9). Reaction 1 is described by a Mars van Krevelen mechanism. For reactions 2–5, a Langmuir–Hinshelwood approach is used. All kinetic equations and parameters are given in Section 3.4.

This work investigates the oxidative dehydrogenation of ethane in a fixed-bed membrane reactor with side injection of reactants, as shown in Figure 8.1b. The reactor consists of a porous tubular membrane and a catalytic fixed bed

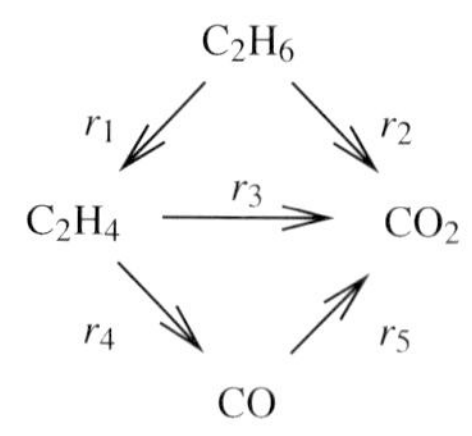

Figure 8.9 Reaction network for the oxidative dehydrogenation of ethane (Klose *et al.*, 2004).

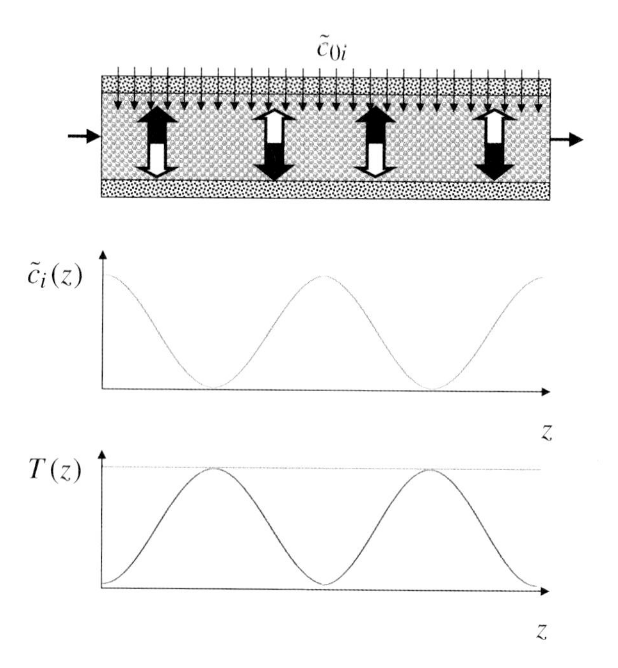

Figure 8.10 Physical interpretation of the stationary pattern formation in a fixed-bed membrane reactor. Meaning of filled arrows: ↑ = charging of energy storage by chemical reaction; ↓ = discharging of energy storage by heat losses. Meaning of open arrows: ↑ = charging of mass storage by mass supply through membrane; ↓ = discharging of mass storage by chemical reaction.

inside the tube. The shell side of the membrane is empty. Reactants can be fed directly to the inlet of the fixed bed with concentrations $\tilde{c}_{i,\text{in}}$ and a temperature T_{in}; alternatively they can enter the fixed bed in a spatially distributed manner via the shell side across the membrane with inlet concentrations $\tilde{c}_{0i}$ and an inlet temperature T_0. Figure 8.10 gives a simple physical explanation for the possible formation of spatial patterns in such a reactor. The fluid flowing through the fixed bed possesses a storage capacity for material of reactants and for heat. A spatial pattern is caused by the alternate charging and discharging of the material storage and the heat storage resulting from the chemical reaction. In sections of high conversion, the material storage is discharged, and the reaction heat released charges the energy storage. In sections of low conversion, the material storage is reloaded by mass transfer from the sweep gas side through the membrane, while heat losses reduce the temperature of the fixed bed and hence discharge the energy storage.

In order to verify the pattern formation in the case of ethane oxidation, a one-dimensional pseudo-homogeneous model of the membrane reactor is used in this Section. Radial concentration and temperature gradients as well as a change in the axial flow rate due to mass transport through the membrane are neglected. The concentration and temperature on the shell side of the membrane are assumed to be constant and identical to the inlet conditions $\tilde{c}_{0i}$ and T_0. The reactants are assumed to behave like ideal gases. The differential equations for the concentration and the temperature inside the fixed bed are obtained from mass and energy balances. They account for accumulation, convection, axial dispersion, chemical reactions, heat transfer through the membrane and mass supply through the membrane. The component mass balances read:

$$\frac{\partial \tilde{c}_i}{\partial t} = -v\frac{\partial \tilde{c}_i}{\partial z} + \sum_{j=1}^{5} \nu_{ij} r_j \rho_{\text{cat}} + \frac{2}{R}\beta(\tilde{c}_{0i} - \tilde{c}_i),\ i = 1, \dots, 5 \tag{8.44}$$

The energy balance leads to:

$$(\rho c_P)_{\text{tot}} \frac{\partial T}{\partial t} = -v(\rho c_P)_f \frac{\partial T}{\partial z} + \lambda \frac{\partial^2 T}{\partial z^2} + \sum_{j=1}^{5} \nu_{ij}(-\Delta_R H)_j r_j \rho_{\text{cat}} + \frac{2}{R}\alpha(T_0 - T) \tag{8.45}$$

In the model, the heat of reaction varies with temperature; the partial molar enthalpy of each component is computed from:

$$h_i = h_{i,\text{ref}} + c_{P,i}(T - T_{\text{ref}}) \tag{8.46}$$

for the values of the enthalpies of formation for each component. The following boundary conditions are used for (8.44) and (8.45):

$$\tilde{c}_i(0,t) = \tilde{c}_{i,\text{in}}(t),\ i = 1, \dots, 5,\ \lambda \left.\frac{\partial T}{\partial z}\right|_{0,t} = v(\rho c_P)_f\,[T(0,t) - T_{\text{in}}(t)] \tag{8.47}$$

$$\left.\frac{\partial T}{\partial z}\right|_{L,t} = 0 \tag{8.48}$$

The set of model equations is completed by a quasi-stationary equation of continuity that determines the flow velocity v.

$$\frac{\partial \rho_f v}{\partial z} = 0,\ v(0,t) = v_{\text{in}}(t) \tag{8.49}$$

The fluid density ρ_f follows from the ideal gas law. In the above equations, t and z are the time and the space coordinates, v is the flow velocity of the gas, $\tilde{c}_1$–$\tilde{c}_5$ stand for the molar concentrations of the components (respectively: ethane, ethene, carbon monoxide, carbon dioxide, oxygen), ν_{ij} are the stoichiometric coefficients, R is the inner radius of the membrane, β is a mass transfer coefficient, $(\rho c_P)_{\text{tot}}$ and $(\rho c_P)_{\text{f}}$ are, respectively, the thermal capacities of the fixed bed and the gas, λ is the axial heat conductivity of the bed and α is a heat transfer coefficient. All model parameters used in the simulations are listed by (Zhang, Mangold, and Kienle , 2006).

8.3.1.1 Analysis of Steady-State Reactor Behavior for Vanishing Heat Dispersion

The analysis of the membrane reactor behavior becomes comparatively simple when the heat dispersion term in Equation 8.45 is neglected. In this case, the steady-state component mass balances and the steady-state energy balance read:

$$v\frac{\partial \tilde{c}_i}{\partial z} = \sum_{j=1}^{5} \nu_{ij} r_j + \frac{2}{R}\beta(\tilde{c}_{0i} - \tilde{c}_i),\ \tilde{c}_i(0) = \tilde{c}_{i,\text{in}},\ i = 1, \ldots, 5 \tag{8.50}$$

$$v(\rho c_P)_f \frac{\partial T}{\partial z} = \sum_{j=1}^{5} \nu_{ij}(-\Delta_R H)_j r_j + \frac{2}{R}\alpha(T_0 - T),\ T(0) = T_{\text{in}} \tag{8.51}$$

Obviously, there is an analogy between the above steady-state equations of an ideal plug flow membrane reactor and the dynamic equations of a CSTR (Nekhamkina *et al.*, 2000a). The dynamic behavior of exothermic CSTRs has been studied in detail and found to be surprisingly complex, even for simple first-order reactions (Uppal and Ray, 1974; Sheplev, Treskov, and Volokitin, 1998). Therefore, an analogous pattern of behavior can be expected for the steady-state solutions of the plug flow membrane reactor considered here. In a first step, spatially homogeneous solutions are investigated, which correspond to the steady-state solutions of a CSTR. The continuation methods in DIVA (Mangold *et al.*, 2000) are used to analyze the dependence of the solutions on the shell side concentrations $\tilde{c}_{0i}$ and the shell side temperature T_0 as the main operation parameters. An example for the results is given in Figure 8.11. Saddle node bifurcations and Hopf bifurcations are found when the temperature and the concentration on the shell side are varied. The curve of saddle node bifurcations in Figure 8.11a borders a region of multiple spatially homogeneous solutions, that is, inside this region there are three different sets of reactor inlet conditions, resulting in three different spatially homogeneous solutions for given shell side conditions. Figure 8.11 also contains information on the stability of the steady-state solutions of Equations 8.50 and 8.51 when moving along the space coordinate z. Stability means in this case that, as z increases, the composition and temperature in the fixed-bed approach asymptotically the homogeneous solution, that is, the steady-state solution of Equations 8.50 and 8.51, if $\tilde{c}_{i,\text{in}}$ and T_{in} are close to but not identical to the homogeneous solution. Obviously, this notion of stability has to be distinguished from the dynamic stability of the homogeneous solutions of Equations 8.44 and 8.45. The curve of Hopf bifurcation points indicates locations in parameter space where spatially periodic solutions emerge.

Only simple period 1 solutions are found when just ethane and oxygen are fed through the membrane. However, the behavior becomes more complex when ethene is also added to the feed, as can be seen from Figures 8.12 and 8.13. Figure 8.12 shows a bifurcation diagram obtained by a continuation of spatially periodic solutions. Under the chosen operation conditions, a sequence of period-doubling bifurcations is found. Selected periodic solutions from the period 1, period 2,

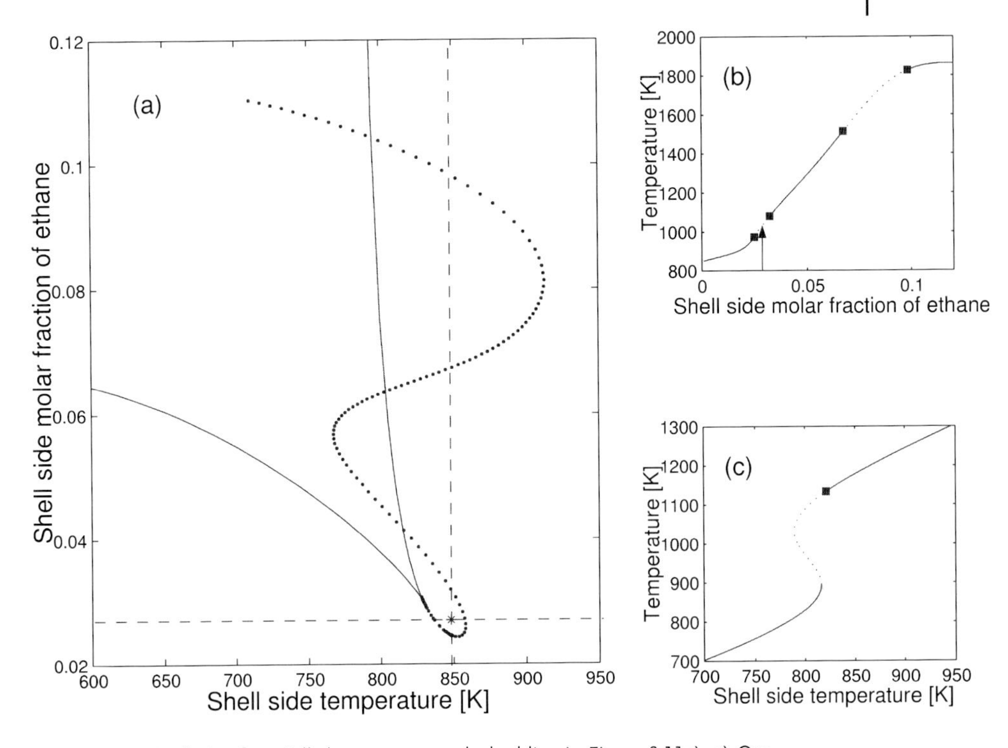

Figure 8.11 Analysis of spatially homogeneous solutions of the ideal plug flow membrane reactor. a) Saddle node bifurcations (solid lines) and Hopf bifurcations (dotted lines), when shell side temperature and shell side mole fraction of ethane are used as bifurcation parameters. b) One parameter continuation with the shell side mole fraction of ethane as continuation parameter and a constant shell side temperature $T_0 = 848\,\text{K}$ (along the vertical dashed line in Figure 8.11a). c) One parameter continuation with the shell side temperature as continuation parameter and a constant shell side mole fraction of ethane $\tilde{y}_{0,C2H6} = 0.04$ (along the horizontal dashed line in Figure 8.11a); solid lines in b) and c) denote stable solutions, dashed lines denote unstable solutions, squares denote Hopf bifurcation points. The asterisk in a) and the arrow in b) indicate the simulation conditions of Figure 8.14.

period 3 and period 4 branches are shown in Figure 8.13a–c. In the vicinity of the period 4 solution branch, even more complex aperiodic solutions are found, as can be seen in Figure 8.13d.

The aperiodic solution in Figure 8.13 is examined more closely by calculating the Lyapunov exponents according to the method given by (Wolf *et al.*, 1985; Benettin *et al.*, 1980; Shimada and Nagashima, 1979), assuming that the reactor has an infinite length. The three leading Lyapunov exponents found are: $\lambda_1 = 0.1765$, $\lambda_2 = 0.0$ and $\lambda_3 = -3.649$, respectively. The positive Lyapunov exponent λ_1 is an indicator of deterministic chaos. The Lyapunov dimension D_L of the aperiodic attractor is calculated according to the Kaplan–Yorke conjecture (Yorke and Kaplan, 1979) from:

$$D_L = 2 + \frac{\lambda_1 + \lambda_2}{|\lambda_3|} \tag{8.52}$$

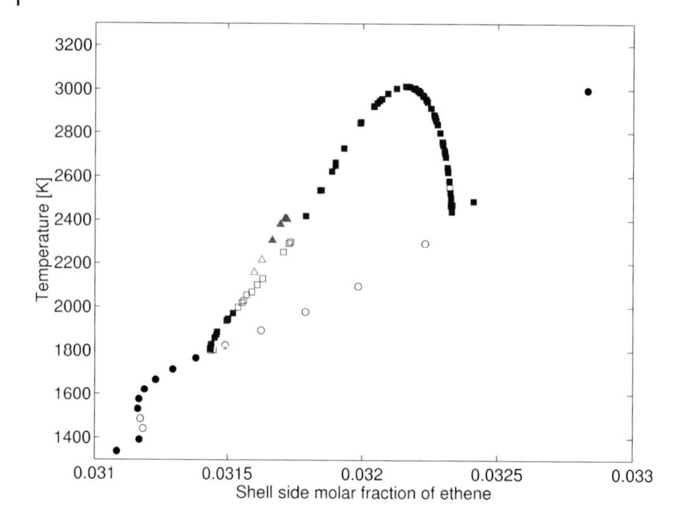

Figure 8.12 Continuation of periodic solutions of the ideal plug flow membrane reactor model. Circles indicate period-1 solutions, squares indicate period-2 solutions, triangles indicate period-4 solutions. Filled symbols denote stable solutions, open symbols denote unstable solutions. Operation conditions on shell side: shell side temperature $T_0 = 820\,\text{K}$, shell side mole fraction of ethane $\tilde{y}_{0,C2H6} = 0.02$, other conditions as in Figure 8.11.

The resulting value of $D_L = 2.048$ is slightly above the lower limit of 2.0 for chaotic attractors. In conclusion, the ideal plug flow reactor model shows a complex steady-state behavior. Some of the pattern solutions found possess maximum temperatures beyond the validity limit of the model. Nevertheless, those solutions indicate operation conditions where interesting phenomena may also occur in more detailed models. As an example, the influence of axial dispersion of heat is studied in the next section.

8.3.1.2 Influence of Heat Dispersion

Figure 8.14 shows steady-state temperature profiles for various values of the axial heat conductivity λ. The operation conditions used are indicated in the bifurcation diagram, Figure 8.11. The steady-state solutions are computed by applying the method of lines to the model (8.44–8.49) using 4000 finite volumes. The resulting algebraic equations are solved by the damped Newton method NLEQ1S (Nowak and Weimann, 1990). For small values of λ, the patterns agree with the results of the ideal plug-flow membrane reactor model obtained by integration along the space coordinate using the integrator DDASAC (Caracotsios and Stewart, 1985). This indicates a reasonable accuracy of the discretized model. For large values of λ the axial heat conduction counteracts the pattern formation and smoothes the temperature profile. The patterns vanish completely when λ exceeds a maximum value, which can be determined by linear analysis (Nekhamkina *et al.*, 2000b).

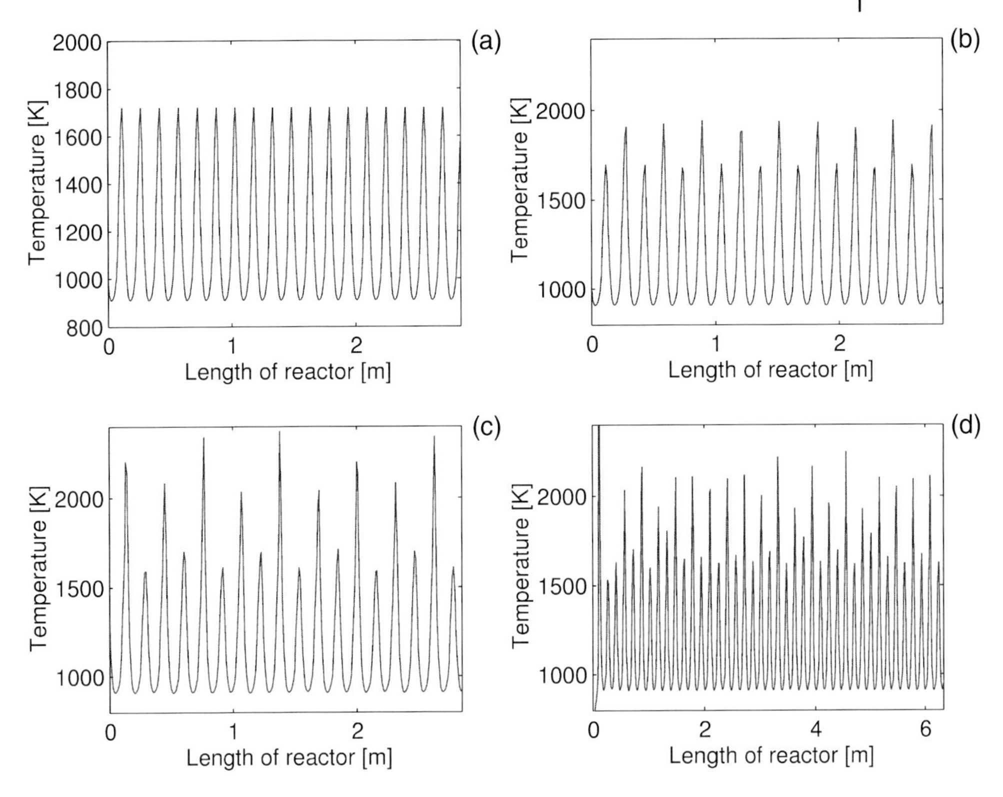

Figure 8.13 Examples of pattern solutions of the ideal plug flow membrane reactor. a) Period-1 oscillation with shell side temperature $T_0 = 820\,K$, shell side mole fractions $\tilde{y}_{0,C2H6} = 0.02$, $\tilde{y}_{0,C2H4} = 0.0313$, other conditions as in Figure 8.11 a). b) Period-2 oscillation with $\tilde{y}_{0,C2H4} = 0.0315$, other conditions as in a). c) Period-4 oscillation with $\tilde{y}_{0,C2H4} = 0.0317$, other conditions as in a). d) Aperiodic solution with $T_0 = 820.8\,K$, $\tilde{y}_{0,C2H4} = 0.0318$, other conditions as in a).

For values of the heat conductivity below this limit, the pattern solution seems to undergo bifurcations: The period of the pattern doubles suddenly for values of λ between 0.0201 and 0.0202 $W\,m^{-1}\,K^{-1}$ (see Figure 8.15).

The dynamic behavior of the system is shown in Figures 8.16 and 8.17. A spatially homogeneous profile is used as initial condition. At the beginning of the simulation, a pattern with a rather short wavelength forms. This pattern moves in the direction of the gas flow towards the reactor outlet. The remaining stationary pattern possesses a longer wave length and a smaller amplitude.

8.3.2
Detailed Membrane Reactor Model

In order to explore the possibility of finding stationary patterns in a laboratory membrane reactor, the following considers a more detailed reactor model. This

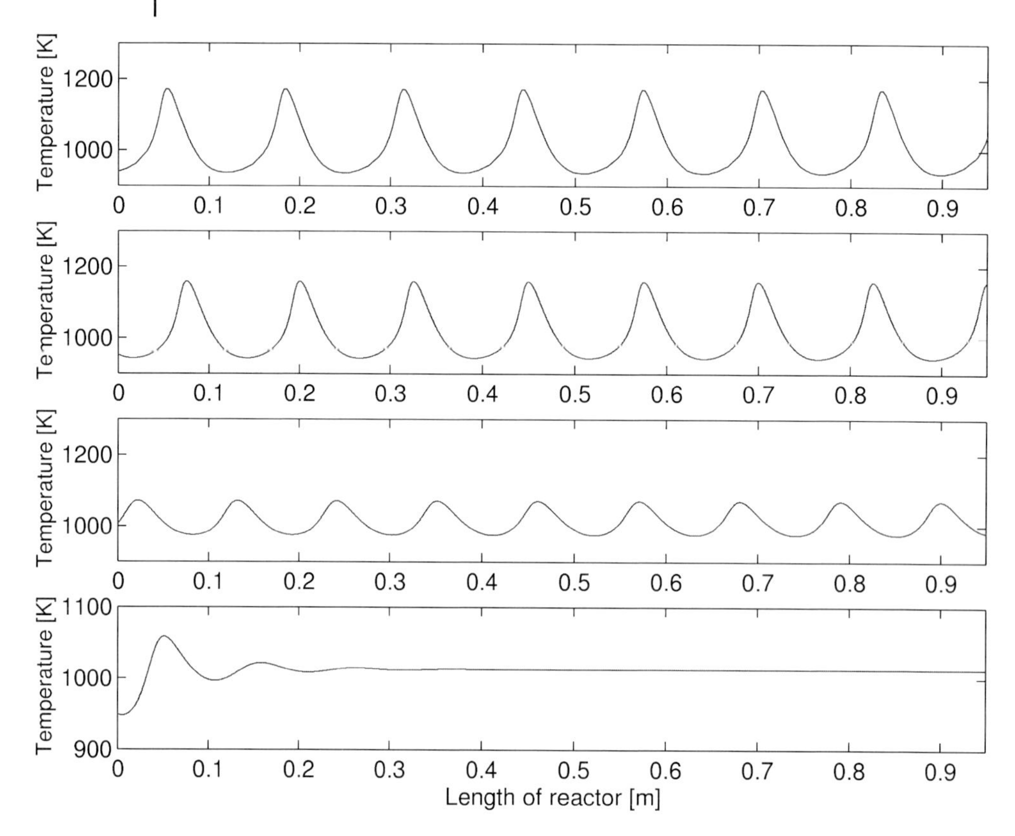

Figure 8.14 Stationary patterns of an ideal plug flow membrane reactor and a membrane reactor model with axial heat dispersion for $\lambda = 0.1$, 0.4 and $1.0\,\mathrm{W\,m^{-1}\,K^{-1}}$ (from top to bottom, respectively). $T_0 = 848\,\mathrm{K}$, $\tilde{y}_{0,C2H6} = 0.027$, other parameters as in Figure 8.11.

model describes the catalytic fixed bed inside a porous membrane, the heat and mass transport through the membrane and the sweep gas side outside the membrane. The model is implemented in the ProMoT process modeling tool, using a model library for membrane reactors available with this software (Mangold, Ginkel, and Gilles, 2004).

8.3.2.1 Main Model Assumptions

The differences between the complete model and simple model are listed below.

- The fixed bed is described in a model as in Equations 8.44–8.49, with the main difference that the dependence of the heat capacities and the total molar mass on the composition of the gases is taken into account.
- The temperature and composition on the sweep gas side are no longer assumed to be constant but are calculated from model equations similar to (8.44–8.49) without the reactive term. Heat loss to the outer reactor wall is described by constant heat transfer parameter, assuming a constant wall temperature.

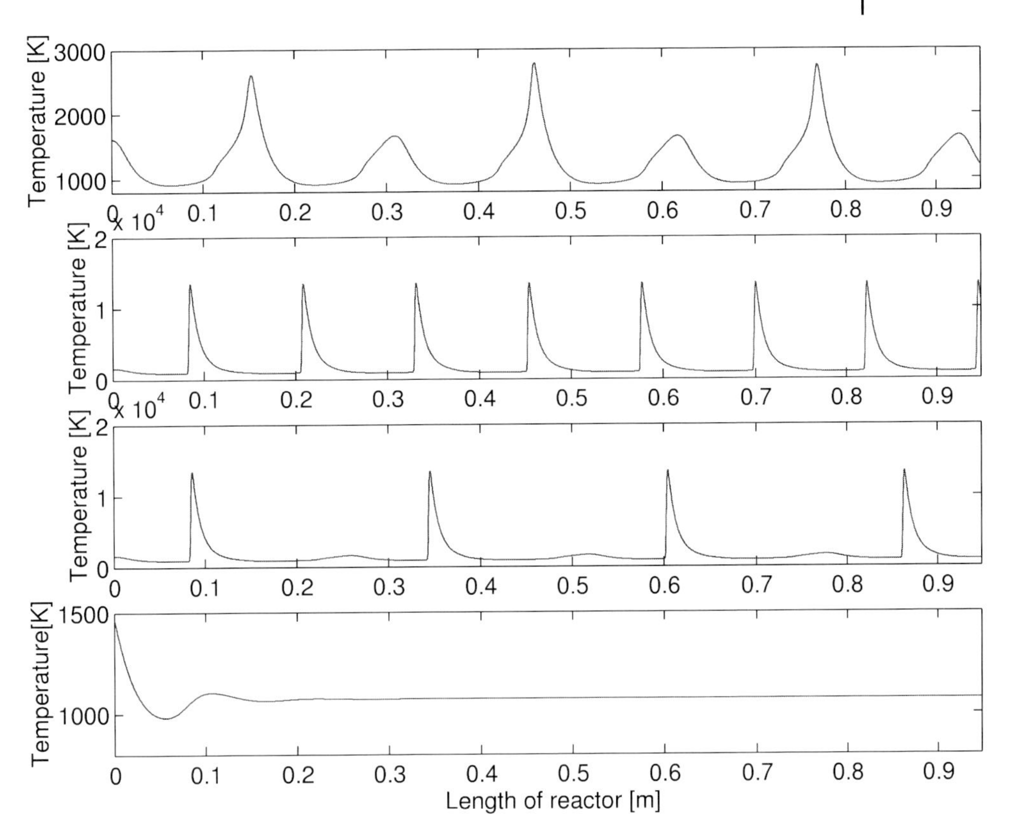

Figure 8.15 Stationary patterns of the temperature for $\lambda = 0.01, 0.0201, 0.0202, 1.0\,\mathrm{W\,m^{-1}\,K^{-1}}$ (from top to bottom, respectively). $T_0 = 820\,\mathrm{K}$, $\tilde{y}_{0,C2H6} = 0.02$, $\tilde{y}_{0,C2H4} = 0.319$, other parameters as in Figure 8.11.

- A more detailed model is used to simulate the mass and heat transfer through the membrane. This model is introduced in the following section.

8.3.2.2 **Model Equations of the Membrane**

The model used here to describe mass transport through the membrane is a simplified version of the model developed by (Hussain, Silalahi, and Tsotsas, 2004). The membrane under consideration consists of four ceramic layers with different porosities. The mass flux through each layer can be described by the dusty gas law (cf. Chapter 2). Neglecting interactions between the components and approximating the partial pressure gradients by the partial pressure differences on both sides of a membrane layer, one obtains for the molar flux of component j through layer i:

$$\dot{n}_{ij} = -\frac{1}{RT_{\mathrm{mem}}d_i}\left(\frac{4}{3}K_{0i}\sqrt{\frac{8RT_{\mathrm{mem}}}{\pi\tilde{M}_j}} + \frac{B_{0i}}{\eta_j}\overline{P}\right)\Delta P_{i,j},\ i = 1,\ldots,4,\ j = 1,\ldots,7 \qquad (8.53)$$

In the above equation, d_i is the thickness of layer i and B_{0i} and K_{0i} are the permeability constant and the Knudsen coefficient of the layer, respectively. $\overline{P}$ is the mean

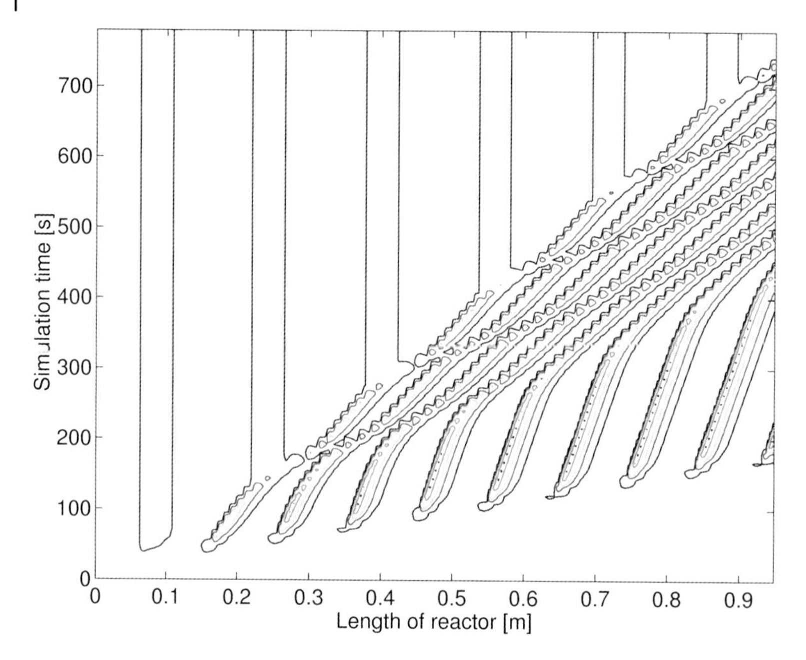

Figure 8.16 Contour plot showing the formation of a sustained pattern starting from a spatially homogeneous initial profile. $T_0 = 848\,\text{K}$, $\tilde{y}_{0,C2H6} = 0.025$ and $\lambda = 0.01\,\text{W}\,\text{m}^{-1}\,\text{K}^{-1}$, all other parameters as in Figure 8.11.

pressure in the membrane, which is set to 1 bar (= 100 kPa) in the simulation. T_{mem} is the temperature in the membrane, which is set to:

$$T_{\text{mem}} = \frac{T_{\text{sw}} + T_{\text{fb}}}{2} \tag{8.54}$$

where T_{SW} is the local temperature on the sweep gas side, and T_{fb} is the local temperature of the fixed bed. The parameters of the membrane layers are taken from (Hussain, Silalahi, and Tsotsas, 2004). Assuming a negligible hold-up of the membrane, the molar fluxes are coupled by:

$$\dot{n}_{j,1} = \dot{n}_{j,2} = \dot{n}_{j,3} = \dot{n}_{j,4} \tag{8.55}$$

The following summation conditions hold for the partial pressure differences:

$$\sum_{i=1}^{4} \Delta P_{i,j} = P_{\text{sw},j} - P_{\text{fb},j},\ j = 1, \ldots, 7 \tag{8.56}$$

The heat transfer from the sweep gas through the membrane to the fixed bed is described by:

$$\dot{q} = \alpha_{\text{mem}} \cdot (T_{\text{sw}} - T_{\text{fb}}) \tag{8.57}$$

The effect of the heat transfer coefficient is discussed in the following section.

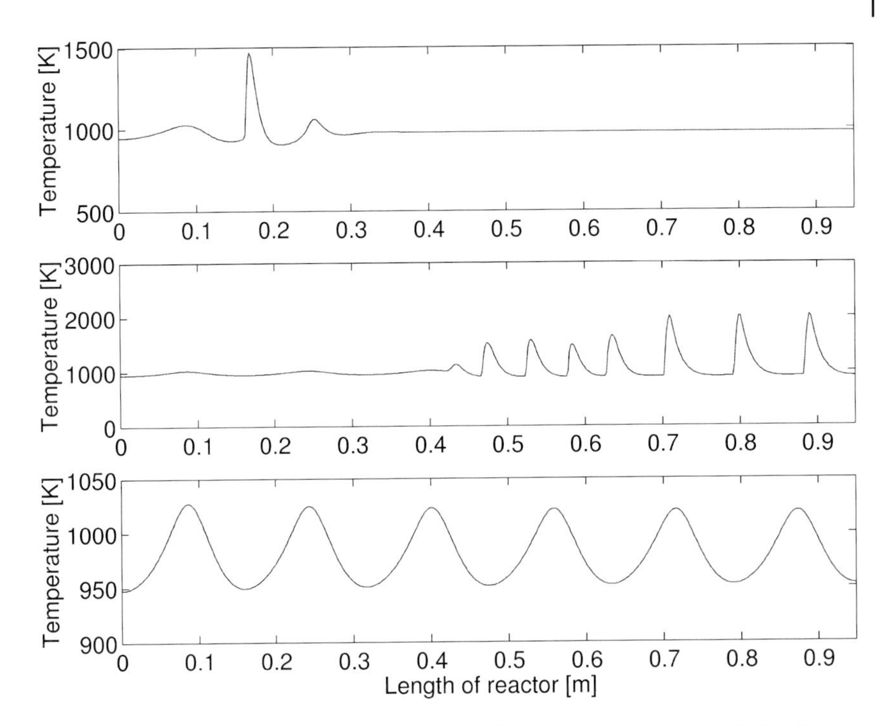

Figure 8.17 Transient pattern at t = 60, 300, 780 s, from top to bottom, respectively, all parameters as in Figure 8.16.

8.3.2.3 Simulation Results

The formation of temperature and concentration patterns is also observed in the more detailed reactor model, as shown in Figure 8.18. As can be seen, the pattern occurs in a temperature and concentration range that can be obtained in a laboratory reactor. In contrast to the simplified model, only patterns with decaying amplitude are found. The main reason for this may be that the concentration of the reactant ethane decreases on the sweep gas side along the reactor coordinate but is constant in the simple model. The pattern turns out to be rather sensitive against the heat transfer coefficient. An increase of $\dot{q}$ reduces the temperature level in the fixed bed, especially in the inlet region, and dampens the temperature oscillations strongly. A decrease of $\dot{q}$ causes a high temperature peak at the reactor inlet but also reduces the oscillation amplitudes in the rear part of the reactor, as can be seen in Figure 8.19.

8.4 Conclusions

This chapter's objective has been to provide some insight in the qualitative nonlinear phenomena that may occur in membrane reactors. The two operation

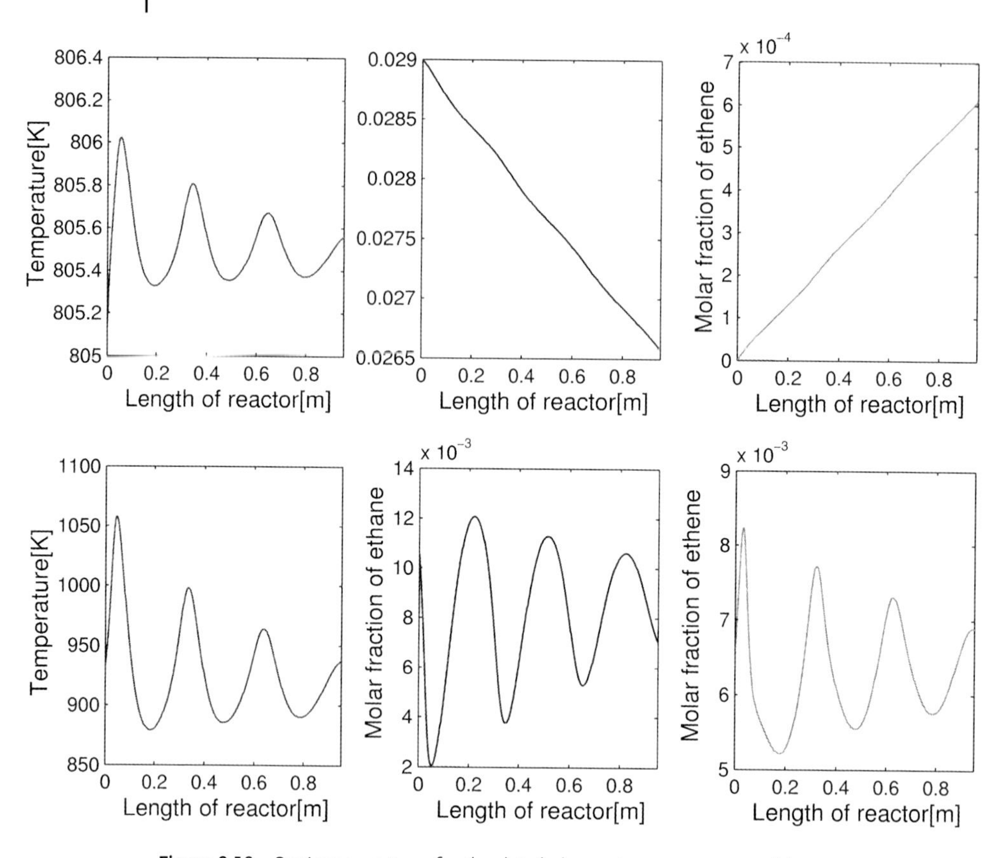

Figure 8.18 Stationary pattern for the detailed membrane reactor model.

modes of selective product removal and of distributed injection of reactants have been considered. In both cases, different nonlinear effects are predominant. When the membrane is used for withdrawing products, then the main concerns are to achieve high conversion and to obtain the target product with high purity. It turns out that total conversion and simultaneous separation requires a membrane that is only permeable for one product. However, in practice the membrane possesses finite permeability for all components. Then its benefit for the reactor performance strongly depends on the order of permeability of the membrane for the different components. For the reaction $A \leftrightarrow B + C$ the behavior is most interesting when A has lowest or highest permeability, because in these cases there is a certain composition for which the change of concentration through reaction and through mass transfer compensate each other. When A has lowest permeability, this composition gives the maximum product concentration achievable when pure A is fed.

When the membrane is used for side injection of reactants to enhance yield, then the stability of the desired solution may become an issue, as shown in the second part of this chapter. Under certain conditions, spatially homogeneous

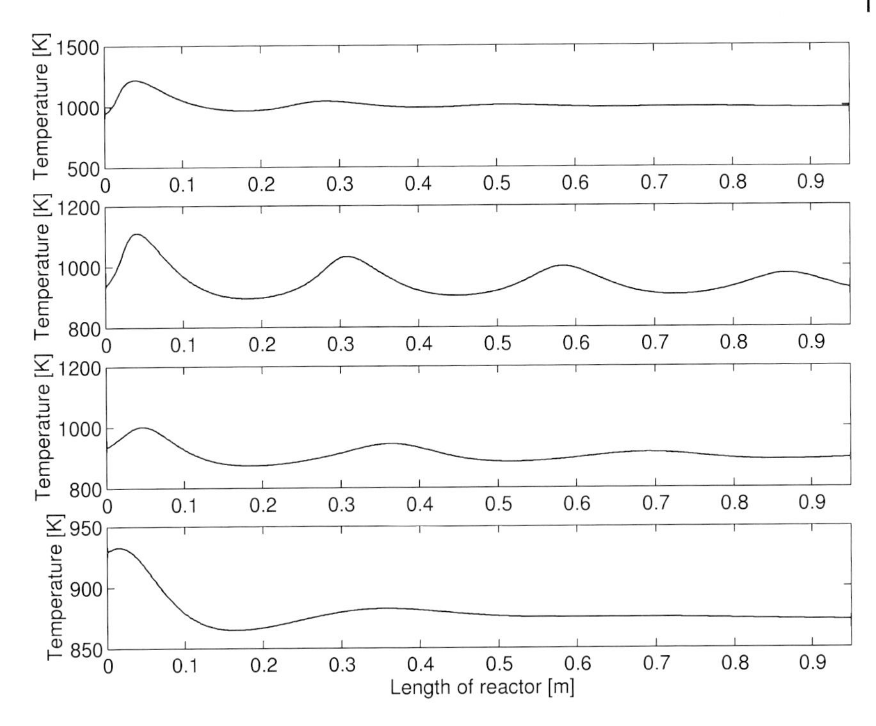

Figure 8.19 Stationary patterns for different heat transfer coefficients: 48, 56, 64 and 72 W m^{-2} K^{-1} (from top to bottom, respectively), all other parameters as in Figure 8.18.

solutions give way to concentration and temperature patterns, which may be quite complex. This can be explained by analogy to the dynamic behavior of CSTRs or physically by the charging and discharging of mass and energy storages along the reactor coordinate. Low axial dispersion and a big reactor length support the effect of pattern formation in membrane reactors.

References

Benettin, G., Galgani, L., Giogilli, A., and Strelcyn, J.-M. (1980) Lyapunov characteristic exponents for smooth dynamical systems and for hamiltonian systems; a method for computing all of them. *Meccanica*, **15**, 9.

Caracotsios, M., and Stewart, W.E. (1985) Sensitivity analysis of initial value problems with mixed odes and algebraic equations. *Comp. Chem. Eng.*, **9**, 359–365.

Fullarton, D., and Schlünder, E.U. (1986) Diffusion distillation – a new separation process for azeotropic mixtures. 1. Selectivity and transfer efficiency. *Chem. Eng. Proc.*, **20**, 255–263.

Glöckler, B., Kolios, G., and Eigenberger, G. (2003) Analysis of a novel reverse-flow reactor concept for autothermal methane steam reforming. *Chem. Eng. Sci.*, **58**, 593–601.

Grüner, S., and Kienle, A. (2004) Equilibrium theory and nonlinear waves for reactive distillation columns and chromatographic reactors. *Chem. Eng. Sci.*, **59**, 901–918.

Grüner, S., Mangold, M., and Kienle, A. (2006) Dynamics of reaction separation processes in the limit of chemical equilibrium. *AIChE J.*, **52** (**3**), 1010–1026.

Huang, Y.-S., Sundmacher, K., Qi, Z., and Schlünder, E.U. (2004) Residue curve maps of reactive membrane separation. *Chem. Eng. Sci.*, **59**, 2863–2879.

Hussain, A., Silalahi, S.H.D., and Tsotsas, E. (2004) Heat transfer in tubular, inorganic membranes. Proceedings of the 4th European Thermal Sciences Conference, Birmingham.

Kienle, A. (2000) Low-order dynamic models for ideal multiplicity of multicomponent distillation processes using nonlinear wave propagation theory. *Chem. Eng. Sci.*, **55**, 1817–1828.

Klose, F., Joshi, M., Hamel, C., and Seidel-Morgenstern, A. (2004) Selective oxidation of ethane over a vox/gamma-al2o3 catalyst–investigation of the reaction network. *Appl. Catal. A*, **260**, 101–110.

Krishna, R., and Wesselingh, J.A. (1997) The Maxwell-Stefan approach to mass transfer. *Chem. Eng. Sci.*, **52**, 861–911.

Mangold, M., Kienle, A., Mohl, K.D., and Gilles, E.D. (2000) Nonlinear computation in DIVA–methods and applications. *Chem. Eng. Sci.*, **55**, 441–454.

Mangold, M., Ginkel, M., and Gilles, E.D. (2004) A model library for membrane reactors implemented in the process modelling tool ProMoT. *Comput. Chem. Eng.*, **28** (**3**), 319–332.

Nekhamkina, O., Rubinstein, B.Y., and Sheintuch, M. (2000a) Spatiotemporal patterns in thermokinetic models of cross-flow reactors. *AIChE J.*, **46**, 1632–1640.

Nekhamkina, O., Nepomnyashchy, A.A., Rubinstein, B.Y., and Sheintuch, M. (2000b) Nonlinear analysis of stationary patterns in convection-reaction-diffusion systems. *Phys. Rev. E*, **61**, 2436–2444.

Nowak, U., and Weimann, L. (1990) A family of newton codes for systems of highly nonlinear equations–algorithm, implementation, application. Technical Report TR 90-10, ZIB, Konrad Zuse Zentrum fur Informationstechnik.

Sheintuch, M., and Nekhamkina, O. (2003) Stationary spatially complex solutions in cross-flow reactors with two reactions. *AIChE J.*, **49**, 1241–1249.

Sheplev, V.S., Treskov, S.A., and Volokitin, E.P. (1998) Dynamics of a stirred tank reactor with first-order reaction. *Chem. Eng. Sci.*, **53**, 3719–3728.

Shimada, I., and Nagashima, T. (1979) A numerical approach to ergodic problem of dissipative dynamical systems. *Prog. Theor. Phys.*, **61**, 1605.

Travnickova, T., Kohout, M., Schreiber, I., and Kubicek, M. (2004) Effects of convection on spatiotemporal solutions in a model of a cross flow reactor. Proceedings of CHISA 2004.

Ung, S., and Doherty, M.F. (1995) Calculation of residue curve maps for mixtures with multiple equilibrium chemical reactions. *Ind. Eng. Chem. Res.*, **34**, 3195–3202.

Uppal, A., and Ray, W.H. (1974) On the dynamic behavior of continuous stirred tank reactors. *Chem. Eng. Sci.*, **29**, 967–985.

Wolf, A., Swift, J.B., Swinney, H.L., and Vastano, J.A. (1985) Determining lyapunov exponents from a time-series. *Phys. D*, **16**, 285–317.

Yorke, J.A., and Kaplan, J.L. (1979) Chaotic behavior of multidimensional difference equations, in *Functional Differential Equations and Approximation of Fixed Points, Vol. 730 of Lecture Notes in Mathematics* (eds H.-O. Peitgen and H.-O. Walther), Springer, pp. 228–237.

Zhang, F., Mangold, M., and Kienle, A. (2006) Stationary spatially periodic patterns in membrane reactors. *Chem. Eng. Sci.*, **61** (**21**), 7161–7170.

9
Comparison of Different Membrane Reactors

Frank Klose, Christof Hamel, and Andreas Seidel-Morgenstern

9.1
General Aspects Regarding Membrane Reactors of the Distributor Type

Membrane reactors of the distributor type clearly possess the potential to enhance selectivities and yields in reaction networks. The main reason is based on the fact that a spatially distributed feeding of the reactants can be beneficial. Hereby the specific reaction rates of the desired and undesired reactions determine the success of the concept. They specify which of the reactants should be introduced in a more concentrated form at the reactor inlets and which reactants should be dosed.

The effect of the membrane acting as a distributor is essentially threefold:

- Distributed dosing allows the altering and adjusting of local concentrations.
- The separated introduction of the reactants causes changes in the local gas velocity profiles and, thus, in the residence times.
- Distributed dosing possesses the potential to smooth the temperature profiles and to decrease local heat accumulation and "hot spots".

All three effects generate differences in overall reactant conversion and product distribution compared to conventional co-feed reactor operation.

An important side effect of the differences in the local concentration profiles is the fact that in membrane reactors the catalysts applied might be in other states. Thus, an optimized operation of membrane reactors requires the development and application of specific catalysts.

Reactant transfer through the membrane can be realized by convection or diffusion in the case of non-selective porous membranes or by ion conduction or solution mechanisms in the case of dense membranes. The membranes can be catalytically inert or they can be themselves catalytically active. Hereby, the provision of sufficient catalytic functionalities inside the membranes or on the membrane surfaces is not trivial and the application of particulate catalysts, as widely studied in this book, appears to be the alternative which is more easily realized.

Membrane Reactors: Distributing Reactants to Improve Selectivity and Yield.
Edited by Andreas Seidel-Morgenstern

ISBN: 978-3-527-32039-4

9.2 Oxidative Dehydrogenation of Ethane in Different Types of Membrane Reactors

Ethylene and propylene are the most important building blocks in organic chemistry. Currently they are industrially produced by steam cracking. Ethylene yields in naphta cracking are in the range of 30%. The cracking of ethane leads to yields above 80% and sets currently the standard (e.g., Weissermel and Arpe, 2003).

Regarding the yield of the oxidative dehydrogenation of ethane to ethylene there are currently no technologies available which allow competion with the steam cracking of ethane. Focusing essentially on conceptional studies in this book this reaction was investigated as a well manageable model reaction in order to evaluate the potential of membrane reactor concepts. Hereby, initially also the option to remove selectively the formed product ethylene in a membrane reactor of the extractor type was considered (Chapter 1). Since currently no membranes are available which are suitable for the effective selective withdrawal of olefins, this option was not further followed.

Oxygen is involved in all main reactions occurring in the network of ethane oxidation. The non-desired deep oxidation reactions are more sensitive to oxygen than the desired olefin formation. This relation in the order of reactions is a general phenomenon in the catalytic oxidation of hydrocarbons over non-noble transition metals and makes axially distributed oxidant dosing meaningful. Thus, it is advantageous to use membranes as an oxidant distributor. Axial oxygen concentration profiles are lowered and axial hydrocarbon profiles are elevated in membrane reactors in comparison to the joint insertion of reactants in conventional packed-bed reactors.

An important aspect regarding membrane reactors is the fact that especially in the inlet sections of the catalyst bed a relative excess of hydrocarbon is present. Thus, it is important to take into account that lowering the oxygen concentration reduces the amount of active vanadium(V) sites on the catalyst and, consequently, the ethane conversion.

Contact time prolongation, which is especially pronounced in the packed-bed membrane reactor (PBMR), enhances ethane conversion but lowers ethylene selectivity. Hereby side reactions are typically negative (e.g., olefin pyrolysis yielding soot), but there can be also positive effects (e.g., due to ethane dehydrogenation).

Flattening the temperature profiles is an important additional benefit of membrane-assisted oxidant dosing. Despite contact time prolongation, the temperature profiles in a pilot-scale packed-bed membrane reactor were significantly less pronounced than in a reference packed-bed reactor operated with identical overall feed amounts. The reduced hot spots decreased ethane conversion but were found to be beneficial for ethylene selectivity.

It was found that all membrane reactor types investigated experimentally demonstrated the beneficial effects mentioned above. However, specific aspects and differences were also observed.

9.2.1 Packed-Bed Membrane Reactor

In order to avoid oxygen limitation in the center of the tubular reactor due to the formation of radial concentration profiles, the PBMR exploiting ceramic membranes in a dead-end configuration was found to operate most efficiently at oxygen to hydrocarbon ratios of unity or slightly above (Chapter 5). To avoid the effect of back-diffusion of products causing a decrease in the reactor performance, high shell to tube side feed ratios should be applied to generate high transmembrane pressure drops. Back-diffusion was observed at low transmembrane pressures differences when sintered metal membranes with larger pores were applied. Furthermore, for the model reaction investigated the PBMR should realize relatively moderate contact times. In comparison to the well established packed-bed reactor (PBR) operation, ethane conversion, ethylene selectivity and ethylene yield were found to be improved in a one-stage membrane reactor and in particular in a three-stage membrane reactor cascade using an increasing oxygen dosing profile (Table 9.1). Furthermore, hot spot formation in the catalyst bed was found to be lower in the PBMR than in the PBR, depending on the operating conditions. The problem of hot spots could be even more overcome by fluidization of the catalyst bed, as applied in the fluidized-bed membrane reactor (FLBMR).

Table 9.1 Performance of different investigated membrane reactor configurations for the ODH of ethane. $T = 600\,°C$, $O_2/C_2H_6 = 1$.

	Conversion C_2H_6	Selectivity C_2H_4	Yield C_2H_4	Weight of catalyst/ total flow rate
Configuration	(%)	(%)	(%)	(kg s/m^3)
PBR (conventional)	55	46	25.6	400
PBMR	50	52	26.2	400
PBMR (three stages; increasing profile)	60	53	31.6	400
FLBMR	43	62	26.7	1750
EMR	46	41	18.9	780

9.2.2 Fluidized-Bed Membrane Reactor

The FLBMR appears to be a promising concept for highly exothermic selective oxidation reactions. Hot spots could be efficiently suppressed by means of the excellent heat exchange properties in a fluidized bed. The exploited thermal homogenization by fluidization led to the highest selectivities with respect to ethylene observed in all membrane reactors studied (62% vs 52% or 53% for the PBMR, see Chapter 6 and Table 9.1). The ethylene yields for the FLBMR and the PBMR were comparable. However, it should be pointed out that higher W/F ratios are necessary in the FLBMR to realize these yields. Similar to the PBMR, oxygen to hydrocarbon ratios above unity are beneficial for increasing the ethane conversion. A drawback in comparison to the PBMR is the higher space consumption of the FLBMR and the demand for additional optimization of the catalyst particles with respect to fluidization and resistance against higher mechanical stresses. Like in the PBMR, the utilization of sintered metal membranes is possible in the FLBMR. Care must be also taken in order to provide sufficient transmembrane pressure drops to avoid the back-diffusion of ethane and ethylene.

9.2.3 Electrochemical Packed-Bed Membrane Reactor

Selective oxygen transport through membranes by oxide anion conduction is realized in the electrochemical membrane reactor (EMR, Chapter 7). This selective transport avoids in the EMR any dilution of the reactants by inert gases such as nitrogen, in contrast to the reactors which apply non-selective membranes (PBMR, FLBMR). In the case study carried out, the subsequent oxygen transfer from the membrane to the catalyst occured via the gas-phase. In the experiments a direct catalyst re-oxidation by accepting oxide anions from the membrane was found to be unlikely. The EMR reached approximately the 20% level with respect to ethylene yield (Table 9.1), which demonstrates the general feasibility of this concept. Oxygen permeation was identified to be still limiting, even though oxygen transfer through the membrane was enforced by electrochemical pumping. The rather low oxygen permeability and the high costs of the membranes can be considered currently as a barrier for a successful scale-up of this type of EMR concept. However, if these limitations can be overcome, the EMR concept offers attractive advantages.

9.3 General Conclusions

Membrane reactors of the distributor type possess an attractive potential to enhance selectivities and yields. The potential is highest when there are pronounced differences in the order of reaction between the desired and undesired reactions.

Results obtained for the oxidative dehydrogenation of ethane confirmed the general potential of the reactor concept. However, the achieved benefits using vanadium oxide catalysts were not sufficient to pass the threshold for industrial exploitation. For reaction systems in which the local concentrations more strongly influence the ratio between the desired and undesired reactions, more pronounced improvements can be expected, rendering membrane reactors of the distributor type attractive for commercialization.

The availability of suitable membranes is another important requirement for successful reactor operation. Hereby a kinetic compatibility between the rates of the reaction and transport processes is essential. This requires a careful selection of the catalysts and membranes. Hereby dead-end configurations could be used efficiently to reduce the need for selective membranes. A general requirement for applications is to avoid the back-transport of products to the feed side.

From the results obtained for the oxidative dehydrogenation of ethane, reactors using particulate catalysts and inert membranes could be recommended. This principle can be realized in packed-bed, fluidized-bed and electrochemical packed-bed membrane reactors. For highly exothermic reactions in particular the fluidized-bed membrane reactor is attractive.

A wider use of catalytic membrane reactors requires several challenging problems in the fields of catalysis, material science and reaction engineering to be tackled. Important problems that must be solved prior to possible application are also related to reactor safety. The reactor principle allows gases to come into contact via a membrane, while, without the membrane, this would create explosive mixtures. Thus, the possibility of membrane breakage must be considered in the design concepts.

With the joint efforts of experts in catalysis, material science and reaction engineering it should be possible to solve the remaining problems, to identify promising reaction systems and to develop successful applications of membrane reactors of the distributor type.

Reference

Weissermel, K., and Arpe, H.-J. (2003) *Industrial Organic Chemistry*, 4th edn, Wiley-VCH Verlag GmbH.

Index

Membrane Reactors: Distributing Reactants to Improve Selectivity and Yield.
Edited by Andreas Seidel-Morgenstern

ISBN: 978-3-527-32039-4